A New Ecology: Systems Perspective

Front cover photo is by B.D. Fath and shows Møns Klint, Denmark.

The back cover photos show (from left to right) (1) a Danish beech forest (Ryget Skov), (2) Krimml Falls in Austria, (3) part of the shore of Namchu Lake in Tibet, (4) Crater Lake, Oregan, USA, and (5) Natron Lake, Tanzania and were taken by S.E. Jørgensen (1 and 4), B.D. Fath (2), and M.V. Jørgensen (3 and 5).

A New Ecology
Systems Perspective

Sven E. Jørgensen
Environmental Chemistry Section
Royal Danish School of Pharmacy
DK-2100 Copenhagen, Denmark

Brian D. Fath
Biology Department
Towson University
Towson, MD 21252, USA

Simone Bastianoni
Department of Chemical and
Biosystems Sciences
University of Siena
53100 Siena, Italy

João C. Marques
Department of Zoology
Institute of Marine Research (IMAR),
University of Coimbra
3004-517 Coimbra, Portugal

Felix Müller
Ecology Centre
University of Kiel
24118 Kiel, Germany

Søren N. Nielsen
Environmental Chemistry Section
Royal Danish School of Pharmacy
DK-2100 Copenhagen, Denmark

Bernard C. Patten
Institute of Ecology
University of Georgia
Athens, GA 30602-2602, USA

Enzo Tiezzi
Department of Chemical and
Biosystems Sciences
University of Siena
53100 Siena, Italy

Robert E. Ulanowicz
Chesapeake Biological Laboratory
P.O. Box 38, 1 Williams Street
Solomons, MD 20688-0038, USA

ELSEVIER

Amsterdam • Boston • Heidelberg • London • New York • Oxford
Paris • San Diego • San Francisco • Singapore • Sydney • Tokyo

Elsevier
Linacre House, Jordan Hill, Oxford OX2 8DP, UK
Radarweg 29, PO Box 211, 1000 AE Amsterdam, The Netherlands

First edition 2007

Copyright © 2007 Elsevier B.V. All rights reserved

No part of this publication may be reproduced, stored in a retrieval system
or transmitted in any form or by any means electronic, mechanical, photocopying,
recording or otherwise without the prior written permission of the publisher

Permissions may be sought directly from Elsevier's Science & Technology Rights
Department in Oxford, UK: phone (+44) (0) 1865 843830; fax (+44) (0) 1865 853333;
email: permissions@elsevier.com. Alternatively you can submit your request online by
visiting the Elsevier web site at http://elsevier.com/locate/permissions, and selecting
Obtaining permission to use Elsevier material

Notice
No responsibility is assumed by the publisher for any injury and/or damage to persons
or property as a matter of products liability, negligence or otherwise, or from any use
or operation of any methods, products, instructions or ideas contained in the material
herein. Because of rapid advances in the medical sciences, in particular, independent
verification of diagnoses and drug dosages should be made

Library of Congress Cataloging in Publication Data
A catalog record is available from the Library of Congress

British Library Cataloguing in Publication Data
A catalogue record is available from the British Library

ISBN: 978-0-444-53160-5

For information on all Elsevier publications
visit our website at books.elsevier.com

Printed and bound in The Netherlands

07 08 09 10 11   10 9 8 7 6 5 4 3 2 1

Working together to grow
libraries in developing countries

www.elsevier.com | www.bookaid.org | www.sabre.org

ELSEVIER   BOOK AID International   Sabre Foundation

CONTENTS

Preface   ix

1. **Introduction: A New Ecology is Needed**   1
   1.1 Environmental management has changed   1
   1.2 Ecology is changing   2
   1.3 Book outline   3

2. **Ecosystems have Openness (Thermodynamic)**   7
   2.1 Why must ecosystems be open?   7
   2.2 An isolated system would die (maximum entropy)   8
   2.3 Physical openness   13
   2.4 The second law of thermodynamics interpreted for open systems   18
   2.5 Dissipative structure   20
   2.6 Quantification of openness and allometric principles   22
   2.7 The cell   30
   2.8 What about the environment?   31
   2.9 Conclusion   32

3. **Ecosystems have Ontic Openness**   35
   3.1 Introduction   35
   3.2 Why is ontic openness so obscure?   36
   3.3 Ontic openness and the physical world   39
   3.4 Ontic openness and relative stability   49
   3.5 The macroscopic openness: Connections to thermodynamics   50
   3.6 Ontic openness and emergence   53
   3.7 Ontic openness and hierarchies   55
   3.8 Consequences of ontic openness: a tentative conclusion   56

4. **Ecosystems have Directionality**   59
   4.1 Since the beginnings of ecology   59
   4.2 The challenge from thermodynamics   60
   4.3 Deconstructing directionality?   62
   4.4 Agencies imparting directionality   63
   4.5 Origins of evolutionary drive   66
   4.6 Quantifying directionality in ecosystems   68
   4.7 Demystifying Darwin   74
   4.8 Directionality in evolution?   76
   4.9 Summary   77

5. Ecosystems have Connectivity — 79

- 5.1 Introduction — 79
- 5.2 Ecosystems as networks — 80
- 5.3 Food webs — 82
- 5.4 Systems analysis — 84
- 5.5 Ecosystem connectivity and ecological network analysis — 86
- 5.6 Network environ analysis primer — 86
- 5.7 Summary of the major insights cardinal hypotheses (CH) from network environ analysis — 92
- 5.8 Conclusions — 101

6. Ecosystems have Complex Dynamics (Growth and Development) — 103

- 6.1 Variability in life conditions — 103
- 6.2 Ecosystem development — 105
- 6.3 Orientors and succession theories — 112
- 6.4 The maximum power principle — 115
- 6.5 Exergy, ascendency, gradients, and ecosystem development — 120
- 6.6 Support for the presented hypotheses — 125
- 6.7 Toward a consistent ecosystem theory — 133
- 6.8 Exergy balances for the utilization of solar radiation — 139
- 6.9 Summary and conclusions — 141

7. Ecosystems have Complex Dynamics – Disturbance and Decay — 143

- 7.1 The normality of disturbance — 143
- 7.2 The risk of orientor optimization — 151
- 7.3 The characteristics of disturbance — 152
- 7.4 Adaptability as a key function of ecosystem dynamics — 156
- 7.5 Adaptive cycles on multiple scales — 160
- 7.6 A case study: Human disturbance and retrogressive dynamics — 164
- 7.7 Summary and conclusions — 166

8. Ecosystem Principles have Broad Explanatory Power in Ecology — 167

- 8.1 Introduction — 167
- 8.2 Do ecological principles encompass other proposed ecological theories?: Evolutionary theory — 168
- 8.3 Do ecological principles encompass other proposed ecological theories?: Island biogeography — 176
- 8.4 Do ecological principles encompass other proposed ecological theories?: Latitudinal gradients in biodiversity — 180
- 8.5 Do ecological principles encompass other proposed ecological theories?: Optimal foraging theory — 184
- 8.6 Do ecological principles encompass other proposed ecological theories?: Niche theory — 187

	8.7	Do ecological principles encompass other proposed ecological theories?: Liebig's law of the minimum	191
	8.8	Do ecological principles encompass other proposed ecological theories?: The river continuum concept (RCC)	194
	8.9	Do ecological principles encompass other proposed ecological theories?: Hysteresis in nature	196
	8.10	Conclusions	198
9.	**Ecosystem Principles have Applications**		**199**
	9.1	Introduction	199
	9.2	Entropy production as an indicator of ecosystem trophic state	200
	9.3	The use of ecological network analysis (ENA) for the simulation of the interaction of the american black bear and its environment	206
	9.4	Applications of network analysis and ascendency to South Florida ecosystems	210
	9.5	The application of eco-exergy as ecological indicator for assessment of ecosystem health	218
	9.6	Emergy as ecological indicator to assess ecosystem health	221
	9.7	The eco-exergy to empower ratio and the efficiency of ecosystems	228
	9.8	Application of eco-exergy and ascendency as ecological indicator to the Mondego Estuary (Portugal)	231
	9.9	Conclusions	241
10.	**Conclusions and Final Remarks**		**243**
	10.1	Are basic ecological properties needed to explain our observations?	243
	10.2	Previous attempts to present an ecosystem theory	243
	10.3	Recapitulation of the ecosystem theory	245
	10.4	Are there basic ecosystem principles?	246
	10.5	Conclusion	248
References			**251**
Index			**273**

PREFACE

The scope of this book is to demonstrate that we do have an ecosystem theory that can be used to describe ecosystem structure and function. It was previously shown in the book, *Integration of Ecosystem Theories: A Pattern* (3rd edition, 2002), that the various contributions to systems ecology are consistent and together form a pattern of ecological processes. My book with Yuri Svirezhev, *Toward a Thermodynamic Theory of Ecosystems* (2004), presented the thermodynamics of this pattern in a mathematical language. This book, *A New Ecology: Systems Perspective*, shows that the basic properties of ecosystems (presented in Chapters 2–7) lead to or are consistent with ten tentative propositions for ecosystems (Chapter 10), which can be used to explain ecological observations (Chapter 8). An ecosystem theory is a prerequisite for wider application of ecological sciences in environmental management because with the theory it becomes feasible to guide conservation or environmental management. Chapter 9 shows how the presented ecosystem theory can be applied to assess ecosystem health, a facet of environmental management. A thermodynamic interpretation of the evolution is under preparation in my other book with Yuri Svirezhev, *A Thermodynamic Theory of the Evolution,* with expected publication in 2007 or early in 2008. The three books *Toward a Thermodynamic Theory of Ecosystems*, this book *A New Ecology: Systems Perspective*, and the coming one, *A Thermodynamic Theory of the Evolution* form a troika that presents a useful ecological theory.

This book has nine authors. The basic outline of the book was formulated during a one-week brainstorming meeting on the Danish island of Møn in June 2005. All nine authors have written parts of the book and have reviewed the contributions of the other authors. The book is therefore a joint effort resulting from close teamwork. I am the first author because the idea to produce a book about ecosystem theory and systems ecology was initiated by me based on a brainstorming meeting with system ecologists. I edited this book with Brian Fath after all the authors had exchanged ideas and reviewed the ten chapters. Brian Fath is therefore considered the second editor of the book. Bai Lian Li (Larry) participated in the brainstorming meeting in Møn and he contributed significantly to the outline of ideas making up the final book. However, due to his engagement with the Eco-summit 2007 in China, he was unable to contribute written material for the book. He is, however, working on a Chinese edition of the book, which we all consider of great importance as China during the last few years has shown an increased interest in environmental problems. This Chinese interest for environmentally sound management is expected to accelerate in the coming years, which makes a Chinese edition of this book even more important. Bernie Patten and Enzo Tiezzi were unable to attend the brainstorming meeting, but they both contributed written material and comments on the chapters (Photos 1 and 2).

Using my 2004 Stockholm Water Prize, I established a foundation to promote ecosystem theory and integrated environmental assessment. The Foundation's grants support brainstorming meetings and travel particularly for young scientists focusing on system ecology, ecological modelling, and lake management. The foundation is named "William

Photo 1: From the brainstorming meeting at Møn, June 2005.

Photo 2: From the brainstorming meeting at Møn, June 2005.

Williams' and Milan Straškraba's Foundation" after two of my close scientist friends who passed away in 2002 and 2000. William Williams has contributed significantly to integrated lake management and Milan Straškraba has played a major role in system ecology in the last two decades of the 20th century. The nine authors express their appreciation to the foundation for the support that has made it possible to publish this book in the hope that it will enhance a wider application of ecosystem theory in ecology to explain observations and to facilitate ecological sound conservation and environmental management.

Sven Erik Jørgensen
Copenhagen, July 2006

1

Introduction: A new ecology is needed

1.1 ENVIRONMENTAL MANAGEMENT HAS CHANGED

The political agenda imposed on ecologists and environmental managers has changed since the early 1990s. Since the Rio Declaration and Agenda 21 the focus has been on sustainability, which inevitably has made ecosystem functioning a core issue. Sustainability Development is, according to the Rio Declaration, defined as follows: *"development that meets the needs of the present without compromising the ability of future generations to meet their own needs."* And, the contrasting parties are invited to, *"act in a way that is economically profitable, socially acceptable, and environmentally compatible."* Already the Rio Declaration emphasized the importance of ecosystems in Principle 7: States shall cooperate in a spirit of global partnership to conserve, protect, and restore the health and integrity of the Earth's ecosystems.

In view of the different contributions to global environmental degradation, states have common but differentiated responsibilities. The developed countries acknowledge the responsibility that they bear in the international pursuit of sustainable development in view of the pressures their societies place on the global environment and of the technologies and financial resources they command.

The Convention of Biodiversity adopted, in 2000, 12 principles—called the Ecosystem Approach—that placed the ecosystem concept even more centrally into environmental management considerations. It is particularly clear from the last 10 of the 12 principles:

(1) The objectives of management of land, water, and living resources are a matter of societal choice.
(2) Management should be decentralized to the lowest appropriate level.
(3) Ecosystem managers should consider the effects (actual or potential) of their activities on adjacent and other ecosystems.
(4) Recognizing potential gains from management, there is usually a need to understand and manage the ecosystem in an economic context. Any such ecosystem-management program should:
   a. Reduce those market distortions that adversely affect biological diversity.
   b. Align incentives to promote biodiversity conservation and sustainable use.
   c. Internalize costs and benefits in the given ecosystem to the extent feasible.

(5) Conservation of <u>ecosystem structure</u> and <u>functioning</u>, in order to maintain <u>ecosystem services</u>, should be a priority target of the ecosystem approach.
(6) Ecosystems must be managed <u>within the limits of their functioning</u>.
(7) The ecosystem approach should be undertaken at the appropriate spatial and temporal scales.
(8) Recognizing the varying temporal scales and lag-effects that characterize ecosystem processes, objectives for ecosystem management should be set for the long term.
(9) Management must recognize that change is inevitable.
(10) The ecosystem approach should seek the appropriate <u>balance between, and integration of</u>, conservation and use of biological diversity.
(11) The ecosystem approach should consider all forms of relevant information, including scientific and indigenous and local knowledge, innovations, and practices.
(12) The ecosystem approach should involve all relevant sectors of society and scientific disciplines.

Also in the book *Ecosystems and Human Well-being, a Report of the Conceptual Framework Working Group of the Millennium Ecosystem Assessment* from 2003, ecosystems are the core topic. In Chapter 2 of the book, it is emphasized that: an assessment of the condition of ecosystems, the provision of services, and their relation to human well-being requires an integrated approach. This enables a decision process to determine which service or set of services is valued most highly and how to develop approaches to maintain services by managing the system sustainably. Ecosystem services are the benefits people obtain from ecosystems. These include provisioning services such a food and water; regulating services such as flood and disease control; cultural services such as spiritual, recreational, and cultural benefits; and supporting services such as nutrient cycling, which maintain the conditions for life on Earth.

Today, environmental managers have realized that maintenance of ecosystem structure and functioning (see Principle 5 above) by an integrated approach is a prerequisite for a successful environmental management strategy, which is able to optimize the ecosystem services for the benefit of mankind and nature. Another question is whether we have sufficient knowledge in ecology and systems ecology today to give the needed information about ecosystem structure, function, and response to disturbance to scientifically pursue the presented environmental management strategy and ecosystem sustainability. In any way, the political demands provide a daunting challenge for ecosystem ecology.

1.2 ECOLOGY IS CHANGING

As a consequence of the changing paradigm direction of environmental management, we need to focus on ecosystem ecology. An ecosystem is according to the Millennium Report (2003) defined as "a dynamic complex of plants, animals and microorganism communities and the nonliving environment, interacting as a functional unit. Humans are an integral part of ecosystems."

A well-defined ecosystem has strong interactions among its components and weak interactions across its boundaries. A useful ecosystem boundary is the place where a number of discontinuities coincide for instance in the distribution of organism, soil type, drainage basin or depth in a water body. At a larger scale, regional and even globally distributed ecosystems can be evaluated based on a commonality of basic structural units.

Three questions are fundamental to pursue for ecosystem-based environmental management:

I: What are the underlying ecosystem properties that can explain their response to perturbations and human interventions?
II: Are we able to formulate at least building blocks of an ecosystem theory in the form of useful propositions about processes and properties? We prefer the word "propositions" and not laws because ecosystem dynamics are so complex that universal laws give way to contextual propensities. The propositions capture these general tendencies of ecosystem properties and processes that can be applied to understand the very nature of ecosystems, including their response to human impacts.
III: Is the ecosystem theory that we can formulate to understand ecosystem properties sufficiently developed to be able to explain ecological observations with practical application for environmental management?

The scope of the book is an attempt to answer these questions to the extent that is currently possible. The authors of this book have realized that an ecosystem theory is a prerequisite for wider application of ecological sciences in environmental management because theory provides a strong guide for environmental management and resource conservation.

1.3 BOOK OUTLINE

Chapters 2–7 present the fundamental properties that explain typical ecosystem processes under "normal" growth and development and their responses to disturbance. These are:

(1) *Ecosystems are open systems*—open to energy, mass, and information. Openness is an absolute necessity because the maintenance of ecosystems far from thermodynamic equilibrium requires an input of energy. This core property is presented in Chapter 2.
(2) *Ecosystems are ontically inaccessible*—meaning that due to their enormous complexity it is impossible to accurately predict in all detail ecosystem behavior. It means that it is more appropriate to discuss the propensity of ecosystems to show a certain pattern or to discuss the direction of responses. This property is presented in Chapter 3.
(3) *Ecosystems have directed development*—meaning they change progressively to increase, in particular, feedback and autocatalysis. It is the observed direction of responses mentioned under point 2. This property is discussed in detail in Chapter 4.
(4) *Ecosystems have network connectivity*—which gives them new and emergent properties. The networks have synergistic properties, which are able to explain the cooperative integration of ecosystem components, which can at least sometimes yield unexpected system relations. This core property is covered in Chapter 5.

(5) *Ecosystems are organized hierarchically*—in the sense that we can understand one level only by understanding interactions with the levels below and above the scale of focus. Often major changes in one level are leveled out in the higher levels, where only minor hierarchical organization changes are observed. The properties associated with the are discussed in great detail by Allen and Starr (1982) in their book *Hierarchy, Perspectives for Ecological Complexity* and in the book *A Hierarchical Concept of Ecosystems* by O'Neill et al. (1986). The scaling theory and the allometric principles are rooted in quantification of openness and are, therefore, presented in Chapter 2. The basic general elements of hierarchy theory are also presented in this chapter. Further examples of the application of hierarchy theory are presented in Chapters 3 and 7.

(6) *Ecosystems grow and develop*—they gain biomass and structure, enlarge their networks, and increase their information content. We can follow this growth and development using holistic metrics such as power, eco-exergy, and ascendency, respectively. For example, incoming solar radiation is first used to cover maintenance of the ecosystem far from thermodynamic equilibrium and afterwards used to move the system further from equilibrium, which increases the power, stored eco-exergy, and ascendency. This growth property is presented in Chapter 6. It is a core property because it explains how ecosystems develop and even evolve. Many ecosystem processes are rooted in the competition for the resources that are needed for growth and can be explained in this light.

(7) *Ecosystems have complex response to disturbance*—but when we understand properties of ecosystems such as adaptation, biodiversity, resistance, and connectedness, to mention a few of the most important properties covered in the book, we can explain and sometimes predict the responses of ecosystems to disturbances. This part of the ecosystem dynamics is presented in Chapter 7.

Chapters 2–7 are directed to answer first question above. The second question is answered throughout these chapters and summarized in Chapter 10, where the presented ecosystem theory is formulated by use of the seven properties and by formulation of ten propositions. The two formulations are completely consistent as discussed in this last chapter of the book.

The last question regarding the applicability of the presented theory to explain ecological observations and to be applied in environmental management is addressed in Chapters 8 and 9. The application of the theory in environmental management has been mostly limited to use of ecological indicators for ecosystem health assessment as described in Chapter 9. The theory has much wider applicability, but the use of ecological indicators has a direct link to ecosystem theory that facilitates testing the theory. Tests of the theory according to its applicability in practical environmental management and to explain ecological observations is crucial for the general acceptance of the ecosystem theory of course; but it does not exclude that it cannot be improved significantly. On the contrary, it is expected that the theory will be considerably improved by persistent and ongoing application because the weaknesses in the present theory will inevitably be uncovered as the number of case studies increases. Discovery of theoretical weaknesses will inspire improvements. Therefore, it is less important that the theory has flaws and lacks important elements than it is that it is sufficiently developed to be directly applied. We, the authors, are of the opinion that we do have an ecosystem theory today that is ready to be applied but which also inevitably will be developed significantly during the next one to two decades due to (hopefully) its wider application.

An ecosystem theory as the one presented in this book may be compared with geographical maps. We had already 2000 years ago geographical maps that could be applied to get an overview of where you would find towns, mountains, forests, etc. These maps were considerably improved and the geographical maps used in the 17th and 18th century were much more accurate and detailed, although they are of course not comparable with the satellite-based maps of today. Our ecosystem theory as presented today may be comparable with the geographical maps of the 18th century. They are, as the more than 200 years old geographical maps, very useful, but they can be improved considerably when new methods, additional information, and additional observations are available. It may take 20 or maybe 50 years before we have the quality of an ecosystem theory comparable with today's geographical maps, but the present level of our ecosystem theory is nevertheless suitable for immediate application. Only through this application we will discover new methods and demand for improvements, both theoretical and practical for science and management, ultimately leading to a more complete and accurate ecosystem theory.

The most fundamental parts of the presented ecosystem theory, particularly the more mathematical aspects, are placed in boxes. It makes it on one hand easy to find the theoretical elements of the entire ecosystem theory but it also facilitates reading for those preferring a less mathematical formulation of an ecosystem theory.

2

Ecosystems have openness (thermodynamic)

> *Without the Sun, everything on Earth dies!*
> (From the plaintive Ukrainian folksong, "Я бачив як вітер…")

2.1 WHY MUST ECOSYSTEMS BE OPEN?

The many 1m-trees that we planted more than 30 years ago in our gardens, which may have been open fields at the time, are today more than 30 m tall. They have increased the structure in the form of stems many times and they have more than a thousand times as many leaves and have grown often more than 1m in height since last spring. The structures of the gardens have changed. Today they have a high biodiversity – not so much due to different plants, but the tall trees and the voluminous bushes with berries attract many insects and birds. The garden today is a much more complex ecosystem. The biomass has increased, the biodiversity has increased and the number of ecological interactions among the many more species has increased.

When you follow the development of an ecosystem over a longer period or even during a couple of spring months, you are witness to one of the many wonders in nature: an inconceivably complex system is developing in front of you. What makes this development of complex (and beautiful) systems in nature possible?

In accordance to classic thermodynamics all isolated systems will move toward thermodynamic equilibrium. All the gradients and structures in the system will be eliminated and a homogenous dead system will be the result. It is expressed thermodynamically as follows: entropy will always increase in a isolated system. As work capacity is a result of gradients in certain intensive variables such as temperature, pressure, and chemical potential, etc. (see Table 2.1), a system at thermodynamic equilibrium can do no work. But our gardens are moving away from thermodynamic equilibrium with almost a faster and faster rate every year. It means that our gardens cannot be isolated. They must be at least non-isolated; but birds and insects and even sometimes a fox and a couple of squirrels enter from outside the garden—from the environment of the garden, maybe from a forest 1000 m away. The garden as all other ecosystems must be open (see also Table 2.2, where the thermodynamic definitions of isolated, closed, and open systems are presented). Gardens are first of all open to energy inputs from the solar radiation, which is absolutely necessary to avoid the system moving toward thermodynamic equilibrium. Without solar radiation the system would die. The energy contained in the solar radiation

Table 2.1 Different forms of energy and their intensive and extensive variables

Energy form	Extensive variable	Intensive variable
Heat	Entropy (J/K)	Temperature (K)
Expansion	Volume (m^3)	Pressure (Pa = kg/s^2 m)
Chemical	Moles (M)	Chemical potential (J/moles)
Electrical	Charge (A·s)	Voltage (V)
Potential	Mass (kg)	(Gravity) (height) (m^2/s^2)
Kinetic	Mass (kg)	0.5 (velocity)^2 (m^2/s^2)

Note: Potential and kinetic energy is denoted mechanical energy.

Table 2.2 Definitions of various thermodynamic systems

System type	Definition
Isolated	No exchange of energy, mass, and information with the environment
Non-isolated	Exchange of energy and information, but no mass with the environment
Closed	Exchange of energy and information, but no mass with the environment
Open	Exchange of energy, mass, and information with the environment

covers the energy needed for maintenance of the plants and animals, measured by the respiration, but when the demand for maintenance energy is covered, additional energy is used to move the system further away from thermodynamic equilibrium. The thermodynamic openness of ecosystems explains why ecosystems are able to move away from thermodynamic equilibrium: to grow, to build structures and gradients.

This openness is in most cases for ecosystems a necessary condition only. For example, a balanced aquarium and also our planet are more non-isolated than open; openness is only incidental. One wonders what would be the elements of sufficient conditions. Openness is obviously not a sufficient condition for ecosystems because all open systems are not ecosystems. If a necessary condition is removed, however, the process or system in question cannot proceed. So openness (or non-isolation) as a necessary condition makes this a pivotal property of ecosystems, one to examine very closely for far-reaching consequences. And if these are to be expressed in thermodynamic terms, ecologists need to be aware that aspects of thermodynamics—particularly entropy and the second law—have for several decades been under some serious challenges in physics, and no longer enjoy the solid standing in science they once held (Capek and Sheehan, 2005). So like a garden, science is open too—ever exploring, changing, and improving. In this chapter, we will not take these modern challenges too much into account.

2.2 AN ISOLATED SYSTEM WOULD DIE (MAXIMUM ENTROPY)

The spontaneous tendency of energy to degrade and be dissipated in the environment is evident in the phenomena of everyday life. A ball bouncing tends to smaller and smaller bounces and dissipation of heat. A jug that falls to the ground breaks (dissipation) into many pieces and the inverse process, which could be seen running a film of the fall backwards, never happens in nature. Except, of course, the jug did come into existence by the same kind of non-spontaneous processes that make the garden grow. It is instructive to ponder how openness or non-isolation operate here, as necessary conditions. Perfume leaves a bottle and dissipates into the room; we never see an empty bottle spontaneously fill, although the laws of probability do allow for this possibility. There is thus a tendency to the heat form and dissipation. The thermodynamic function known as *entropy* (S) is the extensive variable for heat and measure therefore to what extent work has been degraded to heat. Strictly speaking, the entropy concept only applies to isolated systems close to equilibrium, but it is often used in a metaphorical sense in connection with everyday far-from-equilibrium systems. We will follow this practice here as a useful way to consider ecosystems; revisions can come later when thermodynamic ecology is much better understood from theory and greater rigor is possible. Transformations tend to occur spontaneously in the direction of increasing entropy or maximum dissipation. The idea of the passage of time, of the direction of the transformation, is inherent in the concept of entropy. The term was coined by Clausius from $\tau\rho o\pi\eta$ (transformation) and $\varepsilon\nu\tau\rho o\pi\eta$ (evolution, mutation, or even confusion).

Clausius used the concept of entropy and reworded the First and Second Thermodynamic Laws in 1865 in a wider and more universal framework: Die Energie der Welt ist Konstant (the energy of the world is constant) and Die Entropy der Welt strebt einem Maximum zu (The entropy of the world tends toward a maximum). Maximum entropy, which corresponds to the equilibrium state of a system, is a state in which the energy is completely degraded and can no longer produce work. Well, maybe not literally "completely degraded" but rather, let us say, only "degradiented", meaning brought to a point of equilibrium where there is no gradient with its surroundings, therefore no possibility to do work. Energy at 300 K at the earth's surface is unusable, but can do work after it passes to outer space where the temperature is 3 K and a thermal gradient is re-established. Again, it is a common practice to use the term "degraded" in the sense we have, and "completely" for emphasis; for continuity in communication these practices will be followed here.

Entropy is, therefore, a concept that shows us the direction of events. "Time's Arrow", it has been called by Harold Blum (1951). Barry Commoner (1971) notes that sandcastles (order) do not appear spontaneously but can only disappear (disorder); a wooden hut in time becomes a pile of beams and boards: the inverse processes do not occur. The spontaneous direction of an isolated system is thus from order to disorder and entropy, as metaphor, indicates this inexorable process, the process which has the maximum probability of occurring. In this way the concepts of disorder and probability are linked in the concept of entropy. Entropy is in fact a measure of disorder and probability even though for systems like a garden it cannot be measured. Entropy generation can be calculated approximately, however, for reasonably complex systems, and for this one should consult the publications of Aoki (1987, 1988, 1989).

War is a disordering activity, but from such can often arise other levels and kinds of order. For example, a South Seas chieftain once warred on his neighbors and collected their ornately carved wooden thrones as part of the spoils and symbols of their defeat; they came to signify his superiority over his enemies and this enabled him to govern for many years as leader of a well-organized society. This social order, of course, came out of the original disordering activity of warfare, and it was sustained. The captured thrones were stored in a grand thatched building for display on special holidays, a shrine that came to symbolize the chieftain's power and authority over his subjects. One year, a typhoon hit the island and swept the structure and its thrones away in the night. The disordering of the storm went far beyond the scattering of matter, for the social order that had emerged from disorder quickly unraveled also and was swept away with the storm. The remnant society was forced in its recovery to face a hard lesson of the region—"People who live in grass houses shouldn't stow thrones!" In order to understand this order–disorder relationship better, it is useful to describe a model experiment: the mixing of gases.

Suppose we have two gases, one red and one yellow, in two containers separated by a wall. If we remove the wall we see that the two gases mix until there is a uniform distribution: an orange mixture. Well, a uniformly *mixed* distribution, anyway; in a statistical sense the distribution is actually random. If they were originally mixed they would not be expected to spontaneously separate into red and yellow. The "orange" state is that of maximum disorder, the situation of greatest entropy because it was reached spontaneously from a situation of initial order—the maximum of which, by the way, is the uniform distribution. Random, uniform; one must take care in choice of wording. Entropy is a measure of the degree of disorder of the system (notice that the scientific literature presents several definitions of the concept of entropy). The disordered state occurred because it had the highest statistical probability. The law of increasing entropy expresses therefore also a law of probability, of statistical tendency toward disorder. The most likely state is realized, namely the state of greatest entropy or disorder. When the gases mix, the most probable phenomenon occurs: degeneration into disorder—randomness. Nobel Prize winner for physics, Richard Feynman, comments that irreversibility is caused by the general accidents of life. It is not against the laws of physics that the molecules rebound so as to separate; it is simply improbable and would not happen in a million years. Things are irreversible only in the sense that going in one direction is probable whereas going in the other, while it is possible and in agreement with the laws of physics, would almost never happen.

So it is also in the case of our South Sea islanders. Two populations kept separate by distance over evolutionary time could be expected to develop different traits. Let one such set be considered "red" traits, and the other "yellow." Over time, without mixing, the red traits would get redder and the yellow traits yellower—the populations would diverge. If a disordering event like a storm or war caused the islanders to disperse and eventually encounter one another and mix reproductively, their distinctive traits would over a long period of time merge and converge toward "orange." A chieftain governing such a population would not be able to muster the power to reverse the trend by spontaneous means; eugenic management would be required. A tyrant might resort to genocide to develop a genetically pure race of people. Without entropy such an extreme measure, which has over human history caused much misery, would never be needed. Spontaneous de-homogenization could occur, re-establishing the kind of thermodynamic gradient (red vs. yellow) that would again make possible the further ordering work of disordering war. No entropy, no work or war—necessary or sufficient condition?

The principle of increasing entropy is now clearer in orange molecules and people: high-entropy states are favored because they are more probable, and this fact can be expressed by a particular relation as shown by Boltzmann (1905): $S = -k \log p$, where S is entropy, k Boltzmann's constant, and p the probability of an event occurring. The logarithmic dependence makes the probability of zero entropy equal to one. The universality of the law of entropy increase (we speak metaphorically) was stressed by Clausius in the sense that energy is degraded ("de-gradiented") from one end of the universe to the other and that it becomes less and less available in time, until "Wärmetode", or the "thermal death" of the universe. Evolution toward this thermal death is the subject of much discussion. It has been shown (Jørgensen et al., 1995) that the expansion of the universe implies that the thermodynamic equilibrium is moving farther and farther away. In order to extend the theory from the planetary to the cosmic context it is necessary to introduce unknown effects such as gravitation. Current astrophysics suggests an expanding universe that originated in a great primordial explosion (big bang) from a low-entropy state, but the limits of theoretical thermodynamic models do not allow confirmation or provide evidence.

The study of entropy continues: this fundamental concept has been applied to diverse fields such as linguistics, the codification of language and to music and information theory. Thermodynamics has taught us many fascinating lessons, particularly that (I) energy cannot be created or destroyed but is conserved and (II) entropy of isolated systems is always increasing, striking the hours of the cosmic clock, and reminding us that both for man and for energy–matter, time exists and the future is distinct from the past by virtue of a higher value of S.

The second law of thermodynamics, still upheld as one of nature's fundamental laws, addresses the pathways we should avoid in order to keep life on Earth. It shows the universal, inescapable tendency toward disorder (in thermodynamics, the general trend toward an entropy maximum), which is also, again metaphorically, a loss of information and of usable energy availability. This tendency to the Clausius' "thermal death", speaks to the thermodynamic equilibrium, namely the death of biological systems and ecosystems, through the destruction of diversity. There are two ways to achieve such a condition when:

(a) through energy exchanges as heat fluxes, there are no more differences in temperature and nothing more can be done, because no exchange of usable energy is allowed;
(b) a system, becoming isolated, consumes its resources, reaching a great increase in its internal entropy and, at the end, to self-destruction.

For this reason living systems cannot be at the conditions of the thermodynamic equilibrium, but keep themselves as far as possible from that state, self-organizing due to material and energetic fluxes, received from outside and from systems with different conditions of temperature and energy.

To live and reproduce, plants and animals need a continuous flow of energy. This is an obvious and commonly believed truism, but in fact organisms will also readily accept a discontinuous energy inflow, as life in a biosphere, driven by pulsed energy inputs that the periodic motions of the planet provide, demonstrates. The energy of the biosphere that originates in the discontinuous luminous energy of the sun, is captured by plants and passes from one living form to another along the food chain. This radiant pathway that provides us with great quantities of food, fibers, and energy, all of solar origin, has existed for over 4 billion years, a long time if we think that hominids appeared on the earth only 3 million years ago and that known history covers only 10,000 years. The ancestors of today's plants were the blue-green algae, or cyano bacteria, that began to practice photosynthesis, assuming a fundamental role in biological evolution.

All vegetation whether natural or cultivated, has been capturing solar energy for millennia, transforming it into food, fibers, materials and work, and providing the basis for the life of the biosphere. The vast majority of the energy received by the Earth's surface from the sun is dispersed: it is reflected, stored in the soil and water, used in the evaporation of water and so forth. Approximately 1 percent of the solar energy that falls on fertile land and water is fixed by photosynthesis by primary producers in the form of high-energy organic molecules: solar energy stored in chemical bonds available for later use. By biochemical processes (respiration) the plants transform this energy into other organic compounds and work.

The food chain considered in terms of energy flows has a logic of its own: the energy degrades progressively in the different phases of the chain (primary producers and secondary consumers including decomposers), giving back the elementary substances necessary to build again the molecules of living cells with the help of solar energy.

The organization of living beings in mature ecosystems slows the dispersal of energy fixed by plants to a minimum, using it completely for its complex mechanisms of regulation. This is made possible by large "reservoirs" of energy (biomass) and by the diversification of living species. The stability of natural ecosystems, however, means that the final energy yield is zero, except for a relatively small quantity of biomass that is buried underground to form fossils. Relatively small, true, but in absolute terms in some forms enough to power a modern civilization for centuries.

Photosynthesis counteracts entropic degradation insofar as it orders disordered matter: the plant takes up disordered material (low-energy molecules of water and carbon dioxide in disorderly agitation) and puts it in order using solar energy. It organizes the material by building it into complex structures. Photosynthesis is, therefore, the process that by capturing solar energy and decreasing the entropy of the planet paved the way for evolution. Photosynthesis is the green talisman of life, the bio-energetic equivalent of Maxwell's demon that decreases the entropy of the biosphere. On the Earth, living systems need a continuous or discontinuous flow of negative entropy (i.e. energy from outside) and this flow consists of the very solar energy captured by photosynthesis. This input of solar energy is what fuels the carbon cycle. The history of life on the Earth can be viewed as the history of chemotropic life, followed by the photosynthesis and the history of evolution, as the history of a singular planet that learned to capture solar energy and feed on the negative entropy of the universe for the creation of complex self-perpetuating structures (living organisms).

Compared to us, the sun is an enormous engine that produces energy and offers the Earth the possibility of receiving large quantities of negative entropy (organization, life), allowing a global balance that does not contradict the second law of thermodynamics. Every year, the sun sends the Earth 5.6×10^{24} J of energy, over 10,000 times more energy than humans consumes in a year.

2.3 PHYSICAL OPENNESS

An energy balance equation for ecosystems might be written as follows in accordance with the principle of energy conservation:

$$E_{cap} = Q_{evap} + Q_{resp} + \cdots + \Delta E_{bio} \qquad (2.1)$$

Here E_{cap} is external energy captured per unit of time. A part of the incoming energy, solar radiation being the main source for the ecosystems on earth, is captured and a part is reflected unused, determining the albedo of the globe. The more biological structure an ecosystem possesses the more of the incoming energy it is able to capture, i.e. the lower the albedo. The structure acts as an umbrella capturing the incoming solar radiation.

In ecosystem steady states, the formation of biological compounds (anabolism) is in approximate balance with their decomposition (catabolism). That is, in energy terms:

$$\Delta E_{bio} \approx 0 \quad \text{and} \quad E_{cap} \approx Q_{evap} + Q_{resp} + \cdots \qquad (2.2)$$

The energy captured can in principle be any form of energy (electromagnetic, chemical, kinetic, etc.), but for the ecosystems on earth the short-wave energy of solar radiation (electromagnetic energy) plays the major role. The energy captured per unit of time is, however, according to Equation 2.2 used to pay the maintenance cost per unit of time including evapotranspiration and respiration. The overall result of these processes requires that E_{cap} to be greater than 0, which entails openness (or at least non-isolation).

The following reaction chain summarizes the consequences of energy openness (Jørgensen et al., 1999): *source*: solar radiation → *anabolism* (charge phase): incorporation of high-quality energy, with entrained work capacity (and information), into complex bio-molecular structures, entailing antientropic system movement away from equilibrium → *catabolism* (discharge phase): deterioration of structure involving release of chemical bond energy and its degradation to lower states of usefulness for work (heat) → *sink*: dissipation of degraded (low work capacity and high entropy) energy as heat to the environment (and, from earth, to deep space), involving entropy generation and return toward thermodynamic equilibrium. This is how the energy cascade of the planet is usually described. Another way might be to express it in terms of gradient creation and destruction. The high-quality entering energy creates a gradient with baseline background energy. This enables work to be done in which the energy is degradiented and dissipated to space. On arrival there (at approximately 280 K) it locally re-gradients this new environment (at 3 K) but then rapidly disperses into the vacuum of the cosmos at large.

This same chain can also be expressed in terms of matter: *source*: geochemical substrates relatively close to thermodynamic equilibrium → *anabolism*: inorganic chemicals are molded into complex organic molecules (with low probability, it means that the equilibrium constant for the formation process is very low, low entropy, and high distance from thermodynamic equilibrium) → *catabolism*: synthesized organic matter is ultimately decomposed into simple inorganic molecules again; the distance from thermodynamic equilibrium decreases, and entropy increases → *cycling*: the inorganic molecules, returned to near-equilibrium states, become available in the nearly closed material ecosphere of earth for repetition of the matter charge–discharge cycle.

Input environments of ecosystems serve as sources of high-quality energy whose high contents of work and information and low entropy raise the organizational states of matter far from equilibrium. Output environments, in contrast, are sinks for energy and matter lower in work capacity, higher in entropy, and closer to equilibrium. This is one possibility. On the other hand, since output environments also contain equilibrium-avoiding entities (organisms), their energy quality on a local basis might be just as great as that of organisms in input environments. Since, output environments feedback to become portions of input environments living systems operating in the ecosphere, which is energetically non-isolated but materially nearly closed, must seek an adaptive balance between these two aspects of their environmental relations in order to sustain their continued existence. That is, the charge–discharge cycle of the planet wraps output environments around to input environments, which homogenizes gradients and forces gradient-building (anabolic) biological activity.

The expression high-quality energy is used above to indicate that energy can either be applied to do work or it is what is sometimes called "anergy", i.e. energy that cannot do work. The ability to do work can be expressed by:

Work = an extensive variables × a difference in intensive variables

For instance

$$\text{Work} = mg(h_1 - h_2) \qquad (2.3)$$

where m is the mass, g the gravity, h the height, and $(h_1 - h_2)$ the difference in height (see Table 2.1).

The concept exergy was introduced by Rant (1953) to express the work capacity of a system relative to its environment (see details presented in Wall, 1977; Szargut et al., 1988). It was particularly useful when the efficiencies of a power plant or the energy transfer should be expressed. We have therefore:

$$\text{Energy} = \text{exergy} + \text{anergy} \qquad (2.4)$$

$Q_{evap} + Q_{resp}$ in Equations 2.1 and 2.2 represents anergy because it is heat at the temperature of the environment. The temperature of the ecosystem would currently increase, if the ecosystem was not open at both ends, so to say. The heat is exported to the environment. The openness, or actually non-isolation, of ecosystems makes it possible for the systems to capture energy for photosynthesis but also to export the generated heat to maintain an acceptable temperature for the life processes.

Exergy as it is defined technologically cannot be used to express the work capacity of an ecosystem, because the reference (the environment) is the adjacent ecosystem. The Eco-exergy expresses, therefore, the work capacity of an ecosystem compared with the same system as a dead and completely homogeneous system without gradients. See Box 2.1 for definition and documentation of "eco-exergy."

Eco-exergy expresses the development of an ecosystem by its work capacity (see Box 2.1). We can measure the concentrations in the ecosystem, but the concentrations in the reference state (thermodynamic equilibrium; see Box 2.1) can be based on the usual use of chemical equilibrium constants. If we have the process:

$$\text{Component A} \leftrightarrow \text{inorganic decomposition products} \tag{2.6}$$

it has a chemical equilibrium constant, K:

$$K = [\text{inorganic decomposition products}]/[\text{component A}] \tag{2.7}$$

The concentration of component A at thermodynamic equilibrium is difficult to find (see the discussion in Chapter 6), but we can, based on the composition of A, find the concentration of component A at thermodynamic equilibrium from the probability of forming A from the inorganic components.

Box 2.1 Eco-exergy, definition

Eco-exergy was introduced in the 1970s (Jørgensen and Mejer, 1977, 1979; Mejer, 1979; Jørgensen, 1982) to express the development of ecosystems by increase of the work capacity. If we presume a reference environment that represents the system (ecosystem) at thermodynamic equilibrium, which means that all the components are inorganic at the highest possible oxidation state if sufficient oxygen is present (as much free energy as possible is utilized to do work) and homogeneously distributed at random in the system (no gradients), the situation illustrated in Figure 2.1 is valid. As the chemical energy embodied in the organic components and the biological structure contributes far most to the exergy content of the system, there seems to be no reason to assume a (minor) temperature and pressure difference between the system and the reference environment. Under these circumstances we can calculate the exergy content of the system as coming entirely from the chemical energy:

$$\sum (\mu_c - \mu_{co}) N_i \tag{2.5}$$

where μ_c and μ_{co} are the chemical potentials and N in the number of chemical compounds.

This represents the non-flow chemical exergy. It is determined by the difference in chemical potential ($\mu_c - \mu_{co}$) between the ecosystem and the same system at thermodynamic equilibrium. This difference is determined by the concentrations of the considered components in the system and in the reference state (thermodynamic equilibrium), as it is the case for all chemical processes.

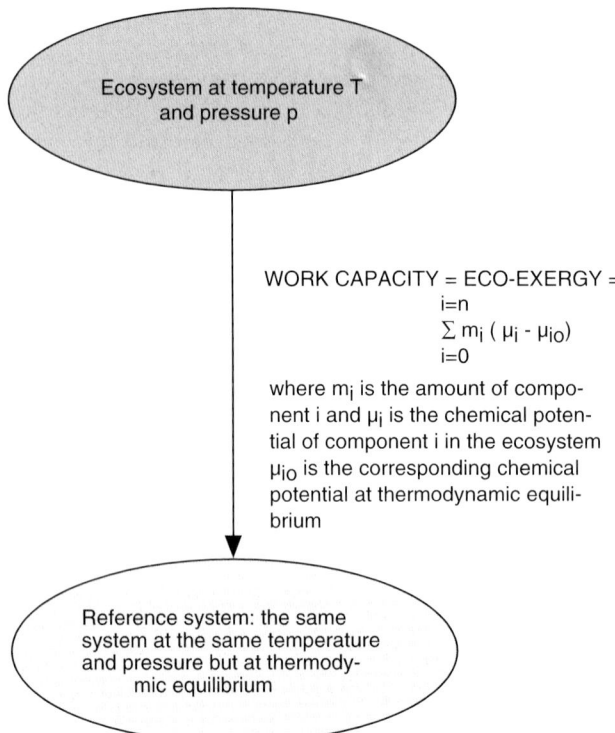

Figure 2.1 The exergy content of the system is calculated in the text for the system relative to a reference environment of the same system at the same temperature and pressure at thermodynamic equilibrium, it means as an inorganic soup with no life, biological structure, information, gradients, and organic molecules.

Eco-exergy is a function of the reference state which is different from ecosystem to ecosystem. Eco-exergy expresses, therefore, the work capacity relative to the same system but at thermodynamic equilibrium. Eco-exergy can furthermore, with the definition given, be applied far from thermodynamic equilibrium. It should be mentioned that eco-exergy cannot be measured, as the total internal energy content of a body or system cannot be measured. Even a small ecosystem contains many microorganisms and it is, therefore, hardly possible by determination of the weight of all components of an ecosystem to assess the eco-exergy of an ecosystem. The eco-exergy of a model of an ecosystem can, however, be calculated as it will be shown in Chapter 6.

We find by these calculations the exergy of the system compared with the same system at the same temperature and pressure but in form of an inorganic soup without any life, biological structure, information, or organic molecules. As $(\mu_c - \mu_{co})$ can be found from the definition of the chemical potential replacing activities by concentrations, we get the following expressions for the exergy:

$$Ex = RT \sum_{i=0}^{n} C_i \ln\left(\frac{C_i}{C_{i,o}}\right) \qquad (2.8)$$

where R is the gas constant (8.317 J/K moles = 0.08207 l·atm/K moles), T the temperature of the environment (and the system; see Figure 2.1), while C_i is the concentration of the ith component expressed in a suitable unit, e.g. for phytoplankton in a lake C_i could be expressed as mg/l or as mg/l of a focal nutrient. $C_{i,o}$ is the concentration of the ith component at thermodynamic equilibrium and n is the number of components. $C_{i,o}$ is of course a very small concentration (except for $i = 0$, which is considered to cover the inorganic compounds), it is therefore possible to use the probability ($p_{i,o}$) (see Chapter 6):

$$\frac{Ex}{V} = RT \sum_{i=0}^{n} p_i \ln\left(\frac{p_i}{p_{i,o}}\right)$$

By using this particular eco-exergy based on the same system at thermodynamic equilibrium as reference, the exergy becomes dependent only on the chemical potential of the numerous biochemical components that are characteristic for life. It is consistent with Boltzmann's statement, that life is a struggle for free energy, that is the work capacity in classic thermodynamics.

As observed above, the total eco-exergy of an ecosystem *cannot* be calculated exactly, as we cannot measure the concentrations of *all* the components or determine all possible contributions to eco-exergy in an ecosystem. Nor does it include the information of interactions. If we calculate the exergy of a fox for instance, the above shown calculations will only give the contributions coming from the biomass and the information embodied in the genes, but what is the contribution from the blood pressure, the sexual hormones, and so on? These properties are at least partially covered by the genes but is that the entire story? We can calculate the contributions from the dominant biological components in an ecosystem, for instance by the use of a model or measurements, that covers the most essential components for a focal problem. The *difference* in exergy by *comparison* of two different possible structures (species composition) is here decisive. Moreover, exergy computations always give only *relative* values, as the exergy is calculated relative to the reference system. These problems will be treated in further details in Chapter 6. For now it is important to realize that it is the metaphorical quality of the exergy concept, and not its measurability, that is most useful to ecologists. Entropy and exergy can both not be measured for ecosystems. It is not always necessary in science to be able make exact measurements. Ecologists rarely do this anyway. Approximations can yield an approximate science, and that is what ecology is. Modeling in particular approximates reality, not duplicates it, or reproduces it exactly because it is impossible due to the high complexity (see also next chapter). Approximate ecology—it can be quite useful and interesting ecology that can be used to quantify (approximately) for instance the influence of anthropogenic impacts on ecosystems. Often concepts and theories, not only measurements, make science interesting. With all the short-comings presented above, eco-exergy gives an approximate, relative measure of how far an ecosystem is from thermodynamic equilibrium and thereby how developed it is. Such assessment of important holistic ecosystem properties is important in systems ecology as well as in environmental management. This explains how eco-exergy has been applied several times successfully to explain ecological observations (see Jørgensen et al., 2002 and Chapter 8) and as indicator for ecosystem health (see Jørgensen et al., 2004 and Chapter 9).

2.4 THE SECOND LAW OF THERMODYNAMICS INTERPRETED FOR OPEN SYSTEMS

If ecosystems were isolated, no energy or matter could be exchanged across their boundaries. The systems would spontaneously degrade their initially contained exergy and increase their entropy, corresponding to a loss of order and organization, and increase in the randomness of their constituents and microstates. This dissipation process would cease at equilibrium, where no further motion or change would be possible. The physical manifestation would ultimately be a meltdown to the proverbial "inorganic soup" containing degradation products dispersed equiprobably throughout the entire volume of the system. All gradients of all kinds would be eliminated, and the system would be frozen in time in a stable, fixed configuration. The high-energy chemical compounds of biological systems, faced suddenly with isolation, would decompose spontaneously (but not necessarily instantaneously) to compounds with high-entropy contents. The process would be progressive to higher and higher entropy states, and would, in the presence of oxygen, end with a mixture of inorganic residues—carbon dioxide, water, nitrates, phosphates, and sulphates, etc. These simpler compounds could never be reconfigured into the complex molecules necessary to carry on life processes without the input of new low-entropy energy to be employed in biosynthesis. An isolated ecosystem could, therefore, in the best case sustain life for only a limited period of time, less than that required from the onset of isolation to reach thermodynamic equilibrium. Observations of properties could not be made, only inferred, because observation requires some kind of exchanges between the system and an observer. There would be no internal processes, because no gradients would exist to enable them. There would only be uninterrupted and uninterruptible stillness and sameness which would never change. The system would be completely static at thermodynamic equilibrium. Thus, in a peculiar way, isolated systems can only be pure abstractions in reality, submitting neither to time passage, change, nor actual observation. They are the first "black holes" of physics, and the antithesis of our systems plus their environments which are the core model for systems ecology. No ecosystem could ever exist and be known to us as an isolated system.

The second law of thermodynamics, though open to question, still retains its status as one of the most fundamental laws of nature. The law has been expressed in many ways. As indicated above: entropy will always increase and exergy will always decrease for an isolated system. Time has one direction. Tiezzi (2003b) concludes that entropy applied to far from thermodynamic equilibrium systems is not a state function since it has intrinsic evolutionary properties, strikingly at variance with classical thermodynamics. Work capacity is constantly lost as heat at the temperature of the environment that cannot do work. It implies that all processes are irreversible. The total reversibility of Newton's Universe (and even of the relativity theories) is no longer valid (Tiezzi, 2003a,b, 2005). The introduction of irreversibility has, however, opened for new emergent possibilities. Without irreversibility there would have been no evolution (Tiezzi, 2005), that is one of the most clear examples of a totally irreversible process. The directionality of ecosystems that will be discussed in Chapter 4, is also a result of the second law of thermodynamics. The second law of thermodynamics and the irreversibility of all processes have given the world new, rich, and beautiful possibilities that a reversible world not could offer.

That is the current dogma, at least, and it is probably true. However, it is useful to at least briefly consider the attributes of a reversible world. Time travel would be possible; this has been amply fantasized in literature. There would be no "evolution" in the sense we understand, but returning to former states could be seen as quite interesting and refreshing, especially if those states were more desirable, let us say further from equilibrium, than their current alternatives. Beauty and rich possibilities—what could be more enriching and beautiful than restoration of former systems, and lives, after wars or other privations, have driven them nearer to equilibrium. Reversibility could produce quite an interesting world, from many perspectives, replacing the humdrum grinding reality of movement toward equilibrium following exergy seeding.

The decrease in entropy or the increase in the eco-exergy in the biosphere depends on its capacity to capture energy from the sun and to retransmit it to space in the form of infrared radiation (positive entropy). If retransmission is prevented, in other words, if the planet were shrouded in an adiabatic membrane (greenhouse effect), all living processes would cease very quickly and the system would decay toward the equilibrium state, i.e. toward thermal death. A sink is just as necessary for life as a source to ensure the temperature that is required for carbon-based life.

Morowitz (1968) continues that all biological processes depend on the absorption of solar photons and the transfer of heat to the celestial sinks. The sun would not be an exergy source if there were not a sink for the flow of thermal energy. The surface of the Earth is at a constant total energy, re-emitting as much energy as it absorbs. The subtle difference is that it is not energy per se that makes life continue but the flow of energy through the system. The global ecological system or biosphere can be defined as the part of the Earth's surface that is ordered by the flow of energy by means of the process of photosynthesis.

The physical chemistry mechanism was elegantly described by Nobel Prize winner Albert Szent-György as the common knowledge that the ultimate source of all our energy and negative entropy is the sun. When a photon interacts with a particle of matter on our globe, it raises an electron or a pair of electrons to a higher energy level. This excited state usually has a brief life and the electron falls back to its basic level in 10^{-7}–10^{-8} s, giving up its energy in one way or another. Life has learned to capture the electron in the excited state, to uncouple it from its partner and to let it decay to its fundamental level through the biological machinery, using the extra energy for vital processes.

All biological processes, therefore, take place because they are utilizing an energy source. With exception of the chemotrophic systems at submarine vents, the ultimate energy source is the solar radiation. Morowitz (1968) notes that it is this tension between photosynthetic construction and thermal degradation that sustains the global operation of the biosphere and the great ecological cycles. This entropic behavior marks the difference between living systems and dead things.

2.5 DISSIPATIVE STRUCTURE

The change in entropy for an *open* system, dS_{system}, consists of an external, exogenous contribution from the environment, $deS = S_{in} - S_{out}$, and an internal, endogenous contribution due to system state, diS, which must always be positive by the second law of thermodynamics (Prigogine, 1980). Prigogine uses the concept of entropy and the second law of thermodynamics far from thermodynamic equilibrium, which is outside the framework of classical thermodynamics, but he uses the concepts only locally.

There are three possibilities for the entropy balance:

$$dS_{system}/dt = deS/dt + diS/dt > 0 \tag{2.9}$$

$$dS_{system}/dt = deS/dt + diS/dt < 0 \tag{2.10}$$

$$dS_{system}/dt = deS/dt + diS/dt = 0 \tag{2.11}$$

The system loses order in the first case. Gaining order (case 2), is *only* possible if $-deS > diS > 0$. Creation of order in a system must be associated with a greater flux of entropy out of the system than into the system. This implies that the system must be open or at least non-isolated.

Case 3, Equation 2.11, corresponds to a stationary situation, for which Ebeling et al. (1990) used the following two equations for the energy (U) balance and the entropy (S) balance:

$$dU/dt = 0 \quad \text{or} \quad deU/dt = -diU/dt = 0 \tag{2.12}$$

and

$$dS_{system}/dt = 0 \quad \text{or} \quad deS/dt = -diS/dt = 0 \tag{2.13}$$

Usually the thermodynamic processes are isothermal and isobaric. This implies that we can interpret the third case (Equations 2.11–2.13) by use of the free energy:

$$deG/dt = T\,diS/dt > 0 \tag{2.14}$$

It means that a "status quo" situation for an ecosystem requires input of free energy or exergy to compensate for the loss of free energy and corresponding formation of heat due to maintenance processes, i.e. respiration and evapotranspiration. If the system is not receiving a sufficient amount of free energy, the entropy will increase. If the entropy of the system will continue to increase, thus, the system will approach thermodynamic equilibrium—the system will die; see Section 2.2. This is in accordance with Ostwald (1931): life without the input of free energy is not possible.

An average energy flow of approximately 10^{17} W by solar radiation ensures the maintenance of life on earth. The surface temperature of the sun is 5800 K and of the earth on average approximately 280 K. This implies that the following export of entropy per unit of time takes place from the earth to the open space:

$$10^{17} \text{ W}(1/5800 \text{ K} - 1/280 \text{ K}) \approx 4 \times 10^{14} \text{ W}/\text{K} \tag{2.15}$$

corresponding to 1 W/m² K.

Prigogine uses the term *dissipative structure* to denote self-organizing systems, thereby indicating that such systems dissipate energy (produce entropy) for the maintenance of their organization (order). The following conclusions are appropriate:

All living systems, because they are subject to the second law of thermodynamics, are inherently dissipative structures. The anabolism combats and compensates for the catabolic deterioration of structure; the two processes operate against one another. Note that the equilibrium "attractor" represents a resting or refractory state, one that is passively devolved to if system openness or non-isolation are compromised (Jørgensen et al., 1999). The term is also commonly used to express the situation when a system is actively pushed or "forced" toward a *steady state.* Though widespread, we do not subscribe to this usage and make a distinction between steady states and equilibria for two reasons:

(1) The state-space system theory we outlined in the conservation chapter of *Ecosystems Emerging* (Patten et al., 1997) precludes anything in system dynamics but a unique input–state–output relationship. Therefore, given an initial state, state-space theory asserts that there exists one and only one sequence of inputs that will put an open system in a given state at a specified final time. For this terminal state to be an "attractor", many input sequences would have to be able to place the system in it, and from many initial states—the attractor would be hard to avoid. This is inconsistent with dynamical state theory.

(2) As observed above, a steady state is a forced (non-zero input) condition; there is nothing "attractive" about it. Without a proper forcing function it will never be reached or maintained. A steady state that is constant may appear equilibrial, but it is really far from equilibrium and maintained by a steady input of energy or matter. We regard equilibrium as a zero-input or resting condition. What are often recognized as local attractors in mathematical models really have no counterparts in nature. Steady states are forced conditions, not to be confused with unforced equilibria which represent states to which systems settle when they are devoid of inputs. The only true natural attractor in reality, and it is global, is the unforced thermodynamic equilibrium.

As an ecosystem is non-isolated, the entropy changes during a time interval, dt can be decomposed into the entropy flux due to exchanges with the environment, and the entropy production due to the irreversible processes inside the system such as diffusion, heat conduction, and chemical reactions. This can also be expressed by use of exergy:

$$Ex/dt = deEx/dt + diEx/dt \qquad (2.16)$$

where $deEx/dt$ represents the exergy input to the system and $diEx/dt$ is the exergy consumed (is negative) by the system for maintenance, etc. Equation 2.16—an exergy version of Equations 2.9 and 2.10—shows among other things that systems can only maintain a non-equilibrium steady state by compensating the internal exergy consumption with a positive exergy influx ($deEx/dt > 0$). Such an influx induces order into the system. In ecosystems the ultimate exergy influx comes from solar radiation, and the order induced is, e.g. biochemical molecular order. If $deEx > -diEx$ (the exergy consumption in the system), the system has surplus exergy input, which may be utilized to construct further order in the system, or as Prigogine (1980) calls it: dissipative structure. The system will thereby move further away from thermodynamic equilibrium. Evolution shows that this situation has been valid for the ecosphere on a long-term basis. In spring and summer ecosystems are in the typical situation that $deEx$ exceeds $-diEx$. If $deEx < -diEx$, the system cannot maintain the order already achieved, but will move closer to the thermodynamic equilibrium, i.e. it will lose order. This may be the situation for ecosystems during fall and winter or due to environmental disturbances.

2.6 QUANTIFICATION OF OPENNESS AND ALLOMETRIC PRINCIPLES

All process rates are in physics described as proportional to a gradient, a conductivity or inverse resistance and to the openness, compare for instance with Fick's laws of diffusion and Ohm's law. The import and export from and to an ecosystem is, therefore, dependent on the differences between the ecosystem and the environment, as well as of openness. For instance, the rate of the reaeration process of a water stream can be expressed by the following equation:

$$R_a = V\, dC/dt = K_a(T)A(C_s - C) \qquad (2.17)$$

or

$$dC/dt = K_a(T)(C_s - C)/d \qquad (2.18)$$

where R_a is the rate of reaeration, K_a a temperature constant for a given stream, A the area $= V/d$, V the volume, d the depth, C_s the oxygen concentration at saturation, and C the actual oxygen concentration. K_a is here the "conductivity" or inverse resistance. The faster the water flow in the stream, the higher is K_a. ($C_s - C$) is the gradient and A, the area, is the openness. Numerous expressions for rates in nature follow approximately the same linear equation.

The surface area of the species is a fundamental property. The surface area indicates quantitatively the size of the boundary to the environment. Flow rates are often formulated in physics and chemistry as area times a gradient, which can be utilized to set up useful relationships between size and rate coefficients in ecology. Loss of heat to the environment must for instance be proportional to the surface area and to the temperature difference, according to the law of heat transfer. The rate of digestion, the lungs, hunting ground, etc. are, on the one hand, determinants for a number of parameters (representing the properties of the species), and on the other hand, they are all dependent on the size of the organism. It is, therefore, not surprising that many rate parameters for plants and animals are highly related to the size, which implies that it is possible to get very good first estimates for most parameters based only on the size. Naturally, the parameters are also dependent on several characteristic features of the species, but their influence is often minor compared with the size, and good estimates are valuable in many ecological models, at least as a starting value in the calibration phase. It is possible, however, to take these variations into account by the use of a form factor = surface/volume. This form factor may vary considerably among species.

The conclusion of these considerations must, therefore, be that there should be many parameters that might be related to simple properties, such as size of the organisms, and that such relationships are based on fundamental biochemistry and thermodynamics (Figures 2.2–2.6).

Above all there is a strong positive correlation between size and generation time, T_g, ranging from bacteria to the biggest mammals and trees (Bonner, 1965). This relationship can be explained by use of the relationship between size (surface) and total metabolic action per unit of body weight mentioned above. It implies that the smaller the organism

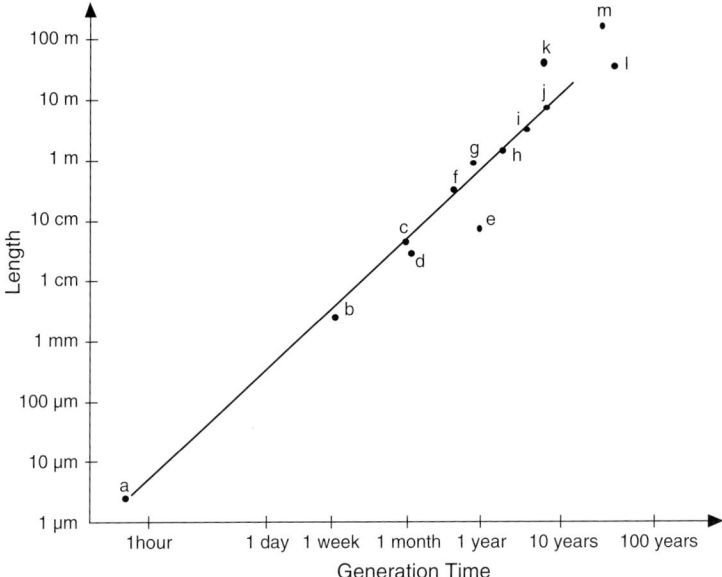

Figure 2.2 Length and generation time plotted on log–log scale: (a) *Pseudomonas*, (b) *Daphnia*, (c) bee, (d) housefly, (e) snail, (f) mouse, (g) rat, (h) fox, (i) elk, (j) rhino, (k) whale, (l) birch, and (m) fir (Peters, 1983). Reproduced from Jørgensen, 2000a.

the greater the metabolic activity. The per capita rate of increase, r, defined by the exponential or logistic growth equations is again inversely proportional to the generation time:

$$dN/dt = rN \qquad (2.19)$$

$$dN/dt = rN(1 - N/K) \qquad (2.20)$$

where N is the population size, r the intrinsic rate of growth, and K the environmental carrying capacity. This implies that r is related to the size of the organism, but, as shown by Fenchel (1974), actually falls into three groups: unicellular, heterotherms, and homeotherms (see Figure 2.3).

The same allometric principles are expressed in the following equations, giving the respiration, food consumption, and ammonia excretion for fish when the weight, W, is known:

$$\text{Respiration} = \text{constant} \times W^{0.80} \qquad (2.21)$$

$$\text{Food consumption} = \text{constant} \times W^{0.65} \qquad (2.22)$$

$$\text{Ammonia excretion} = \text{constant} \times W^{0.72} \qquad (2.23)$$

It is also expressed in the general equation (Odum, 1959, p. 56):

$$m = kW^{-1/3} \qquad (2.24)$$

where k is roughly a constant for all species, equal to approximately 5.6 kJ/g$^{2/3}$ day, and m the metabolic rate per unit weight W.

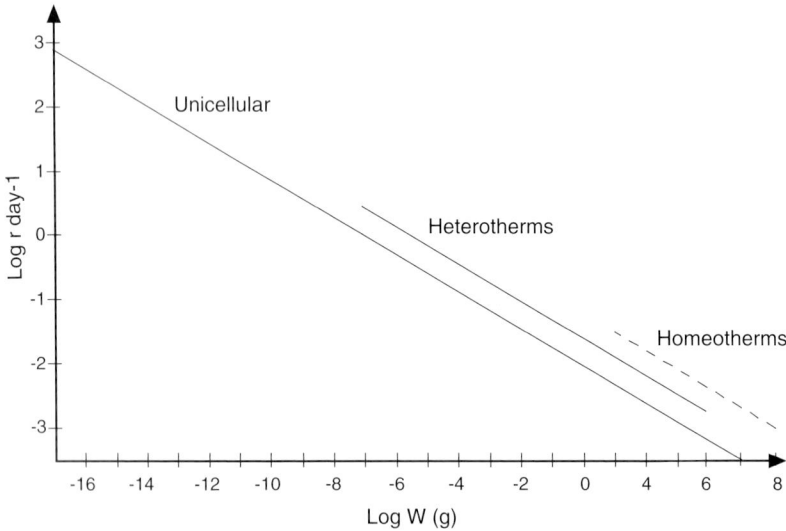

Figure 2.3 Intrinsic rate of natural increase against weight for various animals. After Fenchel (1974). *Source:* Fundamentals of Ecological Modelling by Jørgensen and Bendoricchio.

Chapter 2: Ecosystems have Openness

Similar relationships exist for other animals. The constants in these equations might be slightly different due to differences in shape, but the equations are otherwise the same. All these examples illustrate the fundamental relationship in organisms between size (surface) and biochemical activity. The surface determines the contact with the environment quantitatively, and by that the possibility of taking up food and excreting waste substances.

The same relationships are shown in Figures 2.4–2.6, where biochemical processes involving toxic substances are applied as illustrations. The excretion rate and uptake rate

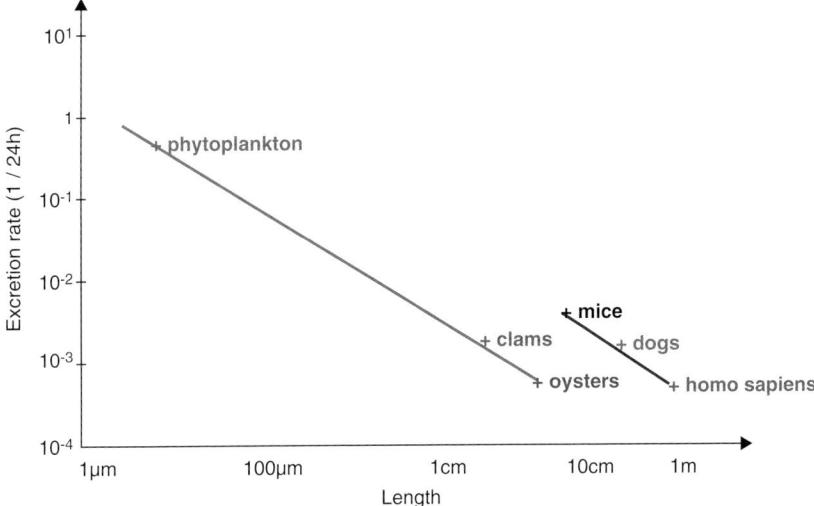

Figure 2.4 Excretion of Cd $(24\,h)^{-1}$ plotted against the length of various animals: (1) *Homo sapiens*, (2) mice, (3) dogs, (4) oysters, (5) clams, and (6) phytoplankton (Jørgensen 1984).

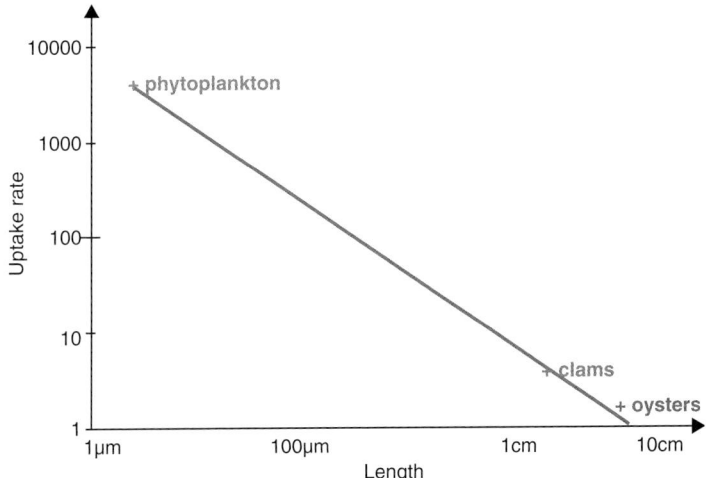

Figure 2.5 Uptake rate ($\mu g/g\ (24\,h)^{-1}$) plotted against the length of various animals (Cd): (1) phytoplankton, (2) clams, (3) oysters. After Jørgensen (1984). *Source:* Fundamentals of Ecological Modelling by Jørgensen and Bendoricchio.

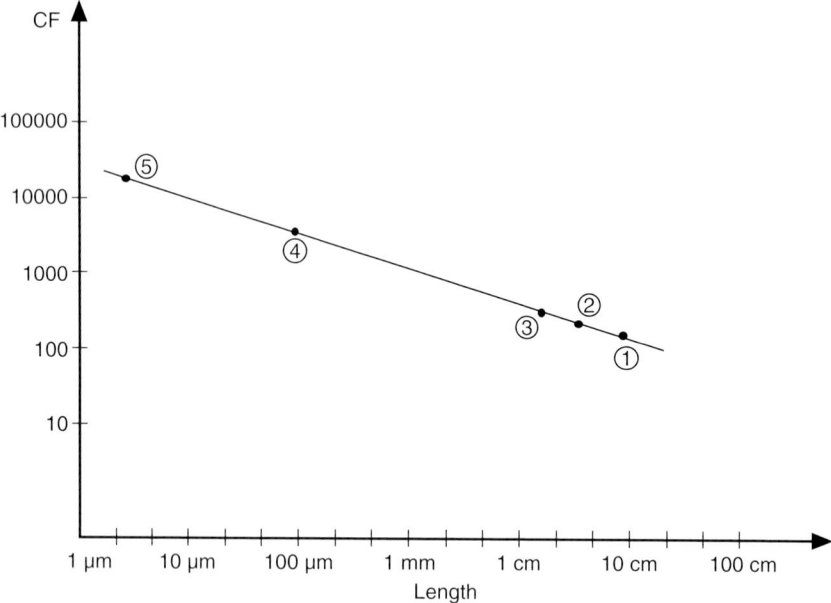

Figure 2.6 Biological concentration factor (BCF) denoted CF for Cd versus length: (1) goldfish, (2) mussels, (3) shrimps, (4) zooplankton, (5) algae (brown-green). After Jørgensen (1984). *Source:* Fundamentals of Ecological Modelling by Jørgensen and Bendoricchio.

(for aquatic organisms) follow the same trends as the metabolic rate. This is of course not surprising, as excretion is strongly dependent on metabolism and the direct uptake dependent on the surface.

These considerations are based on allometric principles (see Peters, 1983; Straškraba et al., 1999), which with other words can be used to assess the relationship between the size of the units in the various hierarchical levels and the process rates, determining the need for the rate of energy supply. All levels in the entire hierarchy of an ecosystem are, therefore, due to the hierarchical organization, characterized by a rate which is ultimately constrained by their size.

Openness is proportional to the area available for exchange of energy and matter, relative to the volume = the inverse space scale (L^{-1}). It may also be expressed as the supply rate = $k \cdot$ gradient $\cdot$ area relative to the rate of needs, which is proportional to the volume or mass. An ecosystem must, as previously mentioned, be open or at least non-isolated to be able to import the energy needed for its maintenance. Table 2.3 illustrates the relationship between hierarchical level, openness, and the four-scale hierarchical properties presented in Simon (1973). The openness is here expressed as the ratio of area to volume.

For the higher levels in the hierarchy approximate values are used. As we move upwards in the hierarchy, the exchange of energy (and matter) becomes increasingly more difficult due to a decreasing openness. It becomes increasingly more difficult to cover needs, which explains why energy density, time scale, and dynamics decrease according to the inverse space scale or openness, or expressed differently as the rates are adjusted

Box 2.2 Basic elements of hierarchy theory

Many of the allometric characteristics described in Section 2.6 are based on correlations between body size and other biological or ecological features of the organisms. These interrelationships are frequently comprehended as basic components of ecological hierarchies and basic objects of scaling procedures. Thus, they are highly correlated to hierarchy theory.

Following Simon (1973), hierarchy is a heuristic supposition to better understand complex systems, and following Nielsen and Müller (2000) hierarchical approaches are prerequisites for the definition of emergent properties in self-organized systems. Hierarchy theory (Allen and Starr, 1982, O'Neill et al., 1986) or the holarchy principle (Kay, 1984) represents an integrative concept of ecosystem-based classification and conception, which is compatible with most of the existing approaches to ecological system analysis. The theory has been developed by Simon (1973), Allen and Starr (1982), and O'Neill et al. (1986) and recently there have been several applications in ecosystem analysis and landscape ecology.

The fundamental unit of hierarchy theory is the holon, a self-regulating open (sub)system (see Figure 2.6). Holons function as autonomous entities and are also components of superior organizational units. They incorporate all inferior subsystems and are parts of higher level systems themselves. Thus, on a specific level of resolution, a biological system consists of interacting entities and is itself a component of a higher organizational unit. Hierarchies are partly ordered sets, in which the subsystems are interacting through asymmetric relationships. These interactions produce an integral activity of the whole, where the variations of the whole complex are significantly smaller than the sums of the variations of the parts. In contrast, the degrees of freedom of single processes are limited by the higher hierarchical level. Controlling functions (constraints) determine the basis for systems organization: microscopic reactions are coordinated at the macroscopic level. O'Neill et al. (1986) defined the interacting constraints of a specific level of an ecosystem as its environmental limits, while the dynamics of lower levels, which generate the behavior of the higher level, are defined as the biotic potential of the system.

The distinction of hierarchical levels has to be determined by the observer as does the definition of the investigated system. Criteria of the levels' differentiation are:

(a) The spatial extent of higher levels is broader than the extent of lower levels. Thus, distinguishing levels is connected with distinguishing spatial scales.
(b) Higher levels change more slowly than lower levels. Significant changes require longer periods on higher levels.
(c) Higher levels control lower levels. Under steady-state conditions they assert the physical, chemical, and biological limits the system of interest can operate within.

(continued)

(d) Higher levels can contain lower levels (nested hierarchies). Accordingly, the spatial and temporal constants of system behavior are important criteria of differentiation. Scale is defined as a holon's spatial and temporal period of integrating, smoothing, and dampening signals before they are converted into messages (Allen and Starr, 1982).
(e) Signals (including fluxes of energy and matter) are filtered in hierarchies. The way a holon converts or ignores signals defines its functional environment and its scale.

All of these assumptions refer to steady-state conditions. The hierarchy of an ecosystem thus continuously develops and its complexity rises during phases of orientor optimization (see Chapters 6, 7, and 9). Whenever phase transitions appear, the hierarchy is broken and the system is enabled to adapt to the changing constraints by forming a new structure.

Table 2.3 Relationship between hierarchical level, openness (area/volume ratio), and approximate values of the Simon's (1973) four scale-hierarchical properties: energy/volume, space scale, time scale, and behavioral frequency

Hierarchical level	Openness[1,3] (A/V, m^{-1})	Energy[2] (kJ/m^3)	Space scale[1] (m)	Time scale[1] (s)	Dynamics[3] (g/m^3 s)
Molecules	10^9	10^9	10^{-9}	$<10^{-3}$	10^4–10^6
Cells	10^5	10^5	10^{-5}	10–10^3	1–10^2
Organs	10^2	10^2	10^{-2}	10^4–10^6	10^{-3}–0.1
Organisms	1	1	1	10^6–10^8	10^{-5}–10^{-3}
Populations	10^{-2}	10^{-2}	10^2	10^8–10^{10}	10^{-7}–10^{-5}
Ecosystems	10^{-4}	10^{-4}	10^4	10^{10}–10^{12}	10^{-9}–10^{-7}

[1]Openness, spatial scale, and time scale are inverse to hierarchical scale.
[2]Energy and matter exchange at each level depend on openness, measured as available exchange area relative to volume. Electromagnetic energy as solar photons comes in small packages (quanta, hv, where h is Planck's constant and v is frequency), which makes only utilization at the molecular level possible. However, cross-scale interactive coupling makes energy usable at all hierarchical levels.
[3]Openness correlates with (and determines) the behavioral frequencies of hierarchical levels.

to make the possible supply of energy sufficient (Figure 2.7). These considerations are consistent with the relationship between size and time scale of levels in the hierarchy, as presented by O'Neill et al. (1986) and Shugart and West (1981).

Exchange of matter and information with the environment of open systems is in principle not absolutely necessary, thermodynamically, as energy input (non-isolation) is sufficient (the system is non-isolated) to ensure maintenance far from equilibrium. However, it often gives the ecosystem some additional advantages, for instance by input of chemical compounds needed for certain biological processes or by immigration of species offering new possibilities for a better ordered structure of the system. All ecosystems are open to exchange of energy, matter, and information with their environment.

Chapter 2: Ecosystems have Openness　　　　　　　　　　　　　　　　　　　　　29

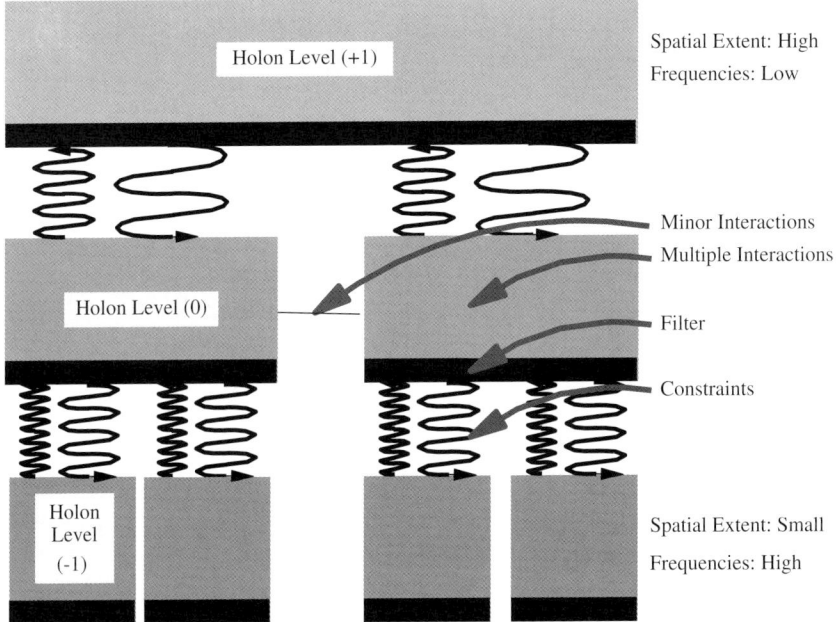

Figure 2.7 A schematic representation of interacting hierarchical levels.

The importance of the openness to matter and information is clearly illustrated in the general relationship between number of species, SD (species diversity), of ecosystems on islands and the area of the islands, A:

$$\text{SD} = C \times A^z \text{ (number)} \tag{2.25}$$

where C and z are constants. The perimeter relative to the area of an island determines how "open" the island is to immigration or dissipative emigration from or to other islands or the adjacent continent. The unit (L^{-1}) is the same as the above used area to volume ratio as a measure of openness.

Different species have very different types of energy use to maintain their biomass. For example, the blue whale uses most (97%) of the energy available for increasing the biomass for growth and only 3% for reproduction. Whales are what we call K-strategists, defined as species having a stable habitat with a very small ratio between generation time and the length of time the habitat remains favorable. It means that they will evolve toward maintaining their population at its equilibrium level, close to the carrying capacity. K-strategists are in contrast to r-strategists which are strongly influenced by any environmental factor. Due to their high growth rate they can, however, utilize suddenly emergent favorable conditions and increase the population rapidly. Many fishes, insects, and other invertebrates are r-strategists. The adult female reproduces more and the proportion going into reproduction can be over 50%.

2.7 THE CELL

The cell is the basic biological unit, as the elementary particles and the elements are the basic units of chemistry. In spite of the enormous variation in the structure and function of different organisms, the fundamental unit, the cell, is with some variations basically the same. Why is the cellular structure the same? First of all, early in evolution the cell demonstrated its functionality. But the use of structural units of small size has also ensured effective transportation by diffusion. Most cells have a diameter between 1 and $20\,\mu m$ (Table 2.4). Cells have, therefore, a relatively high openness (see Table 2.3), that is necessary for the biochemistry of organisms to work. The hierarchical structure, which was presented in Box 2.2 and Figure 2.7 and will be further discussed in Chapters 3 and 7, is a precondition for the needed openness for each level in the hierarchy.

Let us, however, demonstrate the importance of openness by focusing on the cell. The problem is for the cells to have an openness that would match the need for diffusive transportation for the matter needed for the biochemical syntheses that take place in the cells, first of all for the synthesis of proteins.

Protein synthesis takes place in about ten steps from primary gene expression in DNA inside the nucleus to final production of the mature protein at its final destination outside the nucleus but within the plasma membrane. First there is transcription in which the DNA region encoding the gene is transcribed into a complementary messenger RNA (mRNA). Next, in eucaryotes, initial pre-mRNA is spliced and processed to mature mRNA. This is exported across the nuclear envelope into the cytosol. There, codons in ribosomes progressively translate the genetic code into a mature cytosolic protein. This is followed by several steps of sorting and modification involving cytoplasmic ultrastructures such as the endoplasmic reticulum and Golgi apparatus. All the genes of an organism make up its genome. Of these,

Table 2.4 Some differences between prokaryotic and eucaryotic cells

	Prokaryotes	Eucaryotes
Size	$1–10\,\mu m$	$10–100\,\mu m$
Nucleus	None. The chromosomal region is called nucleolus	Nucleus separated from cytoplasm by nuclear envelope
Intracellular organization	Normally, no membrane-separated compartments and no supportive intracellular framework	Distinct compartments, e.g. nucleus, cytosol with cytoskeleton, mitochondria, endoplasmic reticulum, Golgi complex, lysosomes, plastids
Gene structure	No introns, some polycistronic genes	Introns and exons
Cell division	Simple	Mitosis or meiosis
Ribosome	Large 50S subunit and small 30S subunit	Large 60S subunit and small 40S subunit
Reproduction	Parasexual recombination	Sexual recombination
Organization	Mostly single-cellular	Mostly multicellular, with cell differentiation

Source: After Klipp et al. (2005).

only certain ones will be expressed at a given time or for a specific cell type. Some genes which perform basic functions are always required; these are constitutive or housekeeping genes. Others are expressed only under certain conditions (Klipp et al., 2005, pp. 45–47).

Openness in the scenario just given is particularly pronounced at the nuclear and cutoplasmic boundaries, but in fact is expressed all along the way as intracellular structures receive, process, and pass along the various intermediary products in protein synthesis. Is the openness sufficient to ensure uptake of oxygen and nutrients needed for protein synthesis? Matter needed for the biochemistry is proportional to the volume (we presume that the cell is a sphere where d is cell diameter):

$$\text{Volume} = \pi d^3 / 6 \qquad (2.26)$$

The transport from the surface to the cell takes place by a fast active transport and the concentration at the surface is, therefore, 0. The area of the sphere is πd^2. The flux of matter toward the cell is considered constant, which implies that the concentration gradient will decrease with the distance from the cell in the exponent 2:

$$dC/dr = kr^{-2} \qquad (2.27)$$

where r is the distance from the center of the cell (radius). The concentration is 0 at the surface of the cell, i.e. $r = d/2$. The concentration at the distance r from the center of the cell C_r can be found after differentiation of Equation 2.27 to be:

$$C_r = C(1 - d/2r) \qquad (2.28)$$

The diffusion rate, corresponding to the uptake rate is a diffusion coefficient $(D) \times$ the concentration gradient $(dC/dr = Cd/2r^2$ or at the surface $= 2C/d) \times$ the openness = area = πd^2, or therefore $2\pi dDC$, where D is the diffusion coefficient and C the concentration in the environment. The uptake rate relative to the need, denoted UR/N, is found as:

$$UR/N = 12DC/(fd^2) \qquad (2.29)$$

where f is the need per unit of time and volume. The relative uptake rate will be four times smaller, if the diameter is doubled. Relatively small cell sizes are necessary to obtain a sufficient relative uptake rate. This equation demonstrates the importance of the cell size and explains, therefore, indirectly the hierarchical structure, because small cells are the prerequisite for a sufficient supply of nutrients, although there are many additional explanations.

2.8 WHAT ABOUT THE ENVIRONMENT?

Openness is a requisite for moving substance across boundaries, and boundaries imply an inside–outside dichotomy. That is, in departure from thermodynamic equilibrium energy and matter move from outside to inside and dissipation signifies movement in the reverse direction, from interiors to exteriors.

The term "environment" has appeared 44 times previously in this chapter, in a book on ecology, the biological science of environment, and yet we have not once anywhere done anything explicitly with this concept except take it for granted as a reference source and sink from which some older more or less accepted thermodynamics, without its modern challenges, proceeds to operate in the organization of ecosystems. We use the concept of environment, but have not attempted to define it scientifically or explore it in any deep way. There is little in theoretical ecology that elaborates it in substantive scientific terms. It is just a convenient category of "surroundings" that openness requires—some place to derive inputs and exhaust outputs.

Particle scales aside, it is relevant in the context of openness to ask the hard question—"What is environment?" We look around the room or outside the window and see what everyone agrees is "environment." Seeing is only part of it, however; there is also touching and smelling, etc. In other words there are sensory stimuli involved. What about these? Our household pets and the plants in the garden that began this chapter have considerably different sensory apparatus from us. Does that mean environment is relative, something that can only be defined by perception? Or are certain aspects of it accessed differently by different open systems? It is clear from the perspective of reality as a collection of physics' particles, and from mass–energy conservation, that what comes to me at a given moment as visual, auditory, tactile, etc. stimuli cannot also come to you or your dog or plant. At this level it has to be acknowledged that there is a certain uniqueness that attaches to the "environments" of particular open-system receivers of sense-data. Not only is this true for sensory stimuli, but also is true for the masses of matter that enter our bodies as food, and exit as biodegradable products useful as food for other organisms. So environment, it would seem, courses in and out of open systems, and the ultimate particle uniqueness of the substance and signals both seem to confer a central place on the *open system* as the focal arbiter of environment. Afferent input environments coming from the past are originated the moment a unit of high-quality energy or matter crosses the boundary of a receiving open system. This increases the exergy and lowers the entropy of the receiving system. Reciprocally, efferent environments that unfold with the future are founded the moment a unit of energy or matter exits the said open system. This dualistic concept of environment is operationalized in the mathematical theory of *environs* (Patten, 1978, 1982), about which now numerous papers have been published describing the properties of such structures. The dualism is central, and so is the unity of the focal entity–environ triad. Environments and the things with which they are associated can never be separated in environ theory—they are a unit of nature, however intractable, but sometimes with surprising holistic properties.

2.9 CONCLUSION

Openness of systems to exchange of energy and matter implies certain foundational properties in natural organization:

- It is a necessary but not sufficient property for ecosystems.
- It presumes boundaries.

- Boundaries imply bounded entities—something kept in.
- Entities imply an elemental discreteness in natural organization; see Table 2.3.
- Therefore, human sensory apparatus operates correctly in perceiving an essentially particulate world.

Or does it? ... since it is also true that:

- Boundaries imply environments—something kept out.
- Environments are unbounded, thus inherently non-discrete.
- Therefore, human perception is mismatched to environmental reality.

This may be why we know more about disconnected objects than we do about object–environment relationships, and why science is more reductive than holistic. It appears that the property of openness, the subject of this chapter, returns us to the same kind of enigma as the wave–particle duality, giving ecology a deep challenge for its future to unravel.

3

Ecosystems have ontic openness

"…next to music and art, science is the greatest, most beautiful and most enlightening achievement of the human spirit"
(Popper, 1990)

3.1 INTRODUCTION

This chapter's title may mean little to many persons, yet the essence may be understood fairly easily on an intuitive basis. The adjective "ontic", which hardly appears in any dictionary, clearly relates to the term ontology, which is used in philosophy in its widest sense to designate "the way we view the world and how it is composed". Ontic bears the slight difference that it refers to intrinsic properties of the world as we construct it and its behavior, such that it addresses phenomenology as well. Therefore, this chapter complements the concepts of thermodynamic openness addressed in the previous chapter, by including the physical openness available to ecosystem development.

In fact, everybody knows something about openness. We know how it is to be open to another person's opinions, to be open minded, or open to new experiences. We enjoy that surprising things may happen on our (field) trips and journeys (in nature). In fact, any person who has tried to plan exact details for a tour into the wilderness will know how difficult this is. First, we may address the aspect of realizing such a trip and stress that this also implies the acceptance of the fact that unexpected things may or rather will occur. But, second, we have also to address the fact that once an event occurs, it is an outcome of many unexpected events. It is impossible to predict which one and how often such events actually occur. We may expect to bring extra dry socks to use after one incident, an unexpected event. How many persons will be able to foresee exactly how many pairs to bring? Or in other cases we may return with unused socks but found that we needed extra shirts instead. Any of us will know that it is eventually not possible to make such a detailed plan.

In fact, one could have chosen another title to the present chapter: "anything may—but does not—happen". Of which the first part deals with, as we shall see in the following sections, the enormous number of possibilities that exist in general and also in biological systems. The second part indicates that all possibilities have not been realized, partly because it is not physically possible, and partly due to constraints that are described in other chapters of this book.

This chapter is about the ontic openness of ecosystems. It relates directly to the theme of this book and the systemness of ecosystems because ontic openness results, in part, due to the complex web of life constantly combining, interacting, and rearranging, in the natural world to form novel patterns. Furthermore, ontic openness is at least a partial cause of indeterminacy and uncertainty in ecology and thus the reason that we are not able to make exact predictions or measurements with such a high accuracy as for instance in physical experiments. Therefore, when understanding ecosystems from a systems perspective, one cannot overlook the importance of physical openness.

3.2 WHY IS ONTIC OPENNESS SO OBSCURE?

While referring to Section 3.2 of the chapter we have already mentioned that it likely will pose a question to the vast majority of readers, not only the ecologically oriented ones, of: what is the meaning of the title of this chapter? We have tried to foresee this question already by giving a first vague and intuitive explanation. We guess it is likely that only a few readers have met this "phenomenon" before as far as the term ontic openness is concerned. We also expect that very few, if any, of the readers are familiar with texts that deal with the role of ontic openness in an ecological context.

To our knowledge, no such thorough treatment of this topic exists. Rather a number of treatments of more or less philosophical character exist—all of which may be taken into account—and which all together may add up to a composite understanding of what ontic openness may mean and what its importance and consequences to ecological science may be.

Should we attempt to further explain ontic openness very briefly (which is impossible) we would start with openness, and turn the attention to another related word like open-minded. We normally use this word to designate a person that is willing to try out new things, accept novel ideas, maybe a visionary person who is able to think that the world could be different, that matters may be interdependent in other ways than in which we normally think. Many scientists make their breakthrough thanks to such mental openness. Discoveries are often unexpected or unplanned—a phenomenon known in the philosophy of science as *serendipities*. Kuhn also addresses this issue of the scientific procedure when he stresses that paradigm shifts in the evolution of science involves the scientists to come and look at the same object from a different angle or in a different manner.

We now would like, if possible, to remove the psychology element. If we remove the role of *subjectivity*, i.e., that openness relies on one or more person's ability or willingness to see that the surrounding world may be different or could have other possibilities realized than hitherto, then we are really on the right track.

We are now left with an *objective* part of openness. If we can now accept the physical existence of this and that it is a property that penetrates everything, we are getting there. The openness is an objectively existing feature not only of the world surrounding us but also ourselves and our physical lives (e.g., biochemical individuality introduced by Williams, 1998). This is the ontic part of the openness.

Another reason for ontic openness to be not so commonly known among biologist and ecologist is the fact that the progenitors of this concept were dominantly physicists and in particular those in the hard-core areas of quantum mechanics, particle physics, and relativity theory. Furthermore, we typically do not view these areas as being directly relevant to biology or ecology. Also, these theories are not easy to communicate to "outsiders", so even if ecology is considered to be a highly trans-, inter-, and multi-disciplinary science it is perfectly understandable that no one has thought that these hard-core sub-disciplines of physics today could possibly have a message for ecology.

Luckily, one might say, some of the physicists from these areas turned their attention in other directions and started speculating about the consequences of their findings to other areas of natural science such as biology. On several occasions we have found physicists wondering about the distinction between the physical systems and living systems, such as Schrödinger's *What is life*. Living systems are composed of basically the same units, atoms and molecules, and yet they are so different. One physicist, Walter Elsasser, will receive an extra attention in this chapter. Studying his works, in particular from the later part of his productive career, may turn out to be a gold mine of revelations to any person interested in how biology differs from physics and about life itself.

Still not understood or got the idea of what ontic openness is about? Do not worry—you most probably have experienced it and its consequences already. Let us investigate some well-known examples.

Most ecologists have experienced ontic openness already!
Most ecologists will have met ontic openness already—somewhat in disguise—as often our background comes from the gathering of empirical knowledge, an experience we may have achieved through hard fieldwork.

To start, let us consider a hypothetical "test ecologist". Given the information about latitude and a rough characteristic ecosystem type—terrestrial or aquatic—she will be able to decide whether she is expert "enough" in the area to forecast the system state or if she prefers to enlist aid from a person considered to be more knowledgeable in the area. If deciding to be an "expert", then she will for sure be able to tell at least something about the basic properties of the ecosystem, such as a rough estimate of the number and type of species to be expected. Given more details, such as exact geographical position, we may now narrow in on ideas considering our background knowledge. There will be a huge difference in organisms, species composition, production, if we are in the arctic or in the tropics. Likewise, being for instance in the tropics there will be a huge difference between a coral reef in the Pacific Ocean or a mangrove swamp in the Rufiji River Delta. We will be able to begin to form images of the ecosystem in our minds, conceptual models of trophic interactions, community linkages, and functional behavior. Meanwhile, we know very well that to get closer in details with our description we will need additional knowledge, for instance about ecological drivers, such as hydrodynamics, depth, and other external influences, such as human impacts from fisheries, loadings of both organic or inorganic in type, etc.

Nevertheless, given as much information as we possibly can get, and for instance focusing in on a particular geographic position, such as the Mondego River Estuary in Portugal, we will not be able to answer accurately simple questions like: which plant species are present at a certain locality, how are they distributed, or what are their biomass and production? We will more likely be able to give an answer something like that under the given conditions we would consider it to be most likely that some rooted macrophyte will be present and that it would probably be of a type that do not break easily, probably with band-shaped leaves, probably some species of Zostera, etc. We will be able, based on experience and knowledge, to give only an estimate in terms of—what we shall later call the propensity—the system to be of a certain "kind". BUT we will never be completely sure. This is due to ontic openness.

Examples from the world of music
Sometimes, when introducing new concepts, it is useful to make an entrance from an unexpected and totally different angle. In this case, we will consider the world of music—a world with which most people are familiar and have specific preferences. We only know very few people to whom music does not say anything and literally does not "ring a bell".

We consider—in a *Gedanken Experiment*—the situation of an artist set to begin a new composition. To illustrate the universality of the approach we may illustrate the situation by the possible choices in two situations—a small etude for piano or a whole symphony. We shall start by looking at both the situations from a statistical and probabilistic angle. The two situations may look quite different from a macroscopic point of view, but in fact they are not.

In the case of a short piece for piano, a normal house piano has a span of approximately 7 (or 7¼) octaves of 12 notes each giving 84 (or 88) keys in all. If an average chord on the piano has 5 notes in it, then it is theoretically possible to construct 3,704,641,920 or approximately 3.7 billion chords on it (4.7 billion in the case of 88 keys). (Note, that we already here deal with a subset of the $84! = 3.3 \times 10^{126}$ possibilities.) Meanwhile, if the assumption that a chord consists of five notes on average is valid, then it does not take long to reach almost the same level of complexity *sensu lato*. Putting a small piece of music together, assuming that we work in a simple 4/4 and change chords for each quarter, after 16 notes or 4 bars we have reached a level 126×10^{153} of possible ways to construct the music. Many of these possible combinations of notes and chords would not sound as music at all and luckily we are faced with constraints. A physical constraint, such as the human physiology, will serve to limit the number of notes than can be accessed in a single chord (a good piano player will be able to span maybe over one octave per hand, thereby lowering the number of possible variations considerably). Psychological constraints of various kinds do also exist depending on the decisions of the composer or our personal taste—we do want the music to sound "nice".

The situation does not change a lot considering a symphony orchestra although complexity really rises much faster. Considering a relatively small symphony orchestra of say 50 musicians—each having a span of approximately 3 octaves or (36 notes)—even before starting we have 36^{50} or 6.5×10^{77} possibilities of how the first chord may sound. By the second note we have already exceeded any of the above numbers.

Almost no physical constraints exist in this case. The task of the composer is very simple, picking a style of music like the choices between classic or 12-tone music, between piano concerto, opera, or string quartets. The point is now that for each note, for each chord, there are many possibilities of what the composer could write on the sheet, but in fact only

Chapter 3: Ecosystems have ontic openness 39

one ends up being chosen, one "solution" out of an enormous number of possibilities. As we shall see later, the number of possibilities to choose from is so large (*immense*) that it makes no physical sense. Therefore, in the end the choice of the composer is *unique*. The fact that we will anyway be able to determine and talk about such a thing like style is that the composers have had a tendency (see *propensities* later) to choose certain combinations out of the possible.

Let us end this section with a situation most people will know. Considering yourself a skilled person, familiar with the many styles of music, you listen to an unknown piece of music in a radio broadcast. It is a very melodic piece of music in a kind of style you really like and with which you are familiar. You, even without knowing the music, start to hum along with some success, but eventually you will not succeed to be totally right throughout the whole piece. Do not worry it is not you that is wrong, neither is the music—you are just experiencing the ontic openness of someone else, in this case the composer.

3.3 ONTIC OPENNESS AND THE PHYSICAL WORLD

As mentioned above, a number of treatments of this topic exist that all add up to our possible understanding of the importance of ontic openness and what it means in context of our everyday life. Putting them together and taking the statements to a level where we really see them as ontological features, i.e., as ontic, we will be, on one hand forced to reconsider what we are doing, on the other hand, we can look upon the world, and in particular the uncertainties, the emergent properties that we meet, in a much more relaxed manner.

Unfortunately, to ecology and the ecologists, as previously mentioned, the statements that have already been made on openness almost all originate from physicists. In fact, seen from a philosophy of science point of view, this means that the statements are often dominated by arguments deeply rooted in reductionist science, often literally close to an atomistic view. Interesting things happen when the arguments are taken out of the reductionist realm to other levels of hierarchy, i.e., the arguments are taken out of their physical context and extended to biology and eventually—following our purpose of the present book—into ecology.

The basic contributions we think of here may be represented by a number of scientists. A sketch of a few essential ideas that it may be possible to relate to the issue of ontic openness as well as the originators is given in Table 3.1.

In the following sections, we will take a more detailed look at a few of these perspectives. From the table it is evident that we deal with quite recent contributions and some noteworthy overlaps in time. It would, of course, be interesting to know if and how these persons have influenced each other, a thing which may become clear only from close, intensive studies of the time development of their works and biographies. Meanwhile, this would be a tedious task and the possible mutual influence has not been considered in this paper.

It is not possible to measure everything

In the world of physics, the importance of uncertainty and our interference with systems through experiments has been recognized for less than a century. The introduction of concepts such as complementarity and irreversibility has offered solutions to many problems

Table 3.1 A non-exhaustive list of various authors who have addressed the issue of ontic openness of natural, physical, and biological systems

Originator	Era	Idea	Remarks
N. Bohr	1885–1962	Complementarity—the idea that more descriptions are needed	Derived from the wave-particle duality
E. Schrödinger	1887–1961	Order from disorder and order from order	Relates to Elsasser's immense numbers and historical aspects
W. Heisenberg	1901–1976	The principle of uncertainty or indeterminacy, e.g., the simultaneous determination of position and momentum of an electron is not possible	Argued to be valid also for ecosystems by Jørgensen
K.R. Popper	1902–1994	(a) End of fixed probabilities—we need to work with propensities; (b) the open universe	Basic assumption behind Ulanowicz' concept, Ascendency
W.M. Elsasser	1904–1991	Biological systems are heterogeneous and therefore possess immense possibilities which are coped with by agency and history	The combinatorial explosions shaping this phase-space occurs at almost any level of hierarchy
I.A. Prigogine	1917–2003	The understanding of biological systems as dissipative structures and far from equilibrium systems	Assumes that the "Onsager relation" may be extended to the conditions of life (Chapter 6)
C.S. Holling	1930–	The idea that evolution happens through breakdowns that opens up new possibilities through an ordered/cycling process	See creative destruction, Chapter 7; similar to H.T. Odum pulsing paradigm
S.E. Jørgensen	1934–	The Heisenberg uncertainty principle extended to ecosystem measurements	
S.A. Kauffman	1939–	The continuous evolution of biological systems towards the edge of chaos	

Note: At first, the ideas may appear disparate, but in fact all illustrate the necessity to view systems as ontically open.

Chapter 3: Ecosystems have ontic openness

but has simultaneously involved the recognition of limits to the Newtonian paradigm. Below, we deal with some important findings in physics from the 20th century such as the Heisenberg uncertainty principle, the Compton effects, and the relaxation of systems that may have future parallels in ecology.

The Heisenberg principle
The Heisenberg uncertainty relation tells that we cannot know exactly both the position and the velocity of an atom at the same time. At the instant when position is determined, the electron undergoes a discontinuous change in momentum. This change is greater the smaller the wavelength of the light employed. Thus, the more precise the position is determined, the less precise the momentum is known, and vice versa (see Box 3.1).

Box 3.1 The Heisenberg uncertainty principle or principle of indeterminacy

The basic proof shows that the product of position and momentum will always be larger than Planck's constant. This is given explicitly by the following mathematical terms:

$$\Delta s \times \Delta p \geq \frac{1}{2}\hbar = \frac{h}{4\pi}$$

Where, s refers to space, p the momentum, and h the Planck's constant (6.626×10^{-34} J s).

The Compton effect
The Compton effect deals with the change in wavelength of light when scattered by electrons. According to the elementary laws of the *Compton effect*, p_1 and λ_1 stand in the relation:

$$p_1 \times \Delta \lambda_1 \cong h \tag{3.1}$$

$$\Delta E_1 \times \Delta T_1 \cong h \tag{3.2}$$

where p_1 is the momentum of the electron, $\Delta \lambda_1$ the wavelength increase due to the collision, E_1 the energy, and T_1 the time.

Equation 3.1 corresponds to Equation 3.2 and shows how a precise determination of energy can only be obtained at the cost of a corresponding uncertainty in the time (see Box 3.2).

Spin relaxation
Spin relaxation is possible because the spin system is coupled to the thermal motions of the "lattice", be it gas, liquid, or solid. The fundamental point is that the lattice is at thermal equilibrium; this means that the probabilities of spontaneous spin transitions up and down are not equal, as they were for rf-induced transitions (see Box 3.3).

Box 3.2 The Compton effect and directionality

From the uncertainty relation between position and momentum, another relation may be derived. Let v and E be the velocity and energy corresponding to momentum p_x, then:

$$v \Delta p_x \frac{\Delta x}{v} \geq h$$

$$\Delta E \times \Delta t \geq h$$

Where ΔE is the uncertainty of energy corresponding to the uncertainty of momentum Δp_x and Δt the uncertainty in time within which the particle (or the wave packet) passes over a fixed point on the x-axis (Fong, 1962). Thus, irreversibility of time is not taken into account since in the quantum mechanics paradigm time is assumed to be reversible.

We want to point out that if we take as an axiom the irreversibility of time it is an error to calculate the limit:

$$\lim_{\Delta t \to 0} \frac{\Delta s}{\Delta t}$$

because this means that:

$$\forall \varepsilon > 0, \exists \delta > 0 : |\Delta t| < \delta \Rightarrow \left| \frac{\Delta s}{\Delta t} \right| < \varepsilon$$

where:

$$|\Delta t| < \delta \Leftrightarrow |t_1 - t_0| < \delta \Leftrightarrow -\delta < t_1 - t_0 < \delta \Leftrightarrow t_0 - \delta < t_1 < t_0 + \delta$$

Simply speaking it is not possible to think t_1 as approximating t_0 from right, in fact, the state $S(t_0)$ that the functions S reaches when t_1 becomes t_0 from right cannot be the same state $S(t_0)$ that the function assumes as t_1 reaches t_0 from left.

It is well known that if the left and right limits of a function are not identical then the limit does not exist. Hence, we must redefine the time derivative of a function as the left limit, if it exists

$$\lim_{\Delta t \to 0} \frac{\Delta s}{\Delta t}$$

This translates in practice to the statement that in the Cartesian graph it is impossible to cover the t-axis in both sense from left to right and right to left, but in the first manner only.

Box 3.3 Relaxation of systems

Denoting the upward and downward relaxation probabilities by $W_{\alpha\beta}$ and $W_{\beta\alpha}$ (with $W_{\alpha\beta} \neq W_{\beta\alpha}$), the rate of change of N_α is given by:

$$\frac{dN_\alpha}{dt} = N_\beta W_{\beta\alpha} - N_\alpha W_{\alpha\beta}$$

At thermal equilibrium $dN_\alpha/dt = 0$, and denoting the equilibrium population by $N_{0\alpha}$ and $N_{0\beta}$ we see that:

$$\frac{N_{0\beta}}{N_{0\alpha}} = \frac{W_{\alpha\beta}}{W_{\beta\alpha}}$$

The populations follow from Boltzmann's law and so the ratio of the two transition probabilities must also be equal to $\exp(-\Delta E/kT)$. Expressing N_α and N_β in terms of N and n ($n = N_\alpha - N_\beta$) we obtain:

$$\frac{dn}{dt} = -n(W_{\beta\alpha} + W_{\alpha\beta}) + N(W_{\beta\alpha} - W_{\alpha\beta})$$

This may be rewritten as:

$$\frac{dn}{dt} = -\frac{(n - n_0)}{T_1}$$

in which n_0, the population difference at thermal equilibrium, is equal to:

$$n_0 = N\left[\frac{W_{\beta\alpha} - W_{\alpha\beta}}{W_{\beta\alpha} + W_{\alpha\beta}}\right]$$

and $1/T_1$ is expressed by:

$$\frac{1}{T_1} = W_{\alpha\beta} + W_{\beta\alpha}$$

T_1 thus has the dimensions of time and is called the "spin-lattice relaxation time". It is a measure of the time taken for energy to be transferred to other degrees of freedom, i.e., for the spin system to approach thermal equilibrium: Large values of T_1 (minutes or even hours for some nuclei) indicate very slow relaxation (Carrington and McLachlan: Introduction to magnetic resonance).

It is now possible to say something about the width and shape of the resonance absorption line, which certainly cannot be represented by a Dirac δ function.

(*continued*)

> First, it is clear that, because of the spin relaxation, the spin states have a finite lifetime. The resulting line broadening can be estimated from the uncertainty relation:
>
> $$\Delta v \, \Delta t \approx 1$$
>
> and thus we find that the line width due to spin-lattice relaxation will be of the order of $1/T_1$.

Given the remarks made at the start of this section, one may indeed start to wonder and speculate about the relations of these physical systems that obey universal laws when involved at the level of chemistry and biology and how or if these affect living systems at all. This is exactly what the physicist Walter M. Elsasser did and it may be worthwhile to spend a few moments studying his work and conclusions.

What really differs between physics and biology: four principles of Elsasser

The one contributor from Table 3.1 that literally takes the step from physics into biology was Walter M. Elsasser who's "roaming" life is quite impressive. The details of his life are described in a biography[1] by Rubin (1995), who was acquainted with Elsasser in the last 10 years of his life. Most of the information on Elsasser's below is based on this biography and Elsasser's own autobiography (Elsasser, 1978). From these works, one can almost sense that Elsasser's contributions were sparked by ontic openness on his own "body and soul" throughout his career. Rubin (1995) summarized Elsasser's (1987) four basic principles of organisms: (A) ordered heterogeneity, (B) creative selection, (C) holistic memory, and (D) operative symbolism. The first principle is the key reference to ontic openness, while the other points address how this order arises in this "messy" world of immense numbers. In other words, the latter three seem more to be *ad hoc* inventions necessary to elaborate and explain the first.

Background

According to Rubin, Theophile Khan influenced Elsasser's understanding of the overwhelming complexity dominating biological systems as compared with the relative simplicity of physics. Probably, he was also influenced by Wigner from whom he is likely to have picked up group or set theory.

These studies, together with periodical influence from von Neumann, caused him to realize a fundamental difference between physical systems on one side and living systems on the other. Due to his early life education in atomic physics, he considered physical systems as homogenous sets—all atoms and molecules of a kind basically possess the same properties and behavior. At this level, and always near to equilibrium conditions, the world is deterministic and reversible processes dominate.

---

[1] This excellent biography is available on the Internet in several forms. Philosophy of Science students will be provided with a deep insight in how production of a scientist may not necessarily depend on skill or education, but may rather be determined by political and sociological regimens throughout his life.

As opposed to this view, he considered living systems to differ in this fundamental aspect of the homogenous sets. Living systems, he argued, are highly heterogeneous and far more complex than physical systems. Their behavior as opposed to physical systems is non-deterministic and irreversible. This is what we today would designate as far from equilibrium systems or dissipative structures.

The views of Elsasser are at this point derived from studies and knowledge about biological systems at cellular and sub-cellular level, i.e., the boarder between the "dead" physico-chemical systems and the living systems. The "distinction" falls somewhere between the pure chemical oscillations, like in the Beluzov–Zhabotinsky reaction and the establishing of biochemical cycles (autocatalytic cycles or hypercycles of Eigen and Schuster) together with chirality and the coupling to asymmetries introduced by separation of elements and processes by membranes. Part of the living systems indeterminacy is caused by an intrinsic and fundamental (ontic) property of the systems—(ontic) openness.

Ordered heterogeneity
Around the late 1960s, Elsasser directed his attention to the question of what possibly could have happened since the beginning of the universe, i.e., since the Big Bang—the thinking is much along the same line as Jørgensen formulated some decades later where Heisenberg's uncertainty relation is transferred[2] to ecosystems (see later this Section).

Elsasser's starting point was to calculate, roughly at least, how many quantum-level events could have taken place since the Big Bang. Since events at quantum level happens within one billionth of a second he calculates a number to be in order of 10^{25}. Then considering that the number of particles in the form of simple protons that may have been involved in these events to be approximately 10^{85} he calculates the number of possible events to be 10^{110}. Any number beyond this "simply loses its meaning with respect to physical reality" (Ulanowicz, 2006a). Elsasser puts a limit at around 10^{100} (a number known as Googol). Any number beyond this is referred to as an *immense number*. In Elsasser's terminology an immense number is a number whose logarithm itself is large. We claim that such numbers make no sense. And yet, as we saw with the examples from music, any simple everyday event, such as a piece of music, breaks this limit of physical events easily—almost before it is started.

But where does the relevance to ecosystems come in one may ask? Good question—and for once—a very simple answer. The point is that any ecosystem easily goes to a level of complexity where the number of possible events that may occur reaches or exceeds immense numbers. Again, Ulanowicz points out that "One doesn't need Avogadro's number of particles (10^{23}) to produce combinations in excess of 10^{110}, a system with merely 80 or so distinguishable components will suffice" (Ulanowicz, 2006a) as 80! is on the order of 7×10^{118}.

Now, as the vast majority of ecosystems, if not all, exceed this number of components it means that far more possibilities could have been realized, so that out of the phase space of possibilities on a few combinations have been realized. Any state that has occurred is also likely to occur only once—and is picked out of super-astronomical number of

---

[2]This transfer would in the context of philosophy of science be designated as a theoretical reduction—indeed with large epistemic consequences. This is opposed to Elsasser's approach that we here consider within the normal paradigm of physics.

possibilities. The other side of the story, as the title indicates, is that we are also left with a large number of possibilities that have never been and are never going to be realized. In other words, almost all events we may observe around us are literally *unique*. There are simple, repeatable events in nature within the domain of classical probability, but they are sets of a measure zero in comparison with *unique* events.

Meanwhile, we cannot foretell the possibilities of the next upcoming events. If we consider any particular situation, we face a world of unpredictability—a world that is totally ontic open. In fact, taken together, the above means that we should forget about making predictions about ecosystem development or even trying to do this. Luckily, as we shall see later, Karl Popper (1990) advocated a "milder" version of ontic openness.

Whereas up till now we have dealt with heterogeneity at the level of probabilities the following points from Elsasser try to explain how nature copes with this situation.

Creative selection
This point addresses the problems that arise from the immense heterogeneity. How do living systems "decide" among the extraordinary large number of possibilities that exist? Elsasser was precisely aware that living systems were non-deterministic, non-mechanist systems, as opposed to the physical systems that are always identical. As Rubin (1995) states, they "repeat themselves over and over again… but each organism is unique".

Elsasser gives *agency* to the organisms, although judging from this point alone it is not very easy to see where or how the "creativity" arises. Therefore, this point cannot be viewed as isolated from the two additional points below. Selection mechanisms are not ignored in this view that just stresses the intrinsic causes of evolution.

Holistic memory
With memory Elsasser addresses part of what is missing from agency. Again, according to Rubin, the criterion for living system to choose is information stability. Some memory system has to be introduced, as the living systems have to ensure the stability. This point, in addition to agency, also involves history and the ability to convey this history, i.e., heredity to living, organic systems. Although again a part misses on how this information is physically going to be stored, preserved, and conveyed.

Operative symbolism
Lastly, symbolism provides the mechanism for storing this information by introducing DNA as "material carrier of this information". This cannot be seen as isolated from the history of science in the area of genetics. Much of the Elsasser's philosophical work has been written when the material structure and organization of our hereditary material, the chromosomes, was revealed.

The above arguments could be taken as if Elsasser was still basically a true reductionist as we have now got everything reduced into "simple" mechanisms for the conveyance of history. Elsasser was indeed aware of this point and saw the process in a dualistic (not to say dialectic) manner as he stated this mechanism to be holistic in the sense that it had to "involve the entire cell or organism" (see Section 3.6).

Ecology and Heisenberg

According to Jørgensen (1995) "some of the principles of quantum mechanics are (silently and slowly) introduced in ecology" during the last 15 years (this was probably written significantly earlier than 1995!). This is stated to be valid in particular to the area of modeling with the following remarks: "An ecosystem is too complex to allow us to make the number of observations needed to set up a very detailed model—even if we still consider models with a complexity far from that of nature. The number of components (state variables) in an ecosystem is enormous". Taking this argument there is a clear correlation to the ontic openness of Elsasser, for instance through the presentation by Ulanowicz quoted above. Again, the number of components in an ecosystem alone is enough to form a system that is ontic open.

To the empiricist, this means that we have to use our limited resources in time and in particular money in the best possible manner. Who wants to spend unnecessary efforts? Who does not want to be as economically efficient as possible given that research money is always a limiting constraint? Meanwhile, the calculations made by Jørgensen imply a theorem of intrinsic empirical incompleteness. The argument goes as follows (see Box 3.4).

According to Jørgensen the Heisenberg uncertainty principle may now be reformulated, so that it refers to two other measures: uncertainty in time and energy (note the product of the two is consistent with Planck's constant, namely energy times time). The analogous formula reduces to:

$$\Delta t \times \Delta E \geq \frac{1}{2}\hbar \quad (3.3)$$

After all, in the end, the amount of empirical work we can do is dependent on the energy available (not only our own energy) and the time used per measurement.

First, we may now calculate the cumulative amount of energy received by the Earth since its "creation" and the number of measurements that could hypothetically have been made since this "creation". If we consider the amount of energy we could have spent in measuring to be equivalent to the amount of energy received for the past 4.5 billion years, and using 1.731×10^{17} J.s^{-1} as the value for incoming radiation, this gives a total value of

$$\Delta E = \text{No. of years} \times \text{No. of days} \times \text{No. of hours} \times \text{No. of seconds} \times \text{energy s}^{-1}$$
$$(= 4.5 \times 10^9 \times 365.3 \times 24 \times 3600 \times 1.7310 \times 17) \quad (3.4)$$
$$= 2.5 \times 10^{34} \text{ J}$$

Inserting the value of Planck's constant and solving Equation 3.3 we may—again hypothetically—calculate the time necessary for every measurement which will now be

$$\Delta t = \frac{h/4\pi}{\Delta E} = \frac{(6.626 \times 10^{-34})/4\pi}{2.5 \times 10^{34}} = 10^{-67} \text{ s} \quad (3.5)$$

> **Box 3.4** Sampling uncertainties
>
> Given that the amount resources that can be spent on examining an ecosystem is limited to a finite amount of measurement. For this calculation, a limit is set to 10^8, an arbitrarily chosen number, which on one hand seems to be very high in terms of field work, but may be rather realistic when processes such as data logging is involved.
>
> Considering number of dependent variables in the system (n) we need at least the in order to determine the full "phase space" we need make at least m, measurements, where
>
> $$m = 3^{n-1} \qquad (2)$$
>
> This assumes that our knowledge about a given system is so little determined that we have no "a priori" knowledge about the interrelations in the ecosystem, i.e., the physical flows or the regulatory feedbacks in the system. Therefore, we have to assume the worst case—that everything is literally linked to everything. In this case Jørgensen calculates that with the limits of 10^8 number of measurements we can only deal with a system with fewer than 18 components (as $3^{18} = 387.420.489$).
>
> Assuming that our sample is taken from a statistical population with a normal distribution and the standard deviation (σ) of the sample mean ($\bar{x}$) is given by:
>
> $$SD = \frac{\sigma}{\sqrt{\text{No. of samples}}} \qquad (3)$$
>
> Equation (3.3) may be re-organized into
>
> $$\frac{\sigma}{SD \times \sqrt{\text{No. of samples}}} = 1 \qquad (4)$$

Thus, we could possibly make a measurement or sample in 10^{-67} of a second.

If we could have exercised this practice ever since the creation of the Earth, we could have made 4.7×10^{84} measurements.

Returning to Equation 3.2 this means that we will have standard deviation (SD) (accuracy) of

$$SD = \frac{10^{-17}}{\sqrt{4.7 \times 10^{84}}} \approx 10^{-59} \qquad (3.6)$$

or in referring to Equation 3.1 we may never succeed in measuring systems with more than $n = 237$!

To make an intermediate summary there are many ways to express ontic openness. At the same time it has consequences to many relevant aspects of ecology such as the time we use for empirical work as well as the expectations we may have to issues such as accuracy and predictability.

3.4 ONTIC OPENNESS AND RELATIVE STABILITY

After introducing Elsasser's immense numbers and applying Heisenberg's principle to ecology, we end up with a rather pessimistic message to ecologists. In order not to fall totally in despair let us turn to Popper. Although seemingly agreeing mostly with Elsasser, he does present a modified interpretation of the classical probability concept that at the same time offers us a somewhat more optimistic view of what can be done.

Popper, although also a physicist, is best known for his philosophy of science work and the problem of the logics connected to the epistemic of carrying out research like "Logik der Forschung", etc. He is considered to be the father of the research strategy known as falsification.

Popper (1990—reprinted from his lectures in 1930s), in a minor publication: "A world of Propensities", states that he established a common research agenda with Carnap based on "Logik der Forschung". In this agenda, they "agreed to distinguish sharply between, on one hand, probability as it is used in the probabilistic hypotheses of physics, especially of quantum theory, which satisfies the mathematical 'calculus of probabilities', and, on the other hand, the so-called probability of hypotheses, or their degree of confirmation" (Popper, 1990, p. 5, see also Ulanowicz, 1996).

In fact, Popper himself, by addressing our failure to prove anything with a 100 percent certainty, i.e., the total dominance of uncertainty and the higher likelihood of falsification of experiments rather than the opposite, is addressing an openness that is part of the everyday life of all scientists. But again, if this is a property inherent in the systems we work with, then indeed we will be forced to return to the pessimist view presented above. If a true, real feature of the world, then why do science at all? Popper refers to the findings of Heisenberg as "objective indeterminacy", but argues against the solution of translating everything into probabilistic terms. Popper claims that most scientists picking up the probabilities turned it into a question of "lack of knowledge" (the information as entropy approach that is strongly connected to Shannon and von Neumann—our comment) leading to what he calls a subjectivist theory of probability (Popper, 1990, p. 8).

After working with probability theory for more than 35 years, he claims to have come up with "satisfactory and very simple solutions". One of which he refers to as 'the propensity interpretation of probability', a concept that originated back in 1956. Ulanowicz later used this interpretational framework in the development of his Ascendency concept (Box 4.1) that has been proposed to be an indicator of ecosystem development (Ulanowicz, 1986a, 1997).

In his explanation of the propensity interpretation, Popper began with an example of tossing a coin or throwing dice, in which we deal with known equiprobable outcomes—probability of 1/2 or 1/6 of any of the possible outcomes, respectively. Most of us will be familiar with these examples and consider them rather trivial, but what happens in the case when either the coin or the die is manipulated, i.e., loaded.

First of all, it is clear that in this case our assumption of equiprobable outcomes ends. One may introduce a very simple solution to this situation and just continue to work with the new weighted possibilities. We could hope that it would be as simple as that. But the consequence of such a situation on our work is much greater than we may imagine.

At least two major problems originate from the character of the situation: (1) how are the weights determined? (2) What is the consequence to our ability to forecast such a system? In determining the weights, a feasible method may easily be found. We may just continue "normal" coin tossing or dice throwing a considerable number of times, registering the outcome of each event. The point is now that this procedure will eventually take more time (more tosses or throws) in order to reach a reliable result and yet the determined weights will still be connected to a relatively high uncertainty. Popper stated, "instead of speaking of the possibility of an event occurring, we might speak, more precisely, of an inherent propensity to produce, upon repetition, a certain statistical average" (Popper, 1990, p. 11). Each event will happen with a more or less certain probability, a tendency—or as we now know it—a propensity. The immediate effect will be that our chances of successfully predicting a number of sequences will be very small.

We may now consider that the evolving world around us is a composite of events that all have non-fixed probabilities. Assigning weights is further complicated if the weights are not fixed, but rather varying, say on the external conditions in which the event is cast. In fact, adaptation is an inherent property of biological systems, thus, we must consider that the propensities themselves may change with time. This should lead to the understanding that propensities are entailed in the situation not the object. Our ability to predict, or our hopes to do this, will vanish within a short time, just as our abilities to predict the development of music is disappearing after just a few bars of playing as described earlier.

3.5 THE MACROSCOPIC OPENNESS: CONNECTIONS TO THERMODYNAMICS

Although there is possibly a connection between thermodynamic (Chapter 2) and ontic openness, the relation between the two is definitely non-trivial and attempts to distinguish the two will therefore not be included here.

An energy flow can lead to organization (decrease in entropy, e.g., photosynthesis) or destruction (increase in entropy, e.g., a cannon ball, respiration). The same quantity of energy can destroy a wall or kill a man; obviously the loss of information and negentropy is much greater in the second case. Energy and information are never equivalent as demonstrated for instance through Brillouin's refusal of Maxwell's Demon.

The classical example of the mixing of gases in an isolated system shows us that there can be an increase in entropy without energy input from outside. The point is that energy (E) and entropy (S) are both state functions in classical thermodynamics, but energy is intrinsically reversible whereas entropy is not. Entropy has the broken time symmetry (Blum, 1951). In other words, entropy has an energy term plus a time term that energy does not have. Herein lies the physical connection to the concept of exergy dealt with in Chapters 2 and 6.

Energy and mass are conservative quantities, thus it follows that total energy and mass cannot change with time. They may transform to other types of energy and mass but the overall quantities remain the same that is they are reversible. Entropy has an intrinsic temporal parameter. Energy obeys spatial and material constraints; entropy obeys spatial, material, and temporal constraints.

If history and the succession of events are of scientific relevance, the concept of a state function should be revised at a higher level of complexity. The singularity of an event also becomes of particular importance: if a certain quantity of energy is spent to kill a caterpillar, at the same time we lose the information embodied in the caterpillar. But were this the last caterpillar, we could lose its unique genetic information forever. The last caterpillar is different from the nth caterpillar.

The entropy paradox
Stories take place in a setting, the details of which are not irrelevant to the story. What happens in the biosphere, the story of life, depends on the biosphere constraints. Hence it is important to have global models of the biosphere in terms of space, time, matter, energy, entropy, information, and their respective relations.

If we consider the evolutionary transition from anaerobic to aerobic living systems, then the ratio of energy to stored information is clearly different. The information that led to evolution and the organization of the two types of system is not proportional to the flow of energy, due to dissipative losses that also introduces irreversibility.

Thus, entropy breaks the symmetry of time and can change irrespective of changes of energy—energy being a conservative and reversible property, whereas entropy is evolutionary and irreversible *per se*. The flow of a non-conservative quantity, negentropy, makes life go and the occurrence of a negentropic production term is just the point that differs from analysis based on merely conservative terms (energy and matter).

The situation is explained in Figure 3.1 "The death of the deer": mass and energy do not change, whereas entropy does. There is an "entropic watershed", a gradient, between far from equilibrium (living) systems and classical systems (the dead deer or any inorganic, non living system). The essence of the living organism resides in it being a "configuration of processes".

We may conclude that in far from thermodynamic equilibrium systems (biological and ecological) entropy is not a state function, since it has intrinsic evolutionary properties, strikingly at variance with classical thermodynamics.

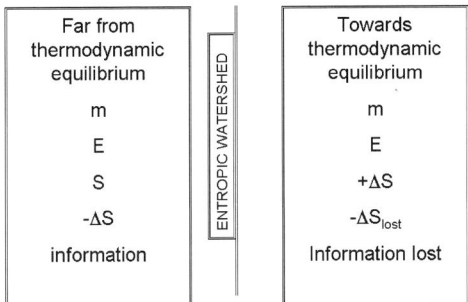

Figure 3.1 The death of the Deer, an example showing the difference between a living, far from equilibrium system compared with the situation after its death where irreversible changes becomes dominant. (After Tiezzi, 2006b.)

It is important to study energy and matter flows, quantities that are intrinsically conserved; it is also important to study entropy flow, an intrinsically evolutionary and non-conserved quantity. But if energy and mass are intrinsically conserved and entropy is intrinsically evolutionary, how can entropy be calculated on the basis of energy and mass quantities (entropy paradox)? This question is still unanswered and all we can do is to note that the eco-dynamic viewpoint is different from that of classical physics and classical ecology.

The probability paradox

The following illustrates that—for even simple far from equilibrium systems—unforeseen consequences to predictability may arise from various aspects of heterogeneity. An event occurs in a stochastic manner because others precede it. Evolutionary events proceed in a manner that depends on time: they show a direction of time; they are irreversible. History determines the environmental and genetic constraints making the future largely unpredictable, as demonstrated several times above. Stochastic or probabilistic elements are unavoidable (although compare the views of Elsasser, Popper, etc.).

Novelty abounds in biological and ecological systems. Ontic openness allows for the emergence of new form and patterns. Previously unobserved events cannot be predictable, while rare and extreme events may or will completely change the dynamics of complex systems.

Figure 3.2 shows the emergence of a probability paradox in the presence of events:

(a) suppose that an oxidation (chemical event), unknown to the observer, arises in the classic "white and black spheres" game: the probability white/black is no more fifty-fifty

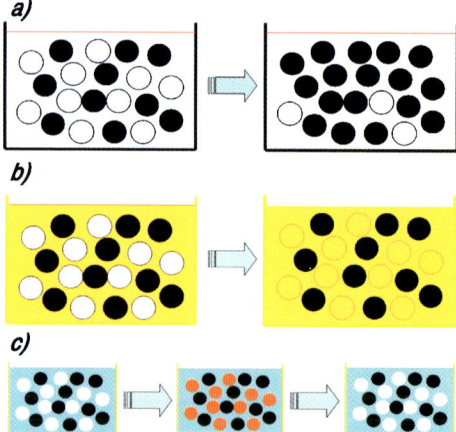

Figure 3.2 Unexpected events that may occur in living systems: (a) oxidation; (b) chameleon effect; (c) oscillating reaction. (After Tiezzi, 2006b.)

(only if the oxidation is changing the white sphere, e.g., to gray, may we know what happened);

(b) suppose that an evolutionary event also occurs, related to the "chameleon" effect (sensible to the environment): again the probability is no more fifty-fifty; moreover the event's interval depends on the "chameleon";

(c) suppose an oscillating event occurs, similar to the Beluzov–Zhabotinsky reaction: the situation is more complex and depends on many parameters. Again the observer has no possibility to predict which sphere will be picked up from the container.

It is possible to conclude that in the far from equilibrium framework a classical probability approach does not apply and new models have to be developed for the Boltzmann's relation $S = k \times \ln W$.

3.6 ONTIC OPENNESS AND EMERGENCE

At all levels of nature we see the emergence of "narrative elements". We are reminded of Scheherazade who interrupts her beautiful story to start another one, even more beautiful. In nature also we have the cosmological history that includes the history of matter, life, humans, and so on till we come to our individual history associated to our consciousness. At all levels we observe events associated with the emergence of novelties, we may associate with the creative power of nature.

These narrative historical aspects are part of complexity. Complex systems share the feature to exhibit a great variety of behaviors. Take an example from chemistry: the Belousov–Zhabotinsky reaction mentioned above. The details are irrelevant here, let us suppose that there are two species of molecules: "red" ones and "blue" ones; moreover they transform one into the other. The behavior of the system depends on the external constraints. Close to equilibrium the collisions are random. There may only appear short living local flashes of color. But far from equilibrium the behavior of this system changes radically. It becomes in succession red then blue then again red. This periodicity indicates the existence of long-range correlations due to the non-equilibrium conditions. "At equilibrium matter is blind, far from equilibrium it begins to see" (Ilya Prigogine[3]).

The fascination of these physical experiments lies in the fact that small variations in a tiny building block of matter manifest themselves as large changes in biological processes. The paradox of modern scientific research in this field lies in the fact that the greater the detail in which we seek "pure" mechanisms or given sub-particles, the more confirmation we have of the validity of quantum mechanics and the more important information we have on the structure of matter. On the other hand, starting from elementary particles, the more we study interactions with biological systems and ecosystems, the more we discover the complexity, irreversibility, and intrinsic aleatory character of nature. In chaos, we

---

[3] From the foreword to Tiezzi (2003a).

rediscover the spontaneity of evolutionary history: a universe in which God plays dice, to invert Einstein's phrase[4].

God was the supreme guarantee of physical determinism. For Einstein, protagonist of the first "heroic" phase of quantum physics, physical determinism applied to any process. However, Max Born[5] once told Einstein that a deterministic universe was innately anathema to him. Born admitted that Einstein might be right, but added that determinism did not seem to hold in physics, much less in other fields. Born criticized Einstein's comment that God does not play dice[6], observing that Einstein's deterministic world needed chance. Born's wife Hedwig had previously written to their "dear friend Albert" that she could not admit a universal law according to which everything was predetermined, including whether or not she vaccinated her child against diphtheria[7].

Both uncertainty equations are related to the complex relation between the observer and the experiment. The first one deals with position and momentum, the second one deals with energy and relaxation time. Both equations assume time reversibility and are valid in a given instant: the momentum is related to the derivative of space with respect to time and the relaxation time is related to the lifetime of the elementary particle in the excited state. Both equations are valid in the quantum physics paradigm and deal with conservative quantities (mass, energy), but not with living systems or evolutionary quantities.

Space and time are categories belonging to different logical types, which should not be confused. By nature, time is evolutionary and irreversible, whereas the space is conservative and reversible. A reversible quantity cannot be differentiated with respect to an irreversible one. It is not possible to compare evolving quantities, such as the life span of the Einstein's twins, in the framework of reversible mechanics. If we deal with evolutionary (living) systems, we may introduce a third concept: Thermodynamic Uncertainty related to the intrinsic irreversible character of time (Tiezzi, 2006a).

Let us say that a thermodynamic uncertainty arises from the experimental existence of the arrow of time and from the experimental evidence that, during the measurements,

---

[4]On 4th December 1926, Einstein wrote to Max Born that although quantum mechanics was worthy of respect, an inner voice told him that it was not yet the right solution because it did not enable us to penetrate the secret of the Great Old Man, who he was sure did not play dice with the world (Science and Life, Letters 1916–1955, letter no. 52 in A. Einstein, H. and M. Born). Max Born considered that there was a profound divergence of viewpoint between Einstein and the following generation, to which Born regarded himself as belonging, though only a few years younger than Einstein. In a previous letter (29th April 1924, no. 48 of the above collection) Einstein observed that the ideas of Niels Bohr on radiation were interesting but he himself did not wish to be led away from rigorous causality. He added that he could not tolerate the idea that an electron exposed to radiation could freely choose when and in which direction to jump. Were this so, he said he would prefer to be a shoemaker or a gambler rather than a physicist. In the introduction to this collection of letters, Werner Heisenberg comments that Einstein agreed with Born on the fact that the mathematical formalism of quantum mechanics, which originated in Göttingen and was subsequently elaborated at Cambridge and Copenhagen, correctly represented the phenomena occurring inside the atom, but that he did not recognize quantum mechanics as a definitive or even exhaustive representation of these phenomena. The theme that God does NOT play dice recurs elsewhere in the Born–Einstein correspondence (e.g., Einstein's letters of 7th September 1944 and 12th October 1953, nos. 81 and 103, respectively).
[5]10th October 1944 (letter no. 84 in Science and Life).
[6]The expression "God plays dice" obviously had an irrational overtone for Einstein, but, as we shall see, not for us.
[7]9th October 1944 (letter no. 82 in Science and Life).

time goes by. Since during the interval of the experiment (measurement) time flows, also the conservative quantities (energy or position) may change leading to a further uncertainty.

Recently astrophysics discovered that the mass of a star is related to the life span of the star itself. The larger is the mass, the less is the life span. This finding may also be related to the uncertainty principle. It seems that there is a sort of uncertainty relation between space and time, where space is related to mass, energy, and the conservative quantities.

3.7 ONTIC OPENNESS AND HIERARCHIES

The above shows the necessity of an extended view of biological systems focusing on the property of heterogeneity and order at the same time. The pertinent mechanisms are encompassed within Elsasser's four principles, but already here the pitfall of a return to reductionism was pointed out. Rather, let us begin with an assumption that includes as given the genetic-level apparatus.

It is easy to see from the composition of nucleic acids or triplet codes that the genome combinatorics will exhibit *immense numbers*. These are also the numbers reached in calculations by Jørgensen et al. (1995) and the many attempts that have followed to calculate an exergy index for ecosystems based on the information content of the genome. Ontic openness is definitely a reality at this level.

Patten later suggested another hierarchical level in addition to the genotype and phenotype levels, namely and the exosomatic envirotype reflecting that an organism's genetic template and physiological manifestation is only realized with respect to its ultimate surroundings and the ecosystems, respectively. Recently, Nielsen (in press) has extended this view by adding a semiotic level above (Figure 3.3). This layer includes all

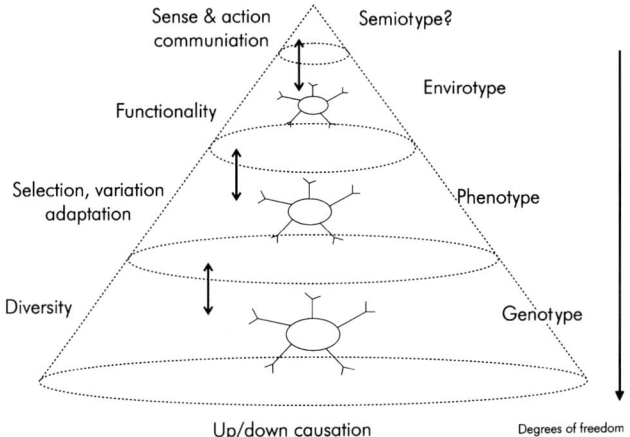

Figure 3.3 A biological hierarchy suggesting that interactions with the environment and finally the semiotics determine the development of the ecosystem (from Nielsen, in press, with permission from Elsevier).

kinds of communicative and cognitive process, i.e., semiotics in a wide sense. This represents the ultimate layer of realizing ecosystem openness.

Thus at each layer of the biological hierarchy we meet a new side of ontic openness. Interactions between hierarchical levels may, as indicated, take place in both upward and downward directions. The traditional view is that as we move up the hierarchy we are narrowing the number of possibilities; therefore, as O'Neill et al. (1986) state, hierarchies are systems of constraints, which only are able to provide system regulations at steady-state conditions. Whenever rare events or system transformations occur the hierarchies are broken, and uncertainty takes place in a broad extent. Emergence due to ontic openness always exists but is just realized in other ways that are not covered by the reductionist view.

3.8 CONSEQUENCES OF ONTIC OPENNESS: A TENTATIVE CONCLUSION

Here we summarize the consequences of ontic openness that will have a deep impact on ecology:

(1) Immense numbers are easily reached.
(2) Possible development and uncertainty.
(3) Uniqueness of the ecosystem.
(4) Agency—how is this uniqueness chosen.
(5) Emergent properties are common.

In the following section we will attempt to address the above points in context of the applying a systems perspective to ecology and ecological theory.

Immense numbers are easily reached

Much of the material given above clearly demonstrates that achieving numbers of interacting elements in ecological systems that are above Googol (10^{100}), and thereby do not in themselves carry any physical meaning, is fairly common if not ubiquitous. A combinatorial view on any level of hierarchy of biological systems is not sufficient in explaining "the meaning of life"—in fact 42 makes more sense and is a better estimate[8].

Possible development and uncertainty

As pointed out by several calculations above, ecosystems have too many distinguishable parts for classical understanding. Even if middle-number systems, possess enough components to exceed the limits, we may for instance consider our capabilities of doing experiments and sampling. In order to accept this, we need to recognize that we do have to live with a high uncertainty, e.g., often expressed in the fact that our standard deviations on any measurement that we make are far beyond the levels accepted by our "colleagues" from physics and chemistry.

---

[8]Meaning of Life given in Douglas Adams', Hitchhikers Guide to the Galaxy.

The uniqueness of ecosystems

This issue could be seen as rather trivial. At each state in its evolution the ecosystem transitions to a new state. The one thing we now can be sure of is that the next state will be just as unique as the previous one, the system will never repeat itself exactly. An event may happen once and never again.

In fact, we did implicitly address this point indirectly in the "Introduction", without putting much attention to it, when we described a situation familiar to most of us: our inability to describe precisely a system without measurements. Meanwhile, not to fall in despair, we may find some satisfaction in the world of propensities. We may not know exactly what happens, but approximately what happens.

Agency of ecosystems

This topic is probably the most problematic. In fact, many of us probably would like the idea that "nature has a life on its own" and this may also correspond nicely with what we observe or have observed. But how to give agency to ecosystems without being accused of romanticism, teleology, idealism, etc. or alternatively getting involved in a debate about intelligent design?

Uniqueness as emergence

Given the sum of the possible conclusions from the above—unexpected things are bound to happen in ecosystems and likewise in ecological research. In fact, when Odum (1969) made his proposal to follow the study of emergent properties of ecosystems as a research strategy he was "only" introducing a suggestion of studying the impact of ontic openness to ecosystems (QED).

The messages of ontic openness to ecology

A view of our world as possessing an essential property such as being ontically open does carry several important messages to ecologists. The ubiquity of emergent properties or unexpected, rare events should, as such, be no surprise to us any longer. Meanwhile, we should not fall in despair; some predictability is still possible, although we should expect accuracy to be small and uncertainty to be high. Probably, understanding the world as propensities rather than fixed possibilities is the way out of this dilemma. The biological world as we see it around us now, its (bio)diversity consists of the part of the openness that was actually realized. It is, together with its individual components, unique and is "locked-in" from many path-dependent evolutionary events. It will never emerge again and as such it should be appreciated a lot more than seems to be the case at the moment.

4

Ecosystems have directionality

"From the way the grass bends, one can know the direction of the wind."
(Chinese Quotation)

*All nature is but art unknown to thee;
All chance, direction which thou canst not see;
All discord, harmony not understood;
All partial evil, universal good;
And, spite of pride, in erring reasons spite,
One truth is clear, Whatever IS, is RIGHT.*
(Alexander Pope, 1773)

4.1 SINCE THE BEGINNINGS OF ECOLOGY

Ecosystems have directionality! This is an extraordinary statement, although the reader might at first wonder why. After all, one observes directional behavior everywhere: A billiard ball, when struck by another ball, will take off in a prescribed direction. Sunflowers turn their heads to face the sun. Copepods migrate up and down in the water column on a daily basis. Yet, despite these obvious examples, scientists have increasingly been trained to regard instances of directionality in nature as having no real basis—epiphenomenal illusions that distract one from an underlying static, isotropic reality.

Before embarking on how ecological direction differs from directionality observed elsewhere, it is worthwhile describing the ecological notion of succession (Odum, 1959). The classical example in American ecology pertains to successive vegetational communities (Cowles, 1899) and their associated heterotrophs (Shelford, 1913)—research conducted on the shores of Lake Michigan. Both Cowles and Shelford had built on the work of the Danish botanist, Eugenius Warming (1909). Prevailing winds blowing against a shore will deposit sand in wave-like fashion. The most recent dunes have emerged closest to the lake itself, while progressively older and higher dunes occur as one proceeds inland. The assumption here, much like the famed ergodic assumption in thermodynamics, is that this spatial series of biotic communities represents as well the temporal evolution of a single ecosystem. The younger, presumably less-mature community consisted of beach grasses and Cottonwood. This "sere" was followed by a Jack pine forest, a xeric Black oak forest, an Oak and hickory moist forest, and the entire progression was thought to "climax" as a Beech-maple forest. The invertebrate and vertebrate communities were observed to segregate more or less among the vegetational

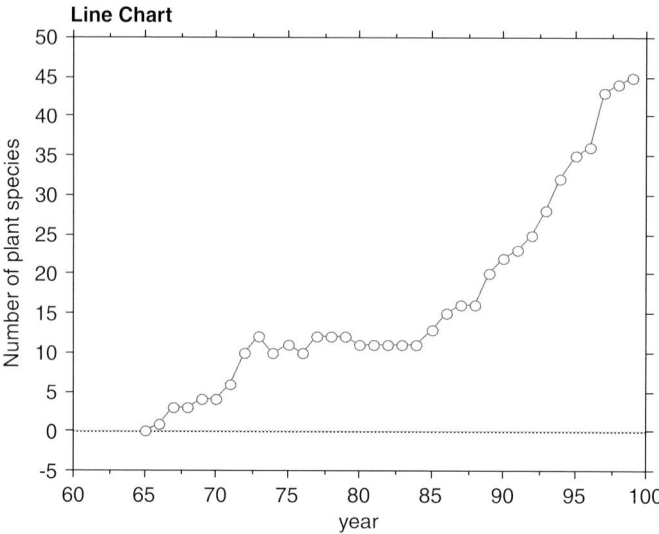

Figure 4.1 Increase over time in the number of plant species found on the newly created island of Surtsey.

zones, although there was more overlap among the mobile heterotrophs than among the sessile vegetation.

Other examples of succession involve new islands that emerge from the sea, usually as the result of volcanic activity. One particular ecosystem that was followed in detail is the sudden emergence in 1963 of the approximately 2.8 km² island, Surtsey, some 33 km south of the large island of Iceland in the North Atlantic. Figure 4.1 depicts the rise in the number of plant species found on the island. (Other measures of succession on Surtsey will be given below).

4.2 THE CHALLENGE FROM THERMODYNAMICS

Now one might well ask how the directionality of these ecosystems differs in any qualitative way from, say the billiard ball mentioned in the opening paragraph of this chapter? For one, the direction of the billiard ball is a consequence of the collision with the other ball, the Newtonian law of momentum and the Newtonian-like law of elasticity. The ball itself remains essentially unchanged after the encounter. Furthermore, if the ball is highly elastic, the encounter is considered reversible. That is, if one takes a motion picture of the colliding balls and the movie is shown to a subject with the projector operating in both the forward and reverse modes, the subject is incapable of distinguishing the original take from its reverse. Reversibility is a key attribute of all Newtonian systems, and until the mid-1960s all Newtonian laws were considered strictly reversible. Early in the 20th century, Aemalie Noether (1918) demonstrated how the property of reversibility was fully equivalent to that of conservation, i.e. all reversible systems are conservative. There is no fundamental change in them, either before or after the event in question.

This pair of fundamental assumptions about how objects behaved set the stage for the first challenge to the Newtonian worldview. In 1820 Sadie Carnot (1824) had been observing the performance of early steam engines in pumping water out of mines. He observed how the energy content (caloric) of the steam used to run the engines could never be fully converted into work. Some of it was always lost forever. This meant that the process in question was irreversible. One could not reverse the process, bringing together the work done by the engine with the dispersed heat and create steam of the quality originally used to run the engine. (See also the discussion of the second law of thermodynamics in Chapter 2).

But the steam, the engine, and the water were all material things, made up of very small particles, according to the atomic hypothesis that had recently been formulated. Elementary particles should obey Newtons laws, which always gave rise to reversible behaviors. Whence, then, the irreversibility? This was a conundrum that for a while placed the atomic hypothesis in jeopardy. The enigma occupied the best minds in physics over the next half century. How it was "resolved" demonstrates volumes about common attitudes toward scientific belief.

Ludwig von Boltzmann (1872) considered the elements of what was called an "ideal gas" (i.e. a gas made up of point masses that did not interact with each other) to obey Newton's laws of motion. He then assumed that the distribution of the momenta of the atoms was normally random. This meant that nearby to any configuration of atoms there were always more equivalent distributions (having same mass and momentum) that were more evenly distributed than there were configurations that were less evenly distributed. Any random walk through the distributions would, therefore, would be biased in the direction of the most probable distribution (the maximum of the normal distribution). Ergo, without violating conservation of mass or momentum at the microlevel, the system at the macrolevel was biased to move in the direction of the most even distribution.

This was a most elegant *model*, later improved by Gibbs (1902). It is worth noting, however that the resolution was a model that was applicable to nature under an exceedingly narrow set of conditions. Nonetheless, it was accepted as validation of the atomic hypothesis and Newtonian reversibility *everywhere*, and it put an end to the controversy. This rush to consensus was, of course, the very antithesis of what later would be exposited as logical positivism—the notion that laws cannot be verified, only falsified. Laws should be the subject of constant and continual scrutiny; and scientists should always strive to falsify existing laws. But when conservation, reversibility, and atomism were being challenged, the response of the community of scholars was precisely the *opposite*—discussion was terminated on the basis of a single model that pertained to conditions that, in relation to the full set of conditions in the universe, amounted to "a set of measure zero"!

Such inconsistencies notwithstanding, the second law does indeed provide a direction for time and introduces history into science. The second law serves as a very significant constraint on the activities of living systems and imparts an undeniable directionality to biology (Schneider and Sagan, 2005).

4.3 DECONSTRUCTING DIRECTIONALITY?

Events in biology have been somewhat the reverse of those in physics. Whereas physics began with directionless laws and was confronted with exceptions, biologists had originally thought that phylogeny took a progressive direction over the eons, culminating in the appearance of humankind at the apex of the natural order—the so-called "natural chain of being." Evolutionary biologists, however, have sought to disabuse other biologists of such directional notions (Gould, 1994). At each turn in its history, a biotic system is subject to random, isotropic influences. What looks in retrospect like a progression has been merely the accumulation of the results of chance influences. Complexity simply accrues until such time as a chance catastrophe prunes the collection back to a drastically simpler composition.

We thus encounter a strong bias at work within the community of scientists to deny the existence of bias in nature (a statement which makes sense only because humanity has been postulated to remain outside the realm of the natural). Physicists and (perhaps by virtue of "physics envy") evolutionary theorists appear keen to deny the existence of direction anywhere in the universe, preferring instead a changeless Eleatic worldview. It is against this background that the notion of direction in ecology takes on such importance.

Directionality, in the form of ecological succession, has been a key phenomenon in ecology from its inception (Clements, 1916). By ecological succession is meant "the *orderly* process of community change" (Odum, 1959) whereby communities replace one another in a given area. Odum (*ibid.*) do not equivocate in saying, "The remarkable thing about ecological succession is that it is directional." In those situations where the process is well known, the community at any given time may be recognized and future changes predicted. That is, succession as a phenomenon appears to be reproducible to a degree.

Of course, it was not long after the ideas of community succession came into play that the opinion arose that its purported direction was illusory. Gleason (1917) portrayed succession in plant communities as random associations of whatever plant species happened to immigrate into the area. Others have pointed out that "seres" of ecological communities almost always differ in terms of the species observed (Cowles, 1899). The ecosystem ecologist takes refuge in the idea that the functional structure nonetheless remains predictable (Sheley, 2002).

The question thus arises as to whether ecological succession is orderly in any sense of the word, and, if so, what are the agencies behind such order? We begin by noting that the directionality of ecosystems is of a different ilk from those mentioned in the opening of this chapter. With regard to all three of those examples, the direction of the system in question was determined by sources *exterior* to the system—by the colliding billiard ball in the first instance, and by the sun as perceived by the sunflower and copepod. It will be argued below, however, that the directionality of an ecosystem derives from an agency active *within* the system itself. Surely, external events do impact the system direction by providing constraints, but any one event is usually incremental in effect. On rare occasions an external event can radically alter the direction and the constitution of the system itself (Prigogine, 1978; Tiezzi, 2006b), but this change is every bit as much a consequence of the system configuration as it is of the external event (Ulanowicz, 2006a). The direction an ecosystem takes is both internal and constitutional. Most change seen elsewhere is neither.

4.4 AGENCIES IMPARTING DIRECTIONALITY

It remains to identify the agency behind any directionality that ecosystems might exhibit. Our natural inclination is such a search would be to look for agencies that conform to our notions of "lawful" behaviors. But such a scope could be too narrow. It would behoove us to broaden our perspective and attempt to generalize the notion of "law" and consider as well the category of "process". A process resembles a law in that it consists of rule-like behaviors, but whereas a law always has a determinate outcome, a process is guided more by its interactions with aleatoric events.

The indeterminacy of such action is perhaps well illustrated by the artificial example of Polya's Urn (Eggenberger and Polya, 1923). Polya's process consists of picking from an urn containing red and blue balls. The process starts with one red ball and one blue ball. The urn is shaken and a ball is drawn at random. If it is a red ball, then the ball is returned to the urn with yet another red ball; if a blue ball is picked, then it likewise is returned with another blue ball. The question then arises whether the ratio of red to blue balls approaches a fixed value. It is rather easy to demonstrate that the law of large number takes over and that after a sufficient number of draws, the ratio changes only within bounds that progressively shrink as the process continues. Say the final ratio is 0.3879175. The second question that arises is whether that ratio is unique? If the urn is emptied and the process repeated, then will the ratio once again converge to 0.3879175? The answer is no. The second time it might converge to 0.81037572. It is rather easy to show in Monte-Carlo fashion that the final ratios of many successive runs of Polya's process are uniformly distributed over the interval from 0 to 1.

One sees in Polya's Urn how direction can evolve out of a stochastic background. The key within the process is the feedback that is occurring between the history of draws and the current one. Hence, in looking for the origins of directionality in real systems, we turn to consider feedback within living systems. Feedback, after all, has played a central role in much of what is known as the theory of "self-organization" (e.g. Eigen, 1971; Maturana and Varela, 1980; DeAngelis et al., 1986; Haken, 1988; Kauffman, 1995). Central to control and directionality in cybernetic systems is the concept of the causal loop. A causal loop, or circuit is any concatenation of causal connections whereby the last member of the pathway is a partial cause of the first. Primarily because of the ubiquity of material recycling in ecosystems, causal loops have long been recognized by ecologists (Hutchinson, 1948).

It was the late polymath, Gregory Bateson (1972) who observed "a causal circuit will cause a non-random response to a random event at that position in the circuit at which the random event occurred." But why is this so? To answer this last question, let us confine further discussion to a subset of causal circuits that are called autocatalytic (Ulanowicz, 1997). Henceforth, autocatalysis will be considered any manifestation of a positive feedback loop whereby the direct effect of every link on its downstream neighbor is positive. Without loss of generality, let us focus our attention on a serial, circular conjunction of three processes—A, B, and C (Figure 4.2) Any increase in A is likely to induce a corresponding increase in B, which in turn elicits an increase in C, and whence back to A.[1]

A didactic example of autocatalysis in ecology is the community that builds around the aquatic macrophyte, *Utricularia* (Ulanowicz, 1995). All members of the genus *Utricularia* are carnivorous plants. Scattered along its feather-like stems and leaves are small bladders, called utricles (Figure 4.3a). Each utricle has a few hair-like triggers at its

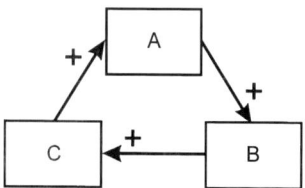

Figure 4.2 Simple autocatalytic configuration of three species.

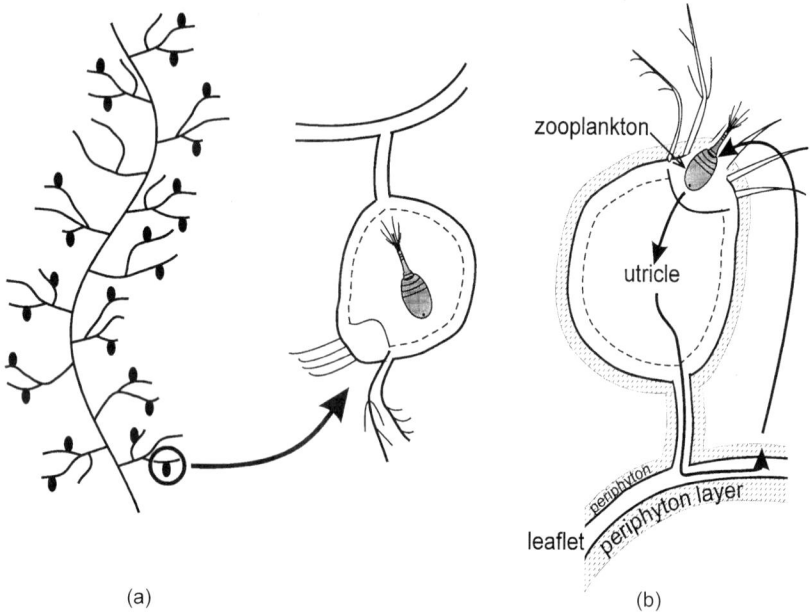

Figure 4.3 The *Utricularia* system. (a) View of the macrophyte with detail of a utricle. (b) The three flow autocatalytic configuration of processes driving the *Utricularia* system.

---

[1] The emphasis in this chapter is on positive feedback and especially autocatalysis. It should be mentioned in passing that negative feedback also plays significant roles in complex ecosystem dynamics (Chapter 7), especially as an agency of regulation and control.

terminal end, which, when touched by a feeding zooplankter, opens the end of the bladder, and the animal is sucked into the utricle by a negative osmotic pressure that the plant had maintained inside the bladder. In nature the surface of *Utricularia* plants is always host to a film of algal growth known as periphyton. This periphyton in turn serves as food for any number of species of small zooplankton. The autocatalytic cycle is closed when the *Utricularia* captures and absorbs many of the zooplankton (Figure 4.3b).

In chemistry, where reactants are simple and fixed, autocatalysis behaves just like any other mechanism. As soon as one must contend with organic macromolecules and their ability to undergo small, incremental alterations, however, the game changes. With ecosystems we are dealing with *open* systems (see Chapter 2), so that whenever the action of any catalyst on its downstream member is affected by contingencies (rather than being obligatory), a number of decidedly non-mechanical behaviors can arise (Ulanowicz, 1997). For the sake of brevity, we discuss only a few:

Perhaps most importantly, autocatalysis is capable of exerting *selection* pressure on its own, ever-changing, malleable constituents. To see this, one considers a small spontaneous change in process B. If that change either makes B more sensitive to A or a more effective catalyst of C, then the transition will receive enhanced stimulus from A. In the *Utricularia* example, diatoms that have a higher P/B ratio and are more palatable to microheterotrophs would be favored as members of the periphyton community. Conversely, if the change in B makes it either less sensitive to the effects of A or a weaker catalyst of C, then that perturbation will likely receive diminished support from A. That is to say the response of this causal circuit is not entirely symmetric, and out of this asymmetry emerges a direction. This direction is not imparted or cued by any externality; its action resides wholly internal to the system. As one might expect from a causal circuit, the result is to a degree tautologous—autocatalytic systems respond to random events over time in such a way as to increase the degree of autocatalysis. As alluded to above, such asymmetry has been recognized in physics since the mid-1960s, and it transcends the assumption of reversibility. It should be emphasized that this directionality, by virtue of its internal and transient nature cannot be considered teleological. There is no externally determined or pre-existing goal toward which the system strives. Direction arises purely out of immediate response by the internal system to a novel, random event impacting one of the autocatalytic members.

To see how another very important directionality can emerge in living systems, one notes in particular that any change in B is likely to involve a change in the amounts of material and energy that are required to sustain process B. As a corollary to selection pressure we immediately recognize the tendency to reward and support any changes that serve to bring ever more resources into B. Because this circumstance pertains to any and all members of the causal circuit, any autocatalytic cycle becomes the epi-center of a *centripetal* flow of resources toward which as many resources as possible will converge (Figure 4.4). That is, an autocatalytic loop *defines itself* as the focus of centripetal flows. One sees didactic example of such centripetality in coral reef communities, which by their considerable synergistic activities draw a richness of nutrients out of a desert-like and relatively inactive surrounding sea. Centripetality is obviously related to the more commonly recognized attribute of system growth (Chapter 6).

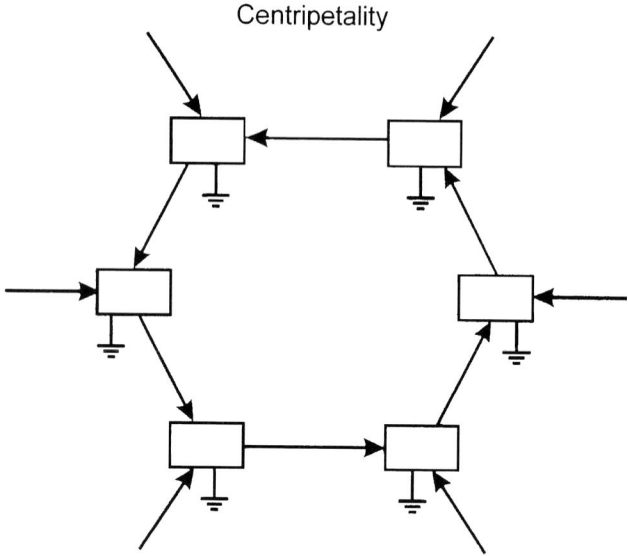

Figure 4.4 The centripetality of an autocatlytic system, drawing progressively more resources unto itself.

4.5 ORIGINS OF EVOLUTIONARY DRIVE

Evolutionary narratives are replete with explicit or implicit references to such actions as "striving" or "struggling", but the origin of such directional behaviors is either not mentioned, or glossed-over. Such actions are simply postulated. But with centripetality we now encounter the roots of such behavior. Suddenly, the system is no longer acting at the full behest of externalities, but it is actively drawing ever more resources unto itself. Bertrand Russell (1960) called this behavior "chemical imperialism" and identified it as the very crux of evolutionary drive.

Centripetality further guarantees that whenever two or more autocatalytic loops exist in the same system and draw from the same pool of finite resources, *competition* among the foci will necessarily ensue, so that another postulated element of Darwinian action finds its roots in autocatalytic behavior. In particular, whenever two loops share pathway segments in common, the result of this competition is likely to be the exclusion or radical diminution of one of the non-overlapping sections. For example, should a new element D happen to appear and to connect with A and C in parallel to their connections with B, then if D is more sensitive to A and/or a better catalyst of C, the ensuing dynamics should favor D over B to the extent that B will either fade into the background or disappear altogether (Figure 4.5). That is, the selection pressure and centripetality generated by complex autocatalysis (a macroscopic ensemble) is capable of influencing the replacement of its own elements. Perhaps the instances that spring most quickly to mind here involve the evolution of obligate mutualistic pollinators, such as yuccas (Yucca) and yucca moths (*Tegeticula, Parategeticula*) (Riley, 1892), which eventually displace other pollinators.

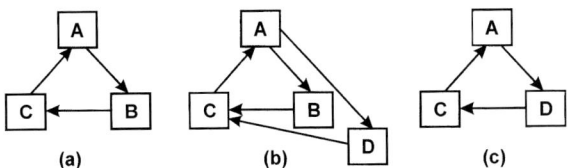

Figure 4.5 Autocatalytic action causing the replacement of element B by a more effective one, D.

It is well-worth mentioning at this point that the random events with which an autocatalytic circuit can interact are by no means restricted to garden-variety perturbations. By the latter are meant simple events that are generic and repeatable. In Chapter 3 it was pointed out how random events can have a complex nature as well and how many such events can be entirely unique for all time. For example, if a reader were to stand on the balcony overlooking Grand Central Station in New York City and photograph a 10×10 m space below, she might count some 90 individuals in the picture. The combinatorics involved guarantee that it is beyond the realm of physical reality that repeating the action at a subsequent time would capture the same 90 individuals in the frame—the habits and routines of those concerned notwithstanding (Elsasser, 1969). Nor are such unique events in any way rare. Even the simplest of ecosystems contains more than 90 distinguishable individual organisms. Unique events are occurring all the time, everywhere and at all levels of the scalar hierarchy. Furthermore, the above-cited selection by autocatalytic circuits is not constrained to act only on simple random events. They can select from among complex, entirely novel events as well.

This ability of an autocatalytic circuit to shift from among the welter of complex events that can impinge upon it opens the door fully to emergence. For in a Newtonian system any chance perturbation would lead to the collapse of the system. With Darwin systems causality was opened up to chance occurrences, but that notion failed to take hold for a long while after Darwin's time, for his ideas had fallen into the shadows by the end of his century (Depew and Weber, 1995). It was not until Fisher and Wright during the late 1920s had rehabilitated Darwin through what is commonly known as "The Grand Synthesis" that evolution began to eclipse the developmentalism that had prevailed in biology during the previous decades. The Grand Synthesis bore marked resemblance to the reconciliation effected in the physical sciences by Boltzmann and Gibbs in that Fisher applied almost the identical mathematics that had been used by Gibbs in describing an ideal gas to the latter's treatment of non-interacting genetic elements. Furthermore, the cardinal effect of the synthesis was similar to the success of Gibbs—it re-established a degree of predictability under a very narrow set of circumstances.

With the recognition of complex chance events, however, absolute predictability and determinism had to be abandoned. There is simply no way to quantify the probability of an entirely unique event (Tiezzi, 2006b). Events must recur at least several times before a probability can be estimated. As compensation for the loss of perfect predictability, emergence no longer need take on the guise of an enigma. Complex and radically chance events are continuously impinging upon autocatalytic systems. The overwhelming majority have no effect whatsoever on the system (which remains *indifferent* to them). A small number impacts the system negatively, and the system must reconfigure itself in countering the effect of the disturbances. An extremely small fraction of the radical events may actually resonate with the autocatalysis and shift it into an entirely new mode of behavior, which can be said to have emerged spontaneously.[2]

Jay Forrester (1987), for example, describes major changes in system dynamics as "shifting loop dominance", by which he means a sudden shift from control by one feedback loop to dominance by another. The new loop could have been present in the background prior to the shift, or it could be the result of new elements entering or arising within the system to complete a new circuit. Often loops can recover from single insults along their circuit, but multiple impacts to several participants, as might occur with complex chance, are more likely to shift control to some other pathway.

One concludes that autocatalytic configurations of flows are not only characteristic of life, but are also central to it. As Popper (1990) once rhapsodically proclaimed, "Heraclitus was right: We are not things, but flames. Or a little more prosaically, we are, like all cells, processes of metabolism; nets of chemical pathways." The central agency of networks of processes is illustrated nicely with Tiezzi's (2006b) comparison of the live and dead deer (just moments after death). The mass of the deer remains the same, as does its form, chemical constitution, energy, and genomic configuration. What the live deer had that the dead deer does not possess is its configuration of metabolic and neuronal processes.

4.6 QUANTIFYING DIRECTIONALITY IN ECOSYSTEMS

It is one thing to describe the workings of autocatalytic selection verbally, but science demands at least an effort at describing how one might go about quantifying and measuring key concepts. At the outset of such an attempt, we should emphasize again the nature of the directionality with which we are dealing. The directionality associated with autocatalysis does not appear in either physical space or, for that matter, in phase space. It is rather more like the directionality associated with time. There direction, or sense, is indicated by changes in a systems-level index—the system's entropy. Increasing entropy identifies the direction of increasing time.

The hypothesis in question is that augmented autocatalytic selection and centripetality are the agencies behind increasing self-organization. Here we note that as autocatalytic configurations displace more scattered interactions, material and energy become increasingly constrained to follow only those pathways that result in greater autocatalytic activities. This tendency is depicted in cartoon fashion in Figure 4.6. At the top is an arbitrary system of four components with an inchoate set of connections between them. In the lower figure one particular autocatalytic feedback loop has come to dominate the system,

---

[2] This emergence differs from Prigogine's "order through fluctuations" scenario in that the system is not constrained to toggle into one of two pre-determined states. Rather, complex chance can carry a system into entirely new modes of behavior (Tiezzi, 2006b). The only criterion for persistence is that the new state be more effective, autocatalytically speaking, than the original.

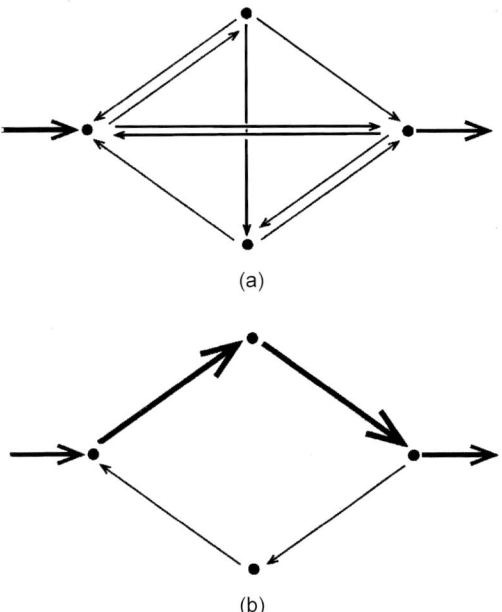

Figure 4.6 Cartoon showing the generic effects of autocatalysis. (a) Inchoate system. (b) Same system after autocatalytic loop has developed.

resulting in fewer effective flows and greater overall activity (as indicated by the thicker surviving arrows). Thus we conclude that quantifying the degree of constraint in an ecosystem must reflect these changes in both the magnitude and intensity of autocatalytic activities. Looked at in obverse fashion, ecosystems with high autocatalytic constraints will offer fewer choices of pathways along which resources can flow.

The appearance of the word "choice" in the last sentence suggests that information theory might be of some help in quantifying the results of greater autocatalysis, and so it is. Box 4.1 details the derivation of a measure called the *System Ascendency*, which quantifies both the total activity of the system as well as the degree of overall constraint extant in the system network. A change in the system pattern as represented in Figure 4.6 will result in a higher value of the ascendency.

In his seminal paper, "The strategy of ecosystem development", Eugene Odum (1969) identified 24 attributes that characterize more mature ecosystems that indicate the direction of ecological succession. These can be grouped into categories labeled species richness, dietary specificity, recycling, and containment. All other things being equal, a rise in any of these four attributes also serves to augment the system ascendency (Ulanowicz, 1986a). It follows as a phenomenological principle "*in the absence of major perturbations, ecosystems have a propensity to increase in ascendency.*" This statement can be rephrased to read that ecosystems exhibit a preferred direction during development: that of increasing ascendency.

Box 4.1 Ascendency: a measure of organization

In order to quantify the degree of constraint, we begin by denoting the transfer of material or energy from prey (or donor) i to predator (or receptor) j as T_{ij}, where i and j range over all members of a system with n elements. The total activity of the system then can be measured simply as the sum of all system processes, $\text{TST} = \sum_{i,j=1}^{n+2} T_{ij}$, or what is called the "total system throughput" (TST). With a greater intensity of autocatalysis, we expect the overall level of system activity to increase, so that T appears to be an appropriate measure. For example, growth in economic communities is reckoned by any increase in gross domestic product, an index closely related to the TST.

In Figure B4.1 is depicted the energy exchanges (kcal/m²/year) among the five major compartments of the Cone Spring ecosystem (Tilly, 1968). The TST of Cone Spring is simply the sum of all the arrows appearing in the diagram. Systematically, this is calculated as follows:

$$\begin{aligned}
\text{TST} &= \sum_{i,j} T_{ij} \\
&= T_{01} + T_{02} + T_{12} + T_{16} + T_{17} + T_{23} + T_{24} + T_{26} + T_{27} + T_{32} \\
&\quad + T_{34} + T_{36} + T_{37} + T_{42} + T_{45} + T_{47} + T_{52} + T_{57} \\
&= 11{,}184 + 635 + 8881 + 300 + 2003 + 5205 + 2309 + 860 \\
&\quad + 3109 + 1600 + 75 + 255 + 3275 + 200 + 370 + 1814 + 167 + 203 \\
&= 42{,}445 \, \text{kcal/m}^2/\text{year}
\end{aligned}$$

where the subscript 0 represents the external environment as a source, 6 denotes the external environment as a receiver of useful exports, and 7 signifies the external environment as a sink for dissipation.

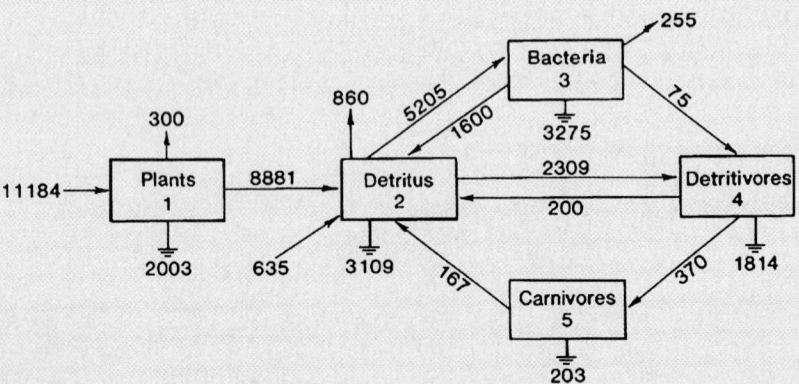

Figure B4.1 Schematic of the network of energy exchanges (kcal/m²/year) in the Cone Spring ecosystem (Tilly, 1968). Arrows not originating from a box represent inputs from outside the system. Arrows not terminating in a compartment represent exports of useable energy out of the system. Ground symbols represent dissipations.

Again, the increasing constraints that autocatalysis imposes on the system channel flows ever more narrowly along fewer, but more efficient pathways—"efficient" here meaning those pathways that most effectively contribute to the autocatalytic process. Another way of looking such "pruning" is to consider that constraints cause certain flow events to occur more frequently than others. Following the lead offered by information theory (Abramson, 1963; Ulanowicz and Norden, 1990), we estimate the joint probability that a quantum of medium is *constrained* both to leave i and enter j by the quotient T_{ij}/T. We then note that the *unconstrained* probability that a quantum has left i can be acquired from the joint probability merely by summing the joint probability over all possible destinations. The estimator of this unconstrained probability thus becomes $\sum_q T_{iq}/T$. Similarly, the unconstrained probability that a quantum enters j becomes $\sum_k T_{kj}/T$. Finally, we remark how the probability that the quantum could make its way by pure chance from i to j, *without* the action of any *constraint*, would vary jointly as the product of the latter two frequencies, or $\sum_q T_{iq} \sum_k T_{kj}/T^2$. This last probability obviously is not equal to the constrained joint probability, T_{ij}/T.

Information theory uses as its starting point a measure of the rareness of an event, first defined by Boltzmann (1872) as $(-k \log p)$, where p is the probability ($0 \leq p \leq 1$) of the given event happening and k is a scalar constant that imparts dimensions to the measure. One notices that for rare events ($p \approx 0$), this measure is very large and for very common events ($p \approx 1$), it is diminishingly small. For example, if $p = 0.0137$, the rareness would be 6.19 k-bits, whereas if $p = 0.9781$, it would be only 0.032 k-bits.

Because constraint usually acts to make things happen more frequently in a particular way (e.g., flow along certain pathways), one expects that, on average, an unconstrained probability would be more rare than a corresponding constrained event. The more rare (unconstrained) circumstance that a quantum leaves i and accidentally makes its way to j can be quantified by applying the Boltzmann formula to the joint probability defined above, i.e., $-k \log(\sum_k T_{kj} \sum_q T_{iq}/T^2)$, and the correspondingly less rare condition that the quantum is constrained both to leave i and enter j becomes $-k \log (T_{ij}/T)$. Subtracting the latter from the former and combining the logarithms yields a measure of the hidden constraints that channel the flow from i to j as $k \log (T_{ij}T/\sum_k T_{kj} \sum_q T_{iq})$.

Finally, to estimate the average constraint at work in the system as a whole, one weights each individual constraint by the joint probability of constrained flow from i to j and sums over all combinations of i and j. That is,

$$\text{AMC} = k \sum_{i,j} \left(\frac{T_{ij}}{T}\right) \log \left(\frac{T_{ij}T}{\sum_k T_{kj} \sum_q T_{iq}}\right)$$

where AMC is the "average mutual constraint" known in information theory as the average mutual information (Rutledge et al., 1976).

(continued)

To illustrate how an increase in AMC actually tracks the "pruning" process, the reader is referred to the three hypothetical configurations in Figure B4.2. In configuration (a) where medium from any one compartment will next flow is maximally indeterminate. AMC is identically zero. The possibilities in network (b) are somewhat more constrained. Flow exiting any compartment can proceed to only two other compartments, and the AMC rises accordingly. Finally, flow in schema (c) is maximally constrained, and the AMC assumes its maximal value for a network of dimension 4.

One notes in the formula for AMC that the scalar constant, k, has been retained. We recall that although autocatalysis is a unitary process, one can discern two separate effects: (a) an extensive effect whereby the activity, T, of the system increases, and (b) an intensive aspect whereby constraint is growing. We can readily unify these two aspects into one measure simply by making the scalar constant k represent the level of system activity, T. That is, we set $k = T$, and we name the resulting product the system *Ascendency*, A, where

$$A = \sum_{i,j} T_{ij} \log\left(\frac{T_{ij}T}{\sum_k T_{kj} \sum_q T_{iq}}\right)$$

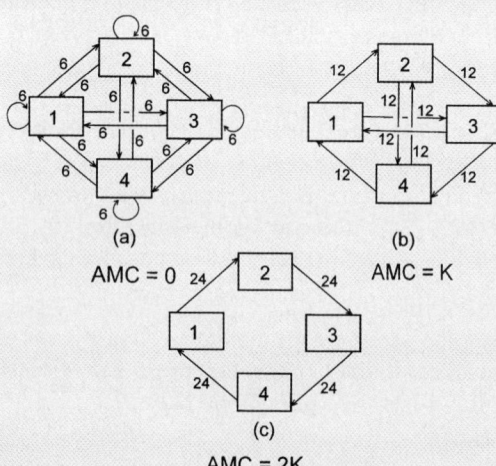

Figure B4.2 Three configurations of processes illustrating how autocatalytic "pruning" serves to increase overall system constraint. (a) A maximally indeterminate four-component system with 96 units of flow. (b) The system in (a) after constraints have arisen that channel flow to only two other compartments. (c) The maximally constrained system with each compartment obligated to support only one other component.

Referring again to the Cone Spring ecosystem network in Figure B4.1, we notice that each flow in the diagram generates exactly one and only one term in the indicated sums. Hence, we see that the ascendency consists of the 18 terms:

$$A = T_{01} \log\left(\frac{T_{01}T}{\sum_k T_{k1} \sum_q T_{0q}}\right) + T_{02} \log\left(\frac{T_{02}T}{\sum_k T_{k2} \sum_q T_{0q}}\right) + \cdots + T_{57} \log\left(\frac{T_{57}T}{\sum_k T_{k7} \sum_q T_{5q}}\right)$$

$$= 20{,}629 - 1481 + 13{,}796 - 94 - 907 + 9817$$
$$+ 4249 + 1004 + 446 + 295 - 147$$
$$+ 142 + 4454 - 338 + 1537 + 2965 + 123 + 236$$
$$= 56{,}725 \text{ kcal-bits/m}^2/\text{year}$$

While ascendency measures the degree to which the system possesses inherent constraints, we wish also to have a measure of the degree of flexibility that remains in the system. To assess the degrees of freedom, we first define a measure of the full diversity of flows in the system. To calculate the full diversity, we apply the Boltzmann formula to the joint probability of flow from i to j, T_{ij}/T, and calculate the average value of that logarithm. The result is the familiar Shannon formula,

$$H = -\sum_{i,j} \left(\frac{T_{ij}}{T}\right) \log\left(\frac{T_{ij}}{T}\right)$$

where H is the diversity of flows. Scaling H in the same way we scaled A, i.e. multiplying H by T, yields the system development capacity, C, as

$$C = -\sum_{i,j} T_{ij} \log\left(\frac{T_{ij}}{T}\right)$$

Now, it can readily be proved that $C \geq A \geq 0$, so that the residual, $(C - A) \geq 0$, as well. Subtracting A from C and algebraically reducing the result yields the residual, Φ, which we call the systems "overhead" as

$$\Phi = C - A = -\sum_{i,j} T_{ij} \log\left(\frac{T_{ij}^2}{\sum_k T_{kj} \sum_q T_{iq}}\right)$$

The overhead gauges the degree of flexibility remaining in the system.

Just as we substituted the values of the Cone Spring flows into the equation for ascendency, we may similarly substitute into this equation for overhead to yield a value of 79,139 kcal-bits/m²/year. Similarly, substitution into the formula for C yields a value of 135,864 kcal-bits/m²/year, demonstrating that the ascendency and the overhead sum exactly to yield the capacity.

The ecologist reading this book is likely to have a healthy appreciation for those elements in nature that do not resemble tightly constrained behavior, as one finds with autocatalysis. In fact, Chapter 3 was devoted in large measure to describing the existence and role of aleatoric events and ontic openness. Hence, increasing ascendency is only half of our dynamical story. Ascendency accounts for how efficiently and coherently the system processes medium. Using the same mathematics as employed above, however, it is also shown in Box 4.1 how one can compute as well an index called the system overhead, Φ, that is complementary to the ascendency and captures how much flexibility the system retains (Ulanowicz and Norden, 1990).

The flexibilities quantified by overhead are manifested as the inefficiencies, incoherencies, and functional redundancies present in the system. Although these latter properties may encumber overall system performance at processing medium, we saw in Chapter 3 how they become absolutely essential to system survival whenever the system incurs a novel perturbation. At such time, the overhead comes to represent the repertoire of potential tactics from which the system can draw to adapt to the new circumstances. Without sufficient overhead, a system is unable create an effective response to the exigencies of its environment. The configurations we observe in nature, therefore, appear to be the results of a dynamical tension between two antagonistic tendencies (ascendency vs. overhead; Ulanowicz, 2006b). The ecosystem needs this tension in order to persist. Should either direction in the transaction atrophy, the system will become fragile either to external perturbations (low overhead) or internal disorder (low ascendency). System fragility is discussed further in Chapter 8.

One disadvantage of ascendency as an index of directionality is that its calculation requires a large amount of data. Currently, the networks accompanying a seres of ecological stages have not yet been assembled. About the closest situation for which data are available is a comparison of two tidal marsh communities, one of which was perturbed by a 6°C rise in temperature caused by thermal effluent from a nearby nuclear power plant, and the other of which remained unimpacted (Homer et al., 1976) Under the assumption that perturbation regresses an ecosystem to an earlier stage, one would expect the unimpacted system to be more "mature" and exhibit a higher ascendency than the heated system.

Homer et al. parsed the marsh gut ecosystem into 17 compartments. They estimated the biomass in each taxon in mgC/m^2 and the flows between taxa in $mgC/m^2/day$. The total system throughputs (T) in the control ecosystem was estimated to be 22,420 $mgC/m^2/day$, and that in the impacted system as 18,050 $mgC/m^2/day$ (Ulanowicz, 1986a,b). How much of the decrease could be ascribed to diminution of autocatalytic activities could not be assessed, suffice it to say that the change was in the expected direction. The ascendency in the heated system fell to 22,433 $mgC-bits/m^2/day$ from a value of 28,337 $mgC-bits/m^2/day$ for the control. The preponderance of the drop could be ascribed to the fall in T, as the corresponding AMC fell by only 0.3%.

4.7 DEMYSTIFYING DARWIN

One possible way around the copious data required to calculate the ascendency might be to search for an indirect measure of the effect of autocatalysis. Along those lines Jørgensen and Mejer (1977) suggested that the directionality in ecosystem succession

Chapter 4: Ecosystems have directionality

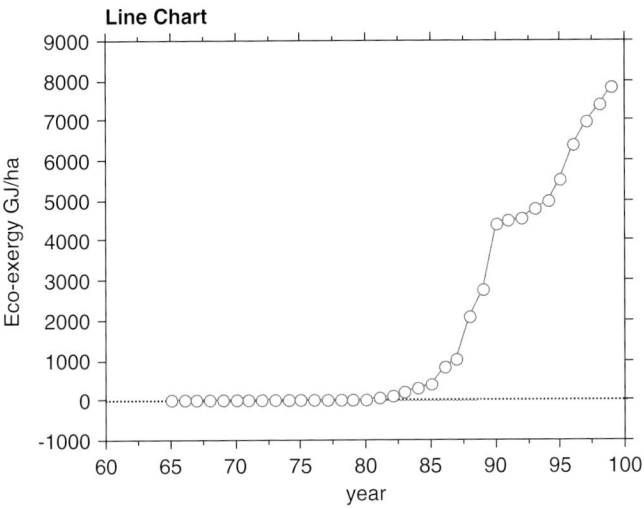

Figure 4.7 Estimated stored exergy among the biota inhabiting Surtsey Island.

might be gauged by the amount of exergy stored among the components of the ecosystem. (Exergy being the net amount of total energy that can be converted directly into work. More to come in Chapter 6.) The working hypothesis is that ecosystems accumulate more stored exergy as they mature. Exergy can be estimated once one knows the biomass densities of the various species, the chemical potentials of components that make up those species and the genetic complexity of those species (Jørgensen et al., 2005, see also Chapter 6). In Figure 4.7, one sees that the stored ecological exergy among the biota of Surtsey Island began to increase markedly after about 1985.

It is perhaps worthwhile at this juncture to recapitulate what has been done: first, we have shifted our focus in ecosystem dynamics away from the normal (symmetrical) field equations of physics and concentrated instead on the origins of asymmetry in any system—the boundary constraints. We then noted how biotic entities often serve as the origins of such constraint on other biota, so that the kernel of ecodynamics is revealed to be the mutual (self-entailing) constraints that occur within the ecosystem itself. We then identified a palpable and measurable entity (the network of material–energy exchanges) on which this myriad of mostly hidden constraints writes its signature. Finally, we described a calculus that could be applied to the network to quantify the effects of autocatalytic selection. Hence, by following changes in the ascendency and overhead of an ecosystem, we are focusing squarely on that which makes ecodynamics fundamentally different from classical dynamics (Ulanowicz, 2004a,b).

The dynamical roots of much of Darwinian narrative having been de-mystified by the directionality inherent in autocatalysis, it is perhaps a bit anti-climatic to note that several other behaviors observed among developing ecosystems also can trace their origins to autocatalysis and its attendant centripetality. Jørgensen and Mejer (1977), as mentioned above, have concluded that ecosystems always develop in the direction of increasing the amount of exergy stored in the system. Maximal exergy storage has proved a useful tool with which to estimate unknown parameters and rates (Jørgensen, 1992a; see also growth and development forms in Chapter 6). Schneider and Kay (1994) hypothesize how systems develop so as to degrade available exergy gradients at the fastest rate possible. This is, however, only correct for the first growth form, growth of biomass because more biomass needs more exergy for respiration to maintain the biomass far from thermodynamic equilibrium. Further details see Chapter 6. Thirdly, the inputs of ecosystems engender many-fold system circulations among the full community—a process called network aggradation (Fath and Patten, 2001). All three behaviors can be traced to autocatalysis and its attendant centripetality (Ulanowicz et al., 2006).

It should be noted in passing how autocatalytic selection pressure is exerted in top-down fashion—contingent action by the macroscopic ensemble on its constituent elements. Furthermore, centripetality is best identified as an agency acting *at* the focal level. Both of these modes of action violate the classical Newtonian stricture called closure, which permits only mechanical actions at smaller levels to elicit changes at higher scales. As noted above, complex behaviors, including directionality, can be more than the ramification of simple events occurring at smaller scales.

Finally, it is worthwhile to note how autocatalytic selection can act to stabilize and regularize behaviors across the hierarchy of scales. Under the Newtonian worldview, all laws are considered to be applicable universally, so that a chance happening anywhere rarely would ramify up and down the hierarchy without attenuation, causing untold destruction. Under the countervailing assumption of ontic-openness, however, the effects of noise at one level are usually subject to autocatalytic selection at higher levels and to energetic culling at lower levels. As a result, nature as a whole takes on habits (Hoffmeyer, 1993) and exhibits regularities; but in place of the universal effectiveness of all natural laws, we discern instead a *granularity* inherent in the real world. That is, models of events at any one scale can explain matters at another scale only in inverse proportion to the remoteness between them. For example, one would not expect to find any connection between quantum phenomena and gravitation, given that the two phenomena are separated by some 42 orders of magnitude, although physicists have searched ardently, but in vain, to join the two. Obversely, the domain within which irregularities and perturbations can damage a system is usually circumscribed. Chance need not unravel a system. One sees demonstrations of systems "healing" in the higher organisms, and even in large-scale organic systems such as the global ecosystem (Lovelock, 1979).

4.8 DIRECTIONALITY IN EVOLUTION?

With the cybernetic narrative of ecosystem development (the New Ecology) now before us, it is perhaps useful to revisit the question of whether the process of biotic evolution might exhibit any form of directionality? Perhaps an unequivocal response is premature, suffice it here to compare the differences in the dynamics of ontogeny, ecosystem development, and evolution. With ontogenetic development, there is no denying the directionality evident in the developing organism. Convention holds that such direction is "programmed" in the genomic material, and no one is going to deny the degree of correspondence between genome and phenome. The question remains, however, as to where does the agency behind such direction reside? It is awkward, to say the least, to treat the genome as some sort of homunculus that directs the development process. Genomic material such as DNA is unlikely to have evolved by random assembly, and outside its network of enzymatic and proteomic reactions it can do nothing of interest (Kauffman, 1993). Its role in ontogeny is probably best described as that of material cause, sensu Aristotle—it is materially necessary, but passive with respect to more efficient (again, sensu Aristotle) agencies that actively read and carry out the anabolic processes. As regards those processes, they form a network that indubitably contains autocatalytic pathways, each with its accompanying directions.

The entire scenario of ontogeny is rather constrained, and noise plays a distinct secondary role. In contrast, the role of genomes is not as prominent in the development of ecosystems (Stent, 1981). While some hysterisis is required of the participating species, the central agencies that provide directions (as argued above) are the autocatalytic loops among the species. The constraints among the species are nowhere near as tight as at the ontogenetic level, and noise plays a much larger role in the direction that a system takes over time.

Evolutionary patterns are not as stereotypical as those in ecological succession. What happens before some cataclysm can be very different from what transpires after the disaster. So evolutionary theorists are probably correct in pointing to random events as playing the larger role over the long run. It appears premature, however, to rule out directional processes altogether. Many species and their genomes survive catastrophes, as do entire autocatalytic ensembles of species at the level of the ecosystem. They provide a degree of history that helps to direct the course of evolution until the next upheaval.

This dynamic is already familiar to us from the workings of Polya's Urn, which we considered earlier. In fact, a reasonable simile would be to consider what might happen if Polya's Urn were upset after some 1000 draws and only a random subset of say 15 balls could be recovered and put back into the Urn to continue the process. Although the subsequent evolution of the ratio of red to blue balls might not converge very closely to what it was before the spill, some remnants of the history would likely keep the ratio from making an extreme jump. Suppose before the spill the ratio had converged rather tightly to 0.739852, and that after the accident ten red balls and five blue balls were recovered. It is exceedingly unlikely that the continuing process would converge to, say 0.25835.

And so it may be on the evolutionary theatre. Not all directions established by ecosystems during one era are necessarily destroyed by a catastrophe that initiates the next. Surviving directions are key to the evolutionary play during the next interval. Thermodynamic and other physical directions notwithstanding, anyone who argues that evolution involves only chance and no directionality is making an ideological statement and not a reasoned "conjecture" because ecosystem have directionality.

4.9 SUMMARY

Ecology, from its very inception, has been concerned with temporal direction. Ecological communities are perforce open systems, and thus are subject to the imperatives of the second law, but there is yet another, internal drive within ecosystems, efforts by evolutionary theorists to deny directionality notwithstanding. Ecosystem dynamics are rooted in configurations of autocatalytic processes, which respond to random inputs in a non-random manner. Autocatalytic processes build on themselves, and in the process give rise to a centripetal pull of energy and resources into the community. Such centripetality is central to the very notion of life and is more basic than even competition, on which conventional evolutionary theory is built. Configurations of processes can select from among complex chance events, any of which can exhibit its own, accidental directionality. Ensuing directionality can be quantified as an increase in an information theoretic measure called *Ascendency*. This directionality opposes the tendency of the second law to disorder systems, but healthy ecosystems need a modicum of *both* trends in order to persist. The resulting dynamic resembles that of a natural dialectic. Finally, although evolution over the longer span might appear adirectional, selection in the nearer ecological timespan always provides the ecosystem with an inherent direction that is an obligate element in a complete description of any particular evolutionary scenario.

5

Ecosystems have connectivity

"Life did not take over the globe by combat, but by networking."
(Margulis and Sagan. Microcosmos)

5.1 INTRODUCTION

The web of life is an appropriate metaphor for living systems, whether they are ecological, anthropological, sociological, or some integrated combination—as most on Earth now are. This phrase immediately conjures up the image of interactions and connectedness both proximate and distal: a complex network of interacting parts, each playing off one another, providing constraints and opportunities for future behavior, where the whole is greater than the sum of the parts. Networks: the term that has received much attention recently due to such common applications as the Internet, "Six Degrees of Separation", terrorist networks, epidemiology, even MySpace®, actually has a long research history in ecology dating to at least Darwin's entangled bank a century and a half ago, through the rise of systems ecology of the 1950s, to the biogeochemical cycling models of the 1970s, and the current focus on biodiversity, stability, and sustainability, which all use networks and network concepts to some extent. It is appropriate that interconnected systems are viewed as networks because of the powerful exploratory advantage one has when employing the tools of network analysis: graph theory, matrix algebra, and simulation modeling, to name a few.

Networks are comprised of a set of objects with direct transaction (couplings) between these objects. Although the exchange is a discrete transfer, these transactions viewed in total link direct and indirect parts together in an interconnected web, giving rise to the network structure. The structural relations that exist can outlast the individual parts that make up the web, providing a pattern for life in which history and context are important. The connectivity of nature has important impacts on both the objects within the network and our attempts to understand it. If we ignore the web and look at individual unconnected organisms, or even two populations pulled from the web, such as one-predator and one-prey, we miss the system-level effects. For example, in a holistic investigation of the Florida Everglades, Bondavalli and Ulanowicz (1999) showed that the American alligator (*Alligator mississippiensis*) has a mutualistic relation with several of its prey items, such that influence of the network trumps the direct, observable act of predation. The connected web of interactions makes this so because each isolated act of predation links together the entire system, such that indirect effects—those mitigated through one or many other objects in the network—can dictate overall relations. While this might seem irrelevant particularly for the individual organisms that end up in the alligator's gut, as a whole the prey population benefits from the presence of the alligator in the web since it also feeds on other organisms in the web which in turn are predators or competitors with the prey.

Such discoveries are not possible without viewing the ecosystem as a connected network. This chapter deals with that connectivity, provides an overview of systems approaches, introduces quantitative methods of ecological network analysis (ENA) to investigate this connectivity and ends with some of the general insight that has been gained from viewing ecosystems as networks. Insight, which at first glance appears surprising and unintuitive, is not that surprising under closer inspection. It only seems so from our current paradigm, which is still largely reductionistic. We hope these examples give further weight for adopting the systems perspective promoted throughout this book.

5.2 ECOSYSTEMS AS NETWORKS

Ecosystems are conceptual and functional units of study that entail the ecological community together with its abiotic environment. Implicit in the concept of any system, such as an ecosystem, is that of a system boundary which demarcates objects and processes occurring within the system from those occurring outside the system. This inside–outside perspective gives rise to two environments, the environment external to the system within which it is embedded, and the environment outside the object of interest but within the system boundaries (the latter has been termed *environ* by Patten, 1978). We typically are not concerned with events occurring wholly outside the system boundary, i.e., those originating and terminating in the environment without entering the system by crossing the system boundary. Furthermore, as open systems, energy–matter fluxes occur across the boundary; these in turn provide the ecosystem with an available source of energy input such as solar radiation and a sink for waste heat. In addition to continuous radiative energy input and output, pulse inputs are important in some ecosystems such as allochthonous organic matter in streams and deltas, and migration in Tundra.

The spatial extent of an ecosystem varies greatly and depends often on the functional processes within the ecosystem boundaries. O'Neill et al. (1986) defined an ecosystem as the smallest unit which can persist in isolation with only its abiotic environment, but this does not give an indication to the area encompassed by the ecosystem. Cousins (1990) has proposed the home range or foraging range of the local dominant top predator arbiter of ecosystem size, which he refers to as an ecosystem trophic module or ecotrophic module. Similar to the watershed approach in hydrology, Power and Rainey (2000) proposed a "resource shed" to delineate the spatial extent of an ecosystem. Taken to the extreme, one could eliminate environment altogether by expanding the boundaries outward indefinitely to subsume all boundary flows, thus making the very concept of environment a paradox (Gallopin, 1981). The idea is not to make the "resource shed" so vast as to include everything in the system boundary, but to establish a demarcation line based on gradients of interior and exterior activities. In fact, in open systems an external reference state is a necessary condition, which frames the ecosystem of interest (Patten, 1978). We give the last word to Post et al. (2005) who stated that different organisms within the ecosystem based on their resource needs and mobility will operate at different temporal and spatial scales, typically leaving the scale context-specific for the research question in hand.

Definitional difficulties aside, one must operationalize an ecosystem so following O'Neill's approach of the smallest unit that could sustain life, the minimum set for a sustainable functioning ecosystem comprises producers and consumers, specifically decomposers (see further below). One visualizes a naturally occurring biotic community to include:

(1) organisms that can draw in and fix external energy into the system, typically primary producers,
(2) additional organisms that feed on this fixed energy, consumers, and
(3) decomposers that close the cycle on material flow as well as provide additional energy pathways.

This biotic community interacts with its abiotic environment acquiring energy, nutrients, water, and physical space to form its place or habitat niche (although habitat is often comprised of other biotic entities). As a result, ecosystems are comprised of many interactions, both biotic and abiotic. This includes interactions between individuals within populations (e.g., mating), interactions between individuals from different species (e.g., feeding), and active and passive interactions of the individuals with their environment (e.g., water and nutrient uptake, excretion, and death). In ecosystem studies two approaches are employed. The first, a "black-box" approach concerns itself entirely with the inputs and outputs to the ecosystem not elucidating the processes that generated them (Likens et al., 1977). The second, generally termed ecological network analysis (ENA), is a detailed accounting of energy–nutrient flows within the ecosystem. In these studies, the focus is usually at the scale of the species or trophospecies (trophic functional groups), and how they interact rather than interactions between individuals of the same species, although these are considered in individual-based models and studies. ENA could even be called reductionistic–holism since it requires fine scale detail of the ecosystem constituents and their interconnections, but uses them to reveal global patterns that shape ecosystem structure and function.

Although interaction networks are ubiquitous, observing them is difficult and this has led to slow recognition of their importance. For example, ecological observations reveal direct transactions between individuals but do not immediately reveal the contextual network in which they play out. Sitting in a forest, one does not readily observe the network, but rather an occasional act of grazing, predation, or death. While watching a wolf take down a deer, it is not apparent what grasses the deer grazed on, now assimilated by the deer, and soon the wolf, not to mention the original source of energy, solar radiation, or nutrients in soil pore water. Since the components form a connected web, it is necessary to study and understand them in relation to the interconnection network, not in isolation or a limited subset of the system.

Each component, in fact, must be connected to others through both its input and output transactions. There are no trivial, isolated components in an ecosystem. Pulling out one species is like pulling one intersection of a spider's web, such that although that one particular facet is brought closer for inspection, the entire web is stretched in the direction of the disturbance. Those sections of the web more closely and strongly connected to the selected node are more affected, but the entire system is warped as each node is embedded within the whole network of webbed interactions. The indicator species approach works because it focuses on those organisms that are deeply embedded in the web (Patten, 2005) and therefore produce a large systemic deformation. The food web is, therefore, in fact, more than just a metaphor; it acknowledges the inherent connectivity of ecosystem interactions.

5.3 FOOD WEBS

Food web ecology has been a driving force in studying the interconnections among species (e.g., MacArthur, 1955; Paine, 1980; Cohen et al., 1990; Polis, 1991; Pimm, 2002). In fact, we typically think of the abundance and distribution of species in an ecological community as being heavily influenced by the interactions with other species (Andrewartha and Birch, 1984), but the species is more than the loci of an envirogram; it is those interactions, that connectivity, with other species and with the environment, which construct the ecosystem. The diversity, stability, and behavior of this complex is governed by such interactions. Here we introduce the standard food web treatment, discuss some of the weakness, while suggesting improvements, and end with an overview of the general insights gained from understanding ecosystem connectivity as revealed by ENA.

A food web is a graph representing the interaction of "who eats whom", where the species are nodes and the arcs are flows of energy or matter. For example, we show a food web diagram typical to what one would find in an introductory biology or ecology textbook (Figure 5.1).

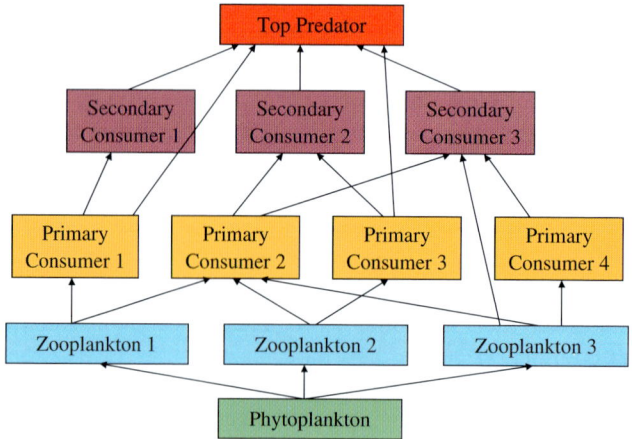

Figure 5.1 Typical ecological food web.

The energy flow enters the primary producer compartments and is transferred "up" the trophic chain by feeding interactions, grazing and then predation, losing energy (not shown) along each step, where after a few steps it has reached a terminal node called a top predator (also known, in Markov chain theory, as an absorbing state). This picture of "who eats whom" has several deficiencies if one wants to understand the entire connectedness as established by the matter–energy flow pattern of the ecosystem:

- First, the diagram excludes any representation of decomposers, identified above as a more fundamental element of ecosystems than more familiar trophic groups like herbivores, carnivores, and omnivores. While decomposers have been an integral part of some ecological research (e.g., microbial ecology, eutrophication models, network analysis, etc.), their role in community food web ecology is just now gaining stature. Prejudices and biases often work to shape science; what food-web ecologist, for example, would *a priori* classify our species (*Homo sapiens*) as detritus feeders as our diet of predominantly dead or not freshly killed organisms (living microbes, parasites, and inquilants in our food aside) in fact rules us to be?
- Second, the diagram shows the top predators as dead-ends for resource flow; if that were the case there would be a continuous accumulation of top predator carcasses throughout the millennia that biological entities called "top-predators" have existed. Nature would be littered with residues of lions, hawks, owls, cougars, wolves, and other "top-predators", even the fiercest of the fierce like *Tyrannosaurus rex* (not to mention other non-grazed or directly eaten materials such as tree trunks, feces, etc.). It would be a different world. Obviously, this is not the case because in reality there is no "top" as far as food resource and energy flow are concerned. The bulk of the energy from "top-predator" organisms, like all others, is consumed by other organisms, although perhaps not as dramatically as in active predation. Although there have been periods in which accumulation rates exceed decomposition rates, resulting in among other things formation of fossil fuels and limestone deposits, but much organic matter is oxidized to carbon dioxide. For our purposes, the relevancy of these flows from top-predators to detritus is that they provide additional connectivity within the ecosystem.
- Third, when decomposers are included in ecosystem models, as there has been some recent effort to do, they are treated as source compartments only. Resource flows out to exploiting organisms, but is not returned as the products and residues of such exploitation. For example, in a commonly studied dataset of 17 ecological food webs (Dunne et al., 2002), 10 included detrital compartments but all of these had in-degrees equal to zero, meaning they received no inputs from other compartments. In reality, all other compartments are the sources for the dead organic material itself (Fath and Halnes, submitted). It is easy enough to correct these flow structures by allowing material from each compartment to flow into the detritus, but this introduces cycling and gives a significantly different picture of the connectance patterns and resulting system dynamics.

The point is that while food webs have been one way to investigate feeding relations in ecology, they are just a starting point for investigating the whole connectivity in ecosystems. Other, more complete, methodologies are needed.

5.4 SYSTEMS ANALYSIS

If the environment is organized and can be viewed as networks of ordered and functioning systems, then it is necessary that we have analysis tools and investigative methodologies that capture this wholeness. Just as one cannot see statistical relationships by visually observing an ecosystem or a mesocosm experiment, one must collect data on the local interactions that can be estimated or measured, then analyze the connectivity and properties that arise from this. In that sense, systems analysis is a tool, similar to statistical analysis, but one that allows the identification of holistic, global properties of organization.

Historically, there are several approaches employed to do just that. One of the earliest was Forrester's (1971) box-and-arrow diagrams. Building on this approach, Meadows et al. (1972) showed the system influence primarily of human population on environmental resource use and degradation. The Forrester approach also later formed the basis for Barry Richmond's STELLA® modeling software first developed in 1985, a widely used simulation modeling package. This type of modeling is based on a simple, yet powerful, principle of modeling that includes Compartments, Connections, and Controls. One of Richmond's main aims with this software was to provide a tool to promote systems thinking. The first chapter of the user manual is an appeal for increased systems thinking (Richmond, 2001). In order to reach an even wider audience, he developed a "Story of the Month" feature which applied systems thinking to everyday situations such as terrorism, climate change, and gun violence. In such scenarios, the key linkage is often not the direct one. System behavior frequently arises out of indirect interactions that are difficult to incorporate into connected mental models. Many societal problems, which may be environmental, economic, or political, stem from the lack of a systems perspective that goes to remote, primary causes rather than stopping at proximate, derivative ones.

Many systems analysis approaches are based on state-space theory Zadeh et al. (1963), which provides a mathematical foundational to understand input–response–output models. Linking multiple states together creates networks of causation Patten et al. (1976), such that input and output orientation and embeddedness of objects influence the overall behavior. Box 5.1 from course material of Patten describes a progression from a simple causal sequence in which one object, through simple connectance, exerts influence over another. Causal chains and networks exhibit indirect causation, followed by a degree of self-control in which feedback ensures that an object's output environ wraps back around to its input environ downstream. Lastly, with holistic causation, systems influence systems. Using network analysis several holistic control parameters have been developed (Patten and Auble, 1981; Fath, 2004; Schramski et al., 2006). Further testing is necessary but these approaches are promising for understanding the overall influence each species has in the system.

Another approach to institutionalize system analysis is Odum's use of energy flow diagrams, which has since spawned the entire industry of emergy (embodied energy) flow analysis for ecosystems, industrial systems, and urban systems (e.g., Odum, 1996; Bastianoni and Marchettini, 1997; Huang and Chen, 2005; Wang et al., 2005; Tilley and Brown, 2006).

> **Box 5.1** Distributed causation in networks
>
> 1. *The causal connective*: $B \to C$
>   There is only a direct effect of B on C.
>
> 2. *The causal chain*: $A \to B(A) \to C$
>   B affects C directly, but A influences C indirectly through B, and C has no knowledge of A.
>
> 3. *The causal network*: $\{A\} \to B(\{A\}) \to C$
> $\{A\}$ is a system, with a full interaction network giving potential for holistic determination.
>
> 4. *Self influence*: $\{A(C)\} \to B(\{A(C)\}) \to C$
>   C is in network $\{A\}$ and exerts indirect causality on itself.
>
> 5. *Holistic influence*: $\{A(B,C)\} \to B(\{A(B,C)\}) \to C$
>   B is also in $\{A\}$ so that B, C and all else in $\{A\}$ influence C indirectly.

The systems analysis approach is also an organizing principle for much of the work at the International Institute for Applied Systems Analysis (IIASA) in Laxenburg, Austria. The institute was established during the Cold War as a meeting ground for East and West scientists and found common ground in the systems approach (www.iiasa.ac.at). Although its focus was not ecology, it has produced several large-scale, interdisciplinary environmental models such as the Regional Air pollution INformation and Simulation (RAINS), population development environment (PDE) models, and lake water quality models.

Another systems approach, food web analysis, is the main ecological approach, but as stated earlier has limited perspective by including only the feeding relations of organisms easily observed and measured, largely ignoring abiotic resources, and operating with a limited analysis toolbox. For example, without the basis of first principles of thermodynamics or graph theory (which are more recently being incorporated) the discipline has been trapped in several "debates" such as "top-down" vs. "bottom-up" control, and interaction strength determination, which have ready alternatives in ENA. Specifically regarding top-down versus bottom-up, Patten and Auble (1981), Fath (2004), and Schramski et al. (2006) all use network analysis to demonstrate and try to quantify the cybernetic and distributed nature of ecosystems.

The latter methodology, ENA, arose specifically to address issues of wholeness and connectivity. It has two major directions, *Ascendency Theory* concerned with ecosystem growth and development, and a system theory of the environment termed *Environ Analysis*. Ascendency theory is summarized elsewhere in this volume (see Box 4.1). After some general remarks on ENA, the remainder of this chapter will sketch connectivity perspectives from the "13 Cardinal Hypotheses" of environ theory.

5.5 ECOSYSTEM CONNECTIVITY AND ECOLOGICAL NETWORK ANALYSIS

The exploration of network connectivity has led to the identification of many interesting, important, and non-intuitive properties. ENA starts with the assumption that a system can be represented as a network of nodes (vertices, compartments, components, etc.) and the connections between them. When there is a flow of matter or energy between any two objects in that system we say there is a direct transaction between them. These direct transactions give rise to both direct and indirect relations between all the objects in the system.

Nobel prize winning economist Wassily Leontief first developed a form of network analysis called input–output analysis (Leontief, 1936, 1951, 1966). Based on system connectivity, it has been applied to many fields. For example, there is a large body of research in the area of social network analysis, which uses the input–output methodology to investigate how individual lives are affected by their web of social connections (Wellman, 1983; Wasserman and Faust, 1994; Trotter, 2000). Input–output analysis has also successfully been applied to study the flow of energy or nutrients in ecosystem models (e.g., Wulff et al., 1989; Higashi and Burns, 1991).

Bruce Hannon (1973) is credited with first applying economic input–output analysis techniques to ecosystems. He pursued this line of research primarily to determine interdependence of organisms in an ecosystem based on their direct and indirect energy flows. Others quickly picked up on this powerful new application and further refined and extended the methodology. Some of the earlier researches in this field include Finn (1976, 1980), Patten et al. (1976), Levine (1977, 1980, 1988); Barber (1978a,b), Patten (1978, 1981, 1982, 1985, 1992), Matis and Patten (1981), Higashi and Patten (1986, 1989), Ulanowicz (1980, 1983, 1986), Ulanowicz and Kemp (1979), Szyrmer and Ulanowicz (1987), and Herendeen (1981, 1989). Both environ analysis and ascendancy theory rely on the input–output analysis basis of ENA.

The analysis itself is computationally not that daunting, but does require some familiarity with matrix algebra and graph theory concepts. The notation and methodology of the two main approaches, *ascendency* and *network environ analysis* (NEA) differ slightly and have been developed in detail elsewhere (see references above), and therefore, we will not repeat here (see Box 5.1 for a very brief introduction to Ascendency). Furthermore, the development of user-friendly software such as ECOPATH (Christensen and Pauly, 1992), EcoNetwork (Ulanowicz, 1999), and more recently WAND by Allesina and Bondavalli (2004) and NEA by Fath and Borrett (2006) are available to perform the necessary computation on network data and will ease the dissemination of these techniques. Following a short NEA primer we sketch the 13 Cardinal Hypotheses (CH) (Patten, in prep) associated with NEA that arise from ecosystem connectivity.

5.6 NETWORK ENVIRON ANALYSIS PRIMER

The details of NEA have been developed elsewhere (see Patten, 1978, 1981, 1982, 1985, 1991, 1992), so below we provide just a general overview for orientation to the discussion below. Ecosystem connections, such as flow of energy of nutrients, provide the framework for the conceptual network. The directed connections between ecosystem compartments provide necessary and sufficient information to construct a network diagram (technically referred to as a digraph) and its associated adjacency matrix—an $n \times n$ matrix with 1s or 0s in each element depending on whether or not the compartments are adjacent. Using this information, structural analysis is possible, which is used to identify the number of indirect pathways and the rate at which these increase with increasing path length. With quantitative information regarding the storages and flows (internal and boundary) of the system compartments, additional functional analyses are possible—primarily referred to as flow, storage, and utility analyses (Table 5.1). The key to the analysis is using the direct adjacency matrix or non-dimensional, normalized matrices in the case of the functional analyses (g_{ij}, p_{ij}, and d_{ij}, respectively) to find indirect pathways or flow, storage, or utility contributions. The network parameters, g_{ij}, p_{ij}, and d_{ij}, in addition to having an important physical characterization in the network, control the integral network organization and structure within the system. Contributions along indirect pathways are revealed through powers of the direct matrix, for example, G has the direct flow intensities, G^2 gives the flow contributions that have traveled 2-step pathways, G^3 those on 3-step pathways, and G^m those on m-step pathways. Given the series constraints, higher order terms approach zero as $m \rightarrow 4$, thereby making it possible to sum the direct and ALL indirect contributions ($m \geq 2$) produce an integral or holistic system evaluation (see Box 5.2). In the case of the functional analyses, integral flow, storage, or utility values are the summation of the direct plus all indirect contributions (N, Q, U, respectively). In this manner it is possible to quantify the total indirect contribution and compare it with the direct flows, the result being that often the direct contribution is less than the indirect, hence leading to the need for a holistic analysis that accounts for and quantifies wholeness and indirectness. This is the primary methodology for investigating system structure, function, and organization using NEA. Below we give two numerical examples that illustrates typical results of a NEA. The next section will give an overview of insights in the resulting, possible effects of networks.

Table 5.1 Overview of network environ analysis

Network Environ Analysis

Path Analysis - enumerates number of pathways in a network

- **Flow Analysis** ($g_{ij} = f_{ij}/T_j$) – identifies flow intensities along indirect pathways
- **Storage Analysis** ($c_{ij} = f_{ij}/x_j$) – identifies storage intensities along indirect pathways
- **Utility Analysis** ($d_{ij} = (f_{ij} - f_{ji})/T_i$) – identifies utility intensities along indirect pathways

> **Box 5.2 Basic notation for network environ analysis**
>
> Flows: f_{ij} = within system flow directed from j to i, comprise a set of transactive flows.
> Boundary transfers: z_j = input to j, y_i = output from i.
> Storages: x_j represent n storage compartments (nodes).
> Throughflow:
> $$T_i^{(in)} = z_i + \sum f_i$$
> $$T_i^{(out)} = \sum f_i + y_i$$
> At steady-state: $T_i^{(in)} = T_i^{(out)} \equiv T_i$.
>
> Non-dimensional, intercompartmental *flows* are given by $g_{ij} = f_{ij}/T_j$
> Non-dimensional, intercompartmental *utilities* are given by $d_{ij} = (f_{ij} - f_{ji})/T_i$.
> Non-dimensional, *storage*-specific, intercompartmental flows are given by
>
> $p_{ij} = c_{ij}\Delta t$, where, for $i \neq j$, $c_{ij} = f_{ij}/x_j$, and for $i = j$, $p_{ii} = 1 + c_{ii}\Delta t$, where $c_{ii} = -T_i/x_i$.
>
> Non-dimensionless integral flow, storage, and utility intensity matrices, N, Q, and U, respectively can be computed as the convergent power series:
>
> $$\begin{aligned} N &= G^0 + G^1 + G^2 + G^3 + \ldots + G^m + \ldots = (I - G)^{-1} \\ Q &= P^0 + P^1 + P^2 + P^3 + \ldots + P^m + \ldots = (I - P)^{-1} \\ U &= D^0 + D^1 + D^2 + D^3 + \ldots + D^m + \ldots = (I - D)^{-1} \end{aligned} \quad (1)$$
>
> The mth order terms, $m = 1, 2, \ldots$, account for interflows over all pathways in the system of lengths m.

Network example 1: aggradation

Using NEA, it is possible to demonstrate how the connections that make up the network are beneficial for the component and the entire ecosystem. Figure 5.2 presents a very simple example, presuming steady-state (input = output) and first order donor determined flows, which is often used in ecological modeling. Figure 5.2a shows the throughflow and exergy storage (based on a retention time of five time units) in the two components with no coupling, i.e., no network connections. Making a simple connection between the two links them physically, and while it changes their individualistic behavior, it also alters the overall system performance. In this case, the throughflow and exergy storage increase because the part of the flow that previously exited the system is no used by the second compartment, thereby increasing the total system throughflow, exergy stored, and average path length. The advantages of integrated systems is also known from industrial ecology in which waste from one industry can be used as raw material for another industry (see, e.g., Gradel and Allenby, 1995; McDonough and Braungart, 2002; Jørgensen, 2006).

Network example 2: Cone Spring ecosystem

For the second example, we use the same Cone Spring ecosystem from the previous chapter, which was used to demonstrate ascendency calculations (this will also help show the

Chapter 5: Ecosystems have connectivity

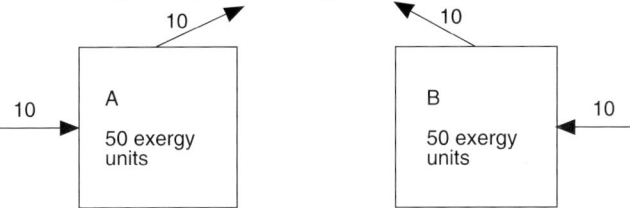

a. No coupling between A and B. The throughflow is 20 and the exergy storage is 100 exergy units

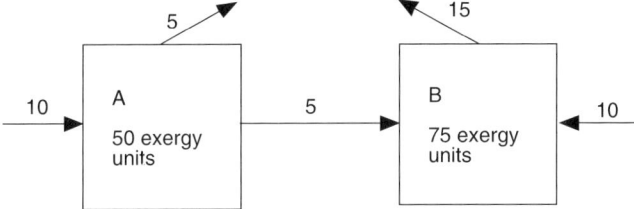

b. A coupling from A to B. The throughflow is now 25 and the exergy storage is 125 exergy units.

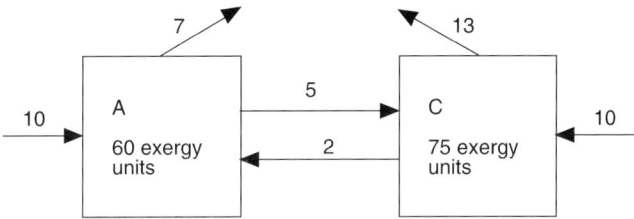

c. A coupling from A to B and a coupling from B to A. The throughflow is now 27 and the exergy storage is 135 exergy units

Figure 5.2 Two compartment system illustrating network aggradation as increased total system throughflow and exergy storage relative to boundary inputs resulting from internal transactional coupling. (a) no coupling. (b) one coupling. (c) cyclic coupling.

similarities and differences between ascendency and environ analysis). Figure B4.1 shows the flows in the system, but since the storage values are not given we limit ourselves here to the flow and utility analyses. From the figure we obtain the following information (note, flows are oriented from columns to rows):

$$A = \begin{bmatrix} 0 & 0 & 0 & 0 & 0 \\ 1 & 0 & 1 & 1 & 1 \\ 0 & 1 & 0 & 0 & 0 \\ 0 & 1 & 1 & 0 & 0 \\ 0 & 0 & 1 & 0 & 0 \end{bmatrix} \quad F = \begin{bmatrix} 0 & 0 & 0 & 0 & 0 \\ 8881 & 0 & 1600 & 200 & 167 \\ 0 & 5205 & 0 & 0 & 0 \\ 0 & 2309 & 75 & 0 & 0 \\ 0 & 0 & 370 & 0 & 0 \end{bmatrix} \quad z = \begin{bmatrix} 11184 \\ 635 \\ 0 \\ 0 \\ 0 \end{bmatrix}$$

$$y = [2303 \quad 3969 \quad 3530 \quad 1814 \quad 203]$$

Compartmental throughflow is the sum either entering OR exiting any compartment. At steady-state these are equal, thus $T_i = z_i + \sum_j f_{ij} = \sum_j f_{ji} + y_i$. Total system throughflow is the sum of all compartmental throughflows: TST $= \sum_i T_i$, which here is 30,626 kcal/m²/y. (Note, in ascendency analysis total system throughput = TST $= \sum_{i,j=1}^{n+2} T_{ij}$, which includes both the input and output terms.) Continuing on with the NEA we present the non-dimensional flow and utility matrices, G and D, respectively:

$$G = \begin{bmatrix} 0 & 0 & 0 & 0 & 0 \\ 0.794 & 0 & 0.308 & 0.084 & 0.451 \\ 0 & 0.453 & 0 & 0 & 0 \\ 0 & 0.201 & 0.014 & 0 & 0 \\ 0 & 0 & 0 & 0.155 & 0 \end{bmatrix}$$

$$D = \begin{bmatrix} 0 & -0.794 & 0 & 0 & 0 \\ 0.773 & 0 & -0.314 & -0.184 & 0.015 \\ 0 & 0.693 & 0 & -0.014 & 0 \\ 0 & 0.885 & 0.0315 & 0 & -0.155 \\ 0 & -0.451 & 0 & 1.00 & 0 \end{bmatrix}$$

Running the analysis gives the integral flow matrix:

$$N = \begin{bmatrix} 1.000 & 0 & 0 & 0 & 0 \\ 0.958 & 1.207 & 0.374 & 0.186 & 0.545 \\ 0.434 & 0.547 & 1.169 & 0.084 & 0.247 \\ 0.199 & 0.251 & 0.092 & 1.039 & 0.113 \\ 0.031 & 0.039 & 0.014 & 0.161 & 1.018 \end{bmatrix}$$

along with the following information: Finn Cycling Index is 0.0919, meaning about 9% of flow is cycled flow, the ratio of direct to indirect flow is 0.9126, meaning that almost half of all total system throughflow traveled along indirect paths (note, this is actually a rare case exception when indirect flow contribution is not a majority), and the network evenness measure (homogenization) is 2.0360, meaning that the values in integral flow matrix, N, are twice as evenly distributed than the values in the direct flow matrix G—which is evident just from eyeballing the two matrices. Another, important analysis possible with the flow data, but not displayed here is the calculation of the actual unit environs, i.e., flow decompositions showing the amount of flow within the system "environ" needed to generate one unit of input or output at each compartment. Unfortunately, this particular example is one in which the powers of the utility matrix, D, do not guarantee convergence (since the maximum eigenvalue of D is greater than 1); and therefore, we cannot present the synergism metrics. Although this is an area of ongoing research, we can speculate about the ecological relationships in such cases by looking at the signs of the utility matrices:

$$\text{sgn}(D) = \begin{bmatrix} 0 & - & 0 & 0 & 0 \\ + & 0 & - & - & + \\ 0 & + & 0 & - & 0 \\ 0 & + & + & 0 & - \\ 0 & - & 0 & + & 0 \end{bmatrix} \quad \text{sgn}(U) = \begin{bmatrix} + & - & + & + & - \\ + & + & - & - & + \\ + & + & + & - & + \\ + & + & - & + & - \\ + & + & - & + & + \end{bmatrix}$$

The direct utility matrix is zero-sum in that for every donor there is a receiver of the flow. In ecological terms with think of a (+, −) relationship as predation or exploitation, but more generally it represents transfer from one compartment to another. Matching compartments pair-wise across the main diagonal gives the relationship type as shown in Table 5.2.

Notice, first that in the integral consideration that all compartments interact with each other, not just directly—there are no zero elements in the matrix. Next, notice that while two of the neutral direct relations became exploitation in one direction or the other, two others became mutualistic relations. Furthermore, note that one direct relations flipped when viewed in light of the system interactions—the relation between compartments 2 and 5 was antagonistic in the direct sense, but mutualistic in the holistic evaluation. The presence of each compartment benefits each other. Lastly, note that overall the integral matrix has more positive signs than negative signs leading to a holistic emergence of network mutualism. See previous literature cited (Patten, 1991, 1992; Fath and Patten, 1998; Fath, 2007) for other examples of utility analysis calculations. Let us now turn to and end with a qualitative interpretation of these network properties.

Table 5.2 Direct and integral relations for Cone Spring ecosystem

Direct	Integral
$(sd_{21}, sd_{12}) = (+, -) \rightarrow$ exploitation	$(sd_{21}, sd_{12}) = (+, -) \rightarrow$ exploitation
$(su_{31}, su_{13}) = (0, 0) \rightarrow$ neutralism	$(su_{31}, su_{13}) = (+, +) \rightarrow$ mutualism
$(su_{41}, su_{14}) = (0, 0) \rightarrow$ neutralism	$(su_{41}, su_{14}) = (+, +) \rightarrow$ mutualism
$(su_{51}, su_{15}) = (0, 0) \rightarrow$ neutralism	$(su_{51}, su_{15}) = (+, -) \rightarrow$ exploitation
$(sd_{32}, sd_{23}) = (+, -) \rightarrow$ exploitation	$(sd_{32}, sd_{23}) = (+, -) \rightarrow$ exploitation
$(sd_{42}, sd_{24}) = (+, -) \rightarrow$ exploitation	$(sd_{42}, sd_{24}) = (+, -) \rightarrow$ exploitation
$(sd_{52}, sd_{25}) = (-, +) \rightarrow$ reverse exploitation	$(sd_{52}, sd_{25}) = (+, +) \rightarrow$ mutualism
$(sd_{43}, sd_{34}) = (+, -) \rightarrow$ exploitation	$(sd_{43}, sd_{34}) = (-, -) \rightarrow$ competition
$(su_{53}, su_{35}) = (0, 0) \rightarrow$ neutralism	$(su_{53}, su_{35}) = (-, +) \rightarrow$ reverse exploitation
$(sd_{54}, sd_{45}) = (+, -) \rightarrow$ exploitation	$(sd_{54}, sd_{45}) = (+, -) \rightarrow$ exploitation

5.7 SUMMARY OF THE MAJOR INSIGHTS CARDINAL HYPOTHESES (CH) FROM NETWORK ENVIRON ANALYSIS

CH-1: network pathway proliferation

After conservative substance enters a system through its boundary it is *transacted*—conservatively transferred—between the living and non-living compartments within the system, being variously transformed and reconfigured by work along the way. The substance that enters as input to a particular compartment always while in the system remains within the output environ of that, and only that, compartment. Thus, environs as partition units within systems (Patten, 1978) defined by different inputs, become entangled within and between the tangible components of the system (this is network enfolding, another of the properties, discussed later). In accordance with second-law requirements, a part (and eventually all) of this substance is continually dissipated back to the environment by the entropy-generating processes that do work and make the system function. At any point in time subsequent to initial introduction, remaining substance continues to be transported around the system, and as it does so it traces out implicit pathways that extend in length by one unit at each transfer step. Pathway numbers increase exponentially with this increasing pathway length, with the result that the interior of the system becomes a complex interconnected network in which all components communicate, indirectly if not directly, with all or virtually all (depending on the connectivity structure) the others. This pathway proliferation is thus one of the sources of an essential holism, which environ theory impresses onto the interiors of systems. And without the openness of semipermeable boundaries, pathways would neither begin nor end, and the interior networks initiating output environs and terminating input environs at boundary points of entry and exodus would never exist.

CH-2: network non-locality

As pathways extend the amount of substance carried along at any given step is less than in the previous step due to dissipation. Therefore, pathways eventually end as they run out of originally introduced material. The rate of decay can be expressed as an exponential function, just as is the rate of pathway proliferation. Dissipation and pathway extension and growth in numbers are in conflict, but early in the transactional sequence following introduction the rate of the former exceeds that of the latter such that the total substance transferred between compartment pairs over the aggregate of pathways of a given length interconnecting them exceeds that of direct intercompartmental transfers. In other words, indirect pathways (those of lengths > 2) deliver more substance from any compartment to any other than a direct link between them. The influences carried by this transferred substance follows the substance itself in its being associated with pathways of particular lengths, and thus the conservative as well as non-conservative causes in the system can be said to be non-local. Indirect effects are dominant in systems, and this is especially true for complex systems, like ecosystems. The limit process that carries introduced conservative substance throughout the system to ultimate dissipation ensures that direct energy–matter links are quantitatively insignificant in comparison with the total. These links, provided by direct interactions such as feeding, serve only to structure the network; they make little contribution to intrasystem determination once this structuring is established. Dominant indirect effects in nature—that is a very different proposition from what we have in ecology at the present time. It is only a hypothesis, but robust in the mathematics of steady-state environ analysis. And each extended pathway that collectively provides its basis begins with openness at the boundary—either reception of input followed by forward passage of material in output environs to ultimate dissipation, or exhaustion from outputs preceded by the traceback of substance in input environs to its boundary points of original introduction.

CH-3: network distributed control
Ecologists from the beginning of the subject have always been concerned with issues of control—allogenic or autogenic—at physiological, population, community, and ecosystem levels of organization. The subject permeates the discipline in many forms. There are no obvious discrete controllers in ecosystems, though there are concepts like "key industry organisms" and "keystone species" that are suggestive of such possibilities. In general, in view of the non-locality property, control in ecosystems would have to be considered as realized by dominantly indirect means. This is the postulate of distributed control, and as with indirectness itself it is clear this has origins in boundary openness (Patten and Auble, 1981; Fath, 2004; Schramski et al., 2006).

CH-4: network homogenization
Another consequence of non-locality is the tendency for intermediate sources and sinks within systems to become blurred. That is, in the limit process that takes introduced energy and matter to ultimate boundary dissipation, there is so much transactional intercompartmental mixing around that causality tends to become evenly spread over the interactive network. In the extreme, this means that all compartments in ecosystems are about equally significant in generating and receiving influences to and from all the others. Originating and terminating at the open boundaries of circumscribing systems, the web of life based on local transactions of energy and matter tends to become quite homogeneous in its unseen ultimate intercomponent relationships (Patten et al., 1990; Fath and Patten, 1999).

CH-5: network internal amplification
It is sometimes observed in the environ mathematics of particular networks that substance introduced into one compartment at the boundary will appear more than once at another compartment, despite boundary dissipation in the interim. This is due to recycling, and it is easily seen how progressively diminishing fractions of a unit of introduced substance can cumulatively produce a sum over time in a limit process that exceeds the original amount. The second law cannot be defeated by this means, but energy cycling (Patten, 1985) following from open boundaries can compensate it and make it appear at least challenged in network organization. This is but one of numerous unexpected properties of networks contributed by cyclic interconnection and system openness.

CH-6: network unfolding

Ever since Raymond Lindeman (1942) pursued Charles Elton's original food cycles, but they came out unintendedly as sequential food chains instead, to which ecologists could better relate, mainstream empirical ecology (e.g., food-web theory, biogeochemical cycling) has had a difficult time returning to meaningful analysis of the concept of cycling. The preoccupation with chains prompted Higashi and Burns (Higashi et al., 1989) to develop a methodology for unfolding an arbitrary network into corresponding isomorphic "macrochains." Emanating from boundary points of input and arrayed pyramidally, these resemble the food pyramids of popular textbook depictions. Because the networks are cyclic, however, the macrochains differ from normal acyclic food chains in being indefinite in extent. Network unfolding refers to the indefinite proliferation of substance-transfer levels in ecosystems. The terminology "transfer pathways" and "transfer levels" is preferred to "food chains" and "trophic levels" because non-trophic as well as trophic processes are involved in any realistic ecosystem. Examples of non-trophic processes include import and export, anabolism and catabolism, egestion and excretion, diffusion and convection, sequestering, immobilization, and so on. Whipple subsequently modified the original unfolding methodology to discriminate the various trophic and non-trophic processes involved (Whipple and Patten, 1993; Whipple, 1998). The transfer levels so discriminated are non-discrete in containing contributions from most, if not all, the compartments in a system, and also they continue to increase in accordance with continuation of the limit process that ultimately dissipates all the introduced substance from the system. Exchange across open borders is at the heart of network unfolding.

CH-7: network synergism

The quantitative methodology of environ theory lends itself to development of certain qualitative aspects of the environmental relation with organisms. Energy and matter are objective quantities, but when cast as resources they engender subjective consequences of having or not having them. A concept introduced in game theory (von Neumann and Morgenstern, 1944, 1947) to describe the usefulness of outcomes or payoffs in games is *utility*. Environ mathematics implements this concept to bridge the gap between objective energy and matter and their subjective value as resources. Utility measures the relative *value* of absolute quantities; it is subjective information extracted from and added onto objective facts (Patten, 1991, 1992). A *zero–sum game* is one in which a winner gains exactly what the loser loses. Each conservative transaction in ecosystems is zero-sum, but it's relative benefit to the gainer and loss to loser may be different. Network synergism concerns how non-zero–sum interactions arise ultimately in conservative flow–storage networks whose proximate transactional linkages are zero-sum (Fath and Patten, 1998). Non-zero–sum interactions tend to be positive such that benefit/cost ratios, which equal one in direct transactions, tend in absolute value to exceed one when non-local indirect effects are taken into account. Such network synergism involves huge numbers of pathways (CH-1), dominant indirect effects (CH-2), and an indefinite transfer-level structure that unfolds as a limit process (CH-6)—all features of utility generation that reflect holistic organization in ecosystems, and the ecosphere. Once again—no open boundaries, no interior networks, no transactional or relational (see immediately below) interactions, and thus no non-zero-sum benefits to components. Life in networks is worth living, it can be said, because the key property of openness as a necessary condition has made possible all subsequent properties derived from it.

CH-8: network mutualism
This property is a qualitative extension of the previous one (Patten, 1991, 1992). Every compartment pair in a transactive network experiences positive (+), negative (−), or neutral *relations* derived from the transactions that directly and indirectly connect them. Ordered pairs of these three signs are nine in number and each pair reflects a qualitative interaction type. For example, the most common types of ecological interactions are (+, −) = predation, (−, −) = competition, (+, +) = mutualism, and (0, 0) = neutralism. Since the signs of benefits and costs in network synergism are + and −, respectively, the shift to |benefit/cost| ratios > 1 in CH-7 carries with it a shift to positive interaction types. This is network mutualism, and it indicates the benefits that automatically accrue to living organisms by their being coupled into transactive networks. Network synergism and mutualism together make nature a beneficial place conducive to life. This is quite different from seeing life only as a Darwinian "struggle for existence"; it is true locally, but not globally. There are built-in, openness-given properties of networks that, on balance, operate to reduce the struggle.

CH-9: network aggradation
As stated above in network example 1, when energy or matter moves across a system boundary, the system moves further from equilibrium and to that extent can be said to aggrade thermodynamically, the opposite of dissipation (boundary exit) and degradation (energy destruction). Aggradation is negentropic, although entropy is still generated and boundary-dissipated by interior aggrading processes. Environ theory appears to solve Schrödinger's *What-is-Life?* riddle (1944) of how antientropic development can proceed against the gradient of second-law degradation and dissipation. It shows a necessary condition for aggradation to be one single interior transaction within the interior system network—*simple adjacent electromagnetic interaction!* Given openness and sustained boundary input and output, there would appear to be no upper bound on this interior aggradation process. Thus, everything in nature that concerns differentiation and diversification of living and non-living structures and processes, and transactional interactions between these both within and across scales, can be seen as incrementally contributing to network aggradation—movement away from equilibrium. Recalling the observation above that solar photons come in small quanta that can only power processes at similarly small scales, and the fact that scales increase bottom-up through interactive coupling, network aggradation would appear to provide, perhaps, an electromagnetic-coupling answer to Schödinger's durable question, "What is life?" Unbounded energy- and matter-based linkage following on boundary openness would be an elegant basis indeed for life in its thermodynamic dimensions—simple and ubiquitous.

CH-10: network boundary amplification
When a compartment within a system brings substance into the system from outside, the importing compartment is favored in development over others that do not do this. The reason is a technical property of both the throughflow- and storage-generating matrices of environ analysis known as *diagonal dominance*. The throughflow case is easiest to explain. Its generating matrix multiplies the system input vector to produce a throughflow vector. Elements of the generating matrix represent the number of times substance introduced at one compartment will appear in another. First introduction by boundary input constitutes a first "hit" to the importing compartment. Non-importing compartments do not receive such first hits. In matrix multiplication of the generating matrix and input vector, importing compartments line up with their corresponding inputs such that first hits are recorded in diagonal positions; that is, input z_i to compartment i appears in the iith position of the generating matrix. This alignment gives the diagonal dominance. Off-diagonal elements represent contributions to i from the other interior compartments, not across the boundary. These do not receive their first-hit from boundary input, but from other interior compartments, and so are correspondingly smaller in numerical value. Storage generation is similar. Elements of storage-generating matrices denote residence times in each compartment of substance derived from other compartments. Diagonal dominance in these generating matrices also associates longer residence times with boundary vs. non-boundary inputs due to the first-hit phenomenon of the throughflow model, and longer residence times result in greater standing stocks. Boundary amplification may offer explanations for many phenomena in ecosystems—edge effects, zonation, ecotones, trophic levels, etc. Take the latter as an example. The transfer levels of network unfolding (CH-7) were seen to be non-discrete due to the mixing around of energy matter in the complex network of indefinitely extending pathways. This negates the mainstream Lindeman (1942) conception of discrete trophic levels. Boundary amplification restores discreteness, however, by giving another argument. Solar photons represent a resource initially outside the ecosystem boundary. Green plants bring them in and thus plant life receives the first-hit advantage and ascends to planetary dominance as a discrete trophic level, the primary producers. In a concentric, onion-like construction of the ecosystem, the resultant living plant biomass represents an untapped resource lying outside the possibilities for use (a functional boundary) until cellulose-digesting animals evolve. When they do, the first-hit advantage establishes them as a second discrete level—herbivores. These organisms initially lie outside the boundary of the next level within, until flesh-eating animals can be developed to employ this resource. Their first-hit advantage produces the third trophic level, carnivores. Omnivores evolve to utilize herbivores and carnivores, and at this point the trophic-dynamic model begins to lose its discreteness. All trophic levels produce dead organic residues, and the procaryotes and eucaryotes were already in place over evolutionary time to utilize these; first-hit boundary amplification establishes them as a discrete tropic category also—decomposers. Boundary amplification is a relatively new property in environ theory. It has the potential to explain the emergence of discrete trophic levels within complex reticular networks, and of course the more general property behind this is system openness.

CH-11: network enfolding

This property refers to the incorporation of indirect energy and matter flows and storages into empirically observed and measured flows and storages. It is another property of coupled systems elucidated by environ mathematics, and it potentially touches many areas of ecology such as chemical stoichiometry, embodied energy ("emergy"), and ecological indicators. Ecologists observe and measure, for example, the chemical composition of organisms or bulk samples. For an entity to have a "composition" means it is a *composite*—made up of materials brought to it from wherever its incoming network reaches in the containing system—directly and indirectly. This has consequences for even a seemingly straightforward concept like a "direct" flow where it turns out to be "macroscopically direct" and must be distinguished from "microscopically direct." To illustrate, in Figure 5.2b the flow f_{21} from compartment 1 to 2 is unambiguously direct. Macroscopically, the link and the process responsible for it (like eating) are direct, and microscopically so are the molecules (food) transferred because it is derived directly from the boundary input z_1. This latter directness is due to the fact that flow f_{21} represents the first transfer of boundary input from the compartment that received it to another; it is uncontaminated by substance from other sources because, in this case, input z_2 cannot reach compartment 1. The situation is different in Figure 5.3. The same flow f_{21} in this figure is still macroscopically adjacent to compartments 1 and 2 (i.e., "direct"), but now it contains composited flow derived from all three inputs. These inputs, the throughflows and storages (not shown) they generate, and also the other three adjacent flows are all complexly enfolded into f_{21}. The enfolding is mutual, and in this case it is universal because all interior network elements are reachable from all the others. The entire system of Figure 5.3 is thus (at steady-state) a composite of itself, which is the ultimate expression of holism. Moreover, this composition property, network enfolding, is true for complex systems generally. If one can imagine empirically sampling this system, f_{21} (and the other interior flows as well) strike the senses as direct. However, they are not since they contain indirect flows from the other sources a few to many times removed, and so are better considered as "adjacent", or perhaps just "observed." In environ mathematics, this

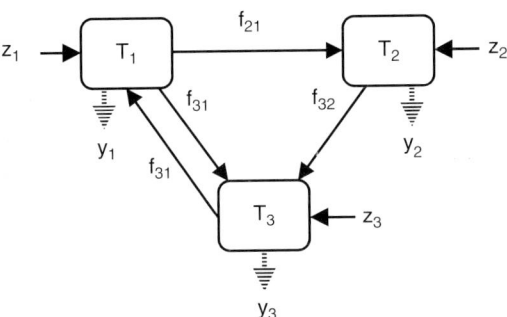

Figure 5.3 Three-compartment model illustrating network enfolding. Ecologically, the compartment labeled T_1 can be taken as producers, T_2 as consumers, and T_3 as decomposers.

embedding or entrainment of newly received inputs into the established flow-stream of the system is reflected in infinite series. Let $f_{21}/T_1 = g_{21}$ define a throughflow-specific dimensionless flow intensity. In effect g_{21} is a probability, thus its powers form a convergent infinite series: $(1 + g_{21} + g_{21}^{(2)} + \cdots + g_{21}^{(m)} + \cdots)$. The parenthesized superscripts denote coefficients derived from matrix, not scalar, multiplication. This power series maps the boundary input z_1 into the portion of throughflow at 2 contributed by this source: $T_{21} = (1 + g_{21} + g_{21}^{(2)} + \cdots + g_{21}^{(m)} + \cdots) z_1$. The first term of the series brings the input into the system: $1 \cdot z_1$. The second term represents the "direct" flow over the link of length 1: $g_{21} z_1$. All other terms represent indirect flows associated with pathways of all lengths 2, 3, …, $m, \ldots$ as $m \to \infty$. The throughflow component T_{21} accordingly contains a plethora of indirect flows: $(g_{21}^{(2)} + \cdots + g_{21}^{(m)} + \cdots) z$. This is one of three elements in the throughflow at compartment 2: $T_2 = T_{21} + T_{22} + T_{23}$. At compartment 1 the throughflow is similarly decomposable: $T_1 = T_{11} + T_{12} + T_{13}$, and at compartment 3: $T_3 = T_{31} + T_{32} + T_{33}$. Each term in these sums has a similar infinite series decomposition to that just given for T_{21}. With this, one can now appreciate there is more than that meets the eye in Figure 5.3. The focal flow f_{21} in question has a decomposition into enfolded elements as follows:

$$\begin{aligned} f_{21} &= g_{21} T_1 = g_{21}(T_{11} + T_{12} + T_{13}) \\ &= g_{21} \big[\big(1 + g_{11} + g_{11}^{(2)} + \cdots + g_{11}^{(m)} + \cdots\big) z_1 + \big(1 + g_{12} + g_{12}^{(2)} + \cdots + g_{12}^{(m)} + \cdots\big) z_2 \\ &\quad + \big(1 + g_{13} + g_{13}^{(2)} + \cdots + g_{13}^{(m)} + \cdots\big) z_3 \big] \end{aligned}$$

This is what an ecologist measuring f_{21} would measure empirically and consider a "direct" flow. One can see, however, that the entire system is embodied in this measurement. This is network enfolding. It gives a strong message about the inherent holism one can expect to be expressed in natural systems and, as stated above, its broad realization is likely to influence many areas of ecology.

CH-12: network environ autonomy
One of the richer consequences of system openness is the extension of "selves" into the broader surroundings that environ theory allows. Input and output environs extend outward from their defining entities, and in a sense reflect and project, respectively, the unique individuality of the latter in and into the world at large. Ownership of this "projection" is never released, however; the defining entities and their paired environs retain a unity that cannot be disassembled, only decomposed, for example by mathematical analysis. Moreover, as the entities themselves, particularly living ones, are unique, so also are the environs they project. A careful reading of the consequences of organisms as open systems having environments that uniquely attach to them is that not only are the organisms autonomous, but so are their environs, although of necessity more diffusely so. Estonian physiologist Jacob von Uexküll (1926) first put forward a view of the organism–environment relationship that is not very far from the one environ theory affords. Uexküll's organisms had an incoming "world-as-sensed" and an outgoing "world-of-action", corresponding to input and output environs, respectively. He held that the world-of-action wrapped around to the world-as-sensed via "function-circles" of the

organisms to produce an organism–environment complex that was a continuous whole, the true functional unit of nature in principle, not the organism by itself. However, Uexküll acknowledged the impossibility of tracing pathways of influence through the general environment from outputs back to inputs, and to that extent the theoretical autonomy of his organism–environment complexes was compromised. In environ theory, however, it is possible to keep track of substances within system boundaries because the systems involved are always models. In this case it can be demonstrated that each input and output environ within a system has its own set of unique characteristics, and these have an integrity and are maintained within the flow–storage stream of the model. The picture that emerges is that environs—partition elements of larger networks—have integrity as consituted though diffuse units, and are indeed autonomous within the systems they occupy. All measures of them that environ theory allows have never revealed two environs the same in any of their characteristics. A given compartment in one environ will have, for a fixed unit of boundary input or output, different flow, throughflow, storage, turnover and output characteristics from the same compartment in other environs. Uexküll may have been more correct than he realized when he wrapped his world-of-action around to his world-as-sensed via his "functions circles" of the organism, and said that the entire existence of the organism is imperilled should these function circles be interrupted.

CH-13: network holoevolution
The modern synthetic theory of biological evolution describes the evolutionary process in terms of two fundamental phenomena—transgenerational descent, and modification of descent over time. However, it only recognizes one kind of descent, genetic descent, and one major modifying process, natural selection that is capable of steering genetic descent non-randomly toward adaptive, fitness-maximizing configurations. Other modifying processes, such as genetic drift and mutation, act at random. Thus, genetic fitness becomes a matter of genes contributed by ancestral organisms to descendants via germ-line inheritance. This "germ track" is separated from a corresponding body or "soma track" of non-heritable, mortal phenotypes by the so-called "Weismann barrier" (1885). In the post-Watson–Crick era of the second half of the 20th Century, this barrier came to mean unidirectional DNA → RNA → protein coding. It ensures that "nothing that happens to the soma can be communicated to the germ cells and their nuclei" (Mayr, 1982, p. 700). This is *genetic determinism*—the doctrine that the structure and function of organisms are exclusively determined by genes. The dogma has always been questioned, but it is now under serious re-examination in systems biology (Klipp et al., 2005) as evidence mounts showing the different ways environment controls gene expression. Environment is underplayed in conventional evolutionary theory, where it appears only as a non-specific agent in natural selection. The environment of environ theory is two-sided, and both sides can be seen to possess potentially heritable elements, enough to support the hypothesis that environment and genomes both code for phenotypes, one from inside, the other from outside. The term *envirotype* has been coined to convey this idea Patten (in prep), and so "holoevolution" (CH-13) postulates joint and balanced contributions to phenotypes, which are mortal, from two evolutionary, potentially immortal, lines of inheritance. These are the conventional genotype, engaged in bottom-up coding within the cell, and a corresponding

external envirotype, manifesting top-down coding from without. Both input and output environs contribute to the heritable qualities of envirotypes, as outlined below:

- *Input-environ-based inheritance.* We can begin with the cell, and then mentally extrapolate outward through higher levels of organization to the organism and beyond. Each level, including that of the whole ecosystem, can be understood to have its own mechanisms of receiving environmental information, generating responses to this, and retaining (inheriting) through natural selection the ability to continue responses that prove beneficial to survival. Consider a cell receiving an energy- or matter-based signal from a near or distant source in its input environ. Biologist Bruce Lipton presents a scenario (http://www.brucelipton.com/newbiology.php) that effectively breaches the Weismann barrier and allows transmission of environmental data directly to the genome. Openness is at the heart of this process because the "cellular brain", as Lipton refers to it, is not located deep inside the cytoplasm or nucleus, but at the cell boundary. It is the cell membrane, or plasmalemma, a crystalline bi-layer of phospholipids and proteins that include a set of "integral membrane proteins" (IMPs) which serve as receptors and effectors. Receptor proteins respond to incoming molecules, or equally electromagnetic energy fields, by changing shape. This enables them to bond with specific effector proteins (enzymes, cytoskeletal elements, or transporters of electrons, protons, ions, or other chemical categories) that carry out behavior. If the requisite effector proteins are not already present in the cytoplasm, the IMP perception units activate expression of appropriate genes in the nucleus to produce new ones. New genes introduced into the DNA $\rightarrow$ RNA $\rightarrow$ protein sequence in the process remain behind to be copied, enabling the response to be repeated if adaptive, or ultimately fall obsolete and become consigned to the genomic set of inactive "junk" genes. Correct activations lead to life-enhancing behaviors, incorrect ones to maladaptation and death. Cellular adaptability thus becomes encoded in response to environmental inputs into new genes that encode new proteins, enabling survival in changing, but history-laden, environments. From the environ perspective, receptor molecules respond to signals transmitted in input environs, and effector molecules transmit the consequences to output environs. This initiates the second phase of environmental inheritance.
- *Output-environ-based inheritance.* When cells or other entities act on their environments the latter are changed as a result. This is "niche construction" Odling-Smee et al. (2003). Its essence is that it alters the machinery of natural selection because selection is, in the first instance, a manifestation of input environs. To the extent, however, that output environs generated by responses of their defining entities wrap around and become elements in those entities' input environs, the process becomes heritable, and epiphenomena such as autoevolution (Lima de Faria, 1988) emerge as distinct possibilities. Metazoan organization as "symbiogenic" aggregates of protozoan antecedents (Margulis, 1981, 1991) is an example. This is based on wrap-around feedback in which unicellular input- and output-environ overlaps are established in multicellular organization and achieve integration and identity. Organized cell communities possess self-similar IMP receptors responsive to the signal content of

hormones and other intercellular regulatory macromolecules. This requires that output-environ elements become input-environ elements. Membrane proteins convert adjacent environmental signals into cellular "awareness", expressed as changes in protein configurations. The movements occasioned by these changes represent useful kinetic energy (exergy) that does the work of achieving further departure from thermodynamic equilibrium, which multicellular organization represents compared to unicells. This is the essence of all antientropic growth and development extending to ecosystems and the ecosphere. Each level has mechanisms peculiar to it for implementing environment-based inheritance and perpetuating all forms of life—operational genotype–phenotype–envirotype complexes—through time. Organisms and their cells below, and communities and ecosystems above, can be said to inherit both their contained genes and attached environments from ancestral forms, and to the extent that these environments manifest holism, the great panoply of life spread over the globe at all levels of organization can be seen as evolving jointly, altogether, in the ecosphere—"holoevolution."

5.8 CONCLUSIONS

As evident throughout this chapter, ecosystems are networks of interacting biota and abiota. Rigorous methodological tools such as input–output analysis and ecological network analysis have been developed to deal with this complexity. As more and more applications of systems and network analyses arise, it is important to remember the common methodological roots of the approaches. In fact, because of its basic assumption about objects connected together as part of a larger system, which is used in several disciplines, the most promising application of network analysis may be as a platform for integrated environmental assessment models to address sustainability issues of combined human–natural systems.

6

Ecosystems have complex dynamics (growth and development)

> *Openness creates gradients*
> *Gradients create possibilities*
> *What you gain in precision,*
> *you lose in plurality*
> (Thermodynamics and Ecological Modelling,
> 2000, S.E. Jørgensen (ed.))

6.1 VARIABILITY IN LIFE CONDITIONS

All known life on earth resides in the thin layer enveloping the globe known as the ecosphere. This region extends from sea level to $\sim 10\,km$ into the ocean depths and approximately the same distance up into the atmosphere. It is so thin that if an apple were enlarged to the size of the earth the ecosphere would be thinner than the peel. Yet a vast and complex biodiversity has arisen in this region. Furthermore, the ecosphere acts as integrator of abiotic factors on the planet accumulating in disproportionate quantities particular elements favored by the biosphere (Table 6.1). In particular, note that carbon is not readily abundant in the abiotic spheres yet is highly concentrated in the biosphere, where nitrogen, silicon, and aluminum, while largely available, are mostly unincorporated.

However, even in this limited domain the conditions for living organisms may vary enormously in time and space.

The climatic conditions:

(1) The temperature can vary from ~ -70 to ~ 55 centigrade.
(2) The wind speed can vary from $0\,km/h$ to several hundred km/h.
(3) The humidity may vary from almost 0–100 percent.
(4) The precipitation from a few millimeter in average per year to several meter per year, which may or may not be seasonally aligned.
(5) Annual variation in day length according to latitude.
(6) Unpredictable extreme events such as tornadoes, hurricanes, earthquakes, tsunamis, and volcanic eruptions.

Table 6.1 Percent composition spheres for five most important elements

Lithosphere		Atmosphere		Hydrosphere		Biosphere	
Oxygen	62.5	Nitrogen	78.3	Hydrogen	65.4	Hydrogen	49.8
Silicon	21.22	Oxygen	21.0	Oxygen	33.0	Oxygen	24.9
Aluminum	6.47	Argon	0.93	Chloride	0.33	Carbon	24.9
Hydrogen	2.92	Carbon	0.03	Sodium	0.28	Nitrogen	0.27
Sodium	2.64	Neon	0.002	Magnesium	0.03	Calcium	0.073

The physical–chemical environmental conditions:

(1) Nutrient concentrations (C, P, N, S, Si, etc.)
(2) Salt concentrations (it is important both for terrestrial and aquatic ecosystems)
(3) Presence or absence of toxic compounds, whether they are natural or anthropogenic in origin
(4) Rate of currents in aquatic ecosystems and hydraulic conductivity for soil
(5) Space requirements

The biological conditions:

(1) The concentrations of food for herbivore, carnivore, and omnivore organisms
(2) The density of predators
(3) The density of competitors for the resources (food, space, etc.)
(4) The concentrations of pollinators, symbiants, and mutualists
(5) The density of decomposers

The human impact on natural ecosystems today adds to this complexity.

The list of factors determining the life conditions is much longer—we have only mentioned the most important factors. In addition, the ecosystems have history or path dependency (see Chapter 5), meaning that the initial conditions determine the possibilities of development. If we modestly assume that 100 factors are defining the life conditions and each of these 100 factors may be on 100 different levels, then 10^{200} different life conditions are possible, which can be compared with the number of elementary particle in the Universe 10^{81} (see also Chapter 3). The confluence of path dependency and an astronomical number of combinations affirms that the ecosphere could not experience the entire range of possible states, otherwise known as non-ergodicity. Furthermore, its irreversibility ensures that it cannot track back to other possible configurations. In addition to these combinations, the formation of ecological networks (see Chapter 5) means that the number of indirect effects are magnitudes higher than the direct ones and they are not negligible, on the contrary, they are often more significant than the direct ones, as discussed in Chapter 5.

What is the result of this enormous variability in the natural life conditions? We have found $\sim 0.5 \times 10^7$ species on earth and it is presumed that the number of species is double or 10^7. They have developed all types of mechanisms to live under the most varied life conditions including ones at the margin of their physiological limits. They have developed defense mechanisms. For example, some plants are toxic to avoid grazing, others have thorns, etc. Animals have developed horns, camouflage pattern, well-developed auditory sense, fast escaping rate, etc. They have furthermore developed integration mechanisms; fitting into their local web of life, often complementing and creating their environmental niche. The multiplicity of the life forms is inconceivable.

The number of species may be 10^7, but living organisms are all different. An ecosystem has normally from 10^{15} to 10^{20} individual organisms that are all different, which although it is a lot, makes ecosystems middle number systems. This means that the number of organisms is magnitudes less than the number of atoms in a room, but all the organisms, opposite the atoms in the rooms, have individual characteristics. Whereas large number systems such as the number of atoms in a room are amenable to statistical mechanics and small number problems such as planetary systems to classical mechanics or individual based modeling, middle number problems contain their own set of challenges. For one thing this variation, within and among species, provides diversity through co-adaptation and co-evolution, which is central both to Darwinian selection and network aggradation.

The competitive exclusion principle (Gause, 1934) claims that when two or more species are competing about the same limited resource only the best one will survive. The contrast between this principle and the number of species has for long time been a paradox. The explanation is rooted in the enormous variability in time and space of the conditions and in the variability of a wide spectrum of species' properties. A competition model, where three or more resources are limiting gives a result very different from the case where one or two resources are limiting. Due to significant fluctuations in the different resources it is prevented that one species would be dominant and the model demonstrates that many species competing about the same spectrum of resources can coexist. It is, therefore, not surprising that there exists many species in an environment characterized by an enormous variation of abiotic and biotic factors.

To summarize the number of different life forms is enormous because there are a great number of both challenges and opportunities. The complexity of ecosystem dynamics is rooted in these two incomprehensible types of variability.

6.2   ECOSYSTEM DEVELOPMENT

Ecosystem development in general is a question of the energy, matter, and information flows to and from the ecosystems. No transfer of energy is possible without matter and information and no matter can be transferred without energy and information. The higher the levels of information, the higher the utilization of matter and energy for further development of ecosystems away from the thermodynamic equilibrium (see also Chapters 2 and 4). These three factors are intimately intertwined in the fundamental nature of complex adaptive systems such as ecosystems in contrast to physical systems, that most often can be described completely by material and energy relations. Life is, therefore, both a material and a non-material (informational) phenomenon. The self-organization of life essentially proceeds by exchange of information.

E.P. Odum has described ecosystem development from the initial stage to the mature stage as a result of continuous use of the self-design ability (E.P. Odum, 1969, 1971a); see the significant differences between the two types of systems listed in Table 6.2 and notice that the major differences are on the level of information. Table 6.2 show what we often call E.P. Odum's successional attributes, but also a few other concepts such as for instance exergy and ecological networks have been introduced in the table.

Table 6.2 Differences between initial stage and mature stage are indicated

Properties	Early stages	Late or mature stage
(A) Energetic		
Production/respiration	$\gg 1$ or $\ll 1$	Close to 1
Production/biomass	High	Low
Respiration/biomass	High	Low
Yield (relative)	High	Low
Specific entropy	High	Low
Entropy production per unit of time	Low	High
Eco-exergy	Low	High
Information	Low	High
(B) Structure		
Total biomass	Small	Large
Inorganic nutrients	Extrabiotic	Intrabiotic
Diversity, ecological	Low	High
Diversity, biological	Low	High
Patterns	Poorly organized	Well organized
Niche specialization	Broad	Narrow
Organism size	Small	Large
Life cycles	Simple	Complex
Mineral cycles	Open	Closed
Nutrient exchange rate	Rapid	Slow
Life span	Short	Long
Ecological network	Simple	Complex
(C) Selection and homeostasis		
Internal symbiosis	Undeveloped	Developed
Stability (resistance to external perturbations)	Poor	Good
Ecological buffer capacity	Low	High
Feedback control	Poor	Good
Growth form	Rapid growth	Feedback controlled
Growth types	r-strategists	K-strategists

Chapter 6: Ecosystems have complex dynamics (growth and development)

The information content increases in the course of ecological development because an ecosystem integrates all the modifications that are imposed by the environment. Thus, it is against the background of genetic information that systems develop which allow interaction of information with the environment. Herein lies the importance in the feedback organism–environment, that means that an organism can only evolve in an evolving environment, which in itself is modifying. The differences between the two stages include entropy and eco-exergy.

The conservation laws of energy and matter set limits to the further development of "pure" energy and matter, while information may be amplified (almost) without limit. Limitation by matter is known from the concept of the limiting factor: growth continues until the element which is the least abundant relatively to the needs by the organisms is used up. Very often in developed ecosystems (for instance an old forest) the limiting elements are found entirely in organic compounds in the living organisms, while there is no or very little inorganic forms left in the abiotic part of the ecosystem. The energy input to ecosystems is determined by the solar radiation and, as we shall see later in this chapter, many ecosystems capture ~75–80 percent of the solar radiation, which is their upper physical limit. The eco-exergy, including genetic information content of, for example, a human being, can be calculated by the use of Equations 6.2 and 6.3 (see also Box 6.3 and Table 6.3). The results are ~40 MJ/g.

A human body of ~80 kg will contain ~2 kg of proteins. If we presume that 0.01 ppt of the protein at the most could form different enzymes that control the life processes and therefore contain the information, 0.06 mg of protein will represent the information content. If we presume an average molecular weight of the amino acids making up the enzymes of ~200, then the amount of amino acids would be $6 \times 10^{-8} \times 6.2 \times 10^{23}/200 \approx 2 \times 10^{17}$, that would give an eco-exergy that is (10^{-5} moles/g, T = 300 K, 20 different amino acids):

$$= 8.314 \times 80,000 \times 300 \times 10^{-5} \times 2 \times 10^{17} \ln 20 = 1.2 \times 10^{12} \text{ GJ}$$

It corresponds to 1.5×10^7 GJ/g. These are back of the envelope calculations and do not represent what is expected to be the information content of organisms in the future; but it seems possible to conclude that the development of the information content is very, very far from reaching its limit, in contrast to the development of the material and energy relations (see Figure 6.1).

Information has some properties that are very different from mass and energy.

(1) *Information unlike matter and energy can disappear without trace.* When a frog dies the enormous information content of the living frog may still be there a microseconds after the death in form of the right amino-acid sequences but the information is useless and after a few days the organic polymer molecules have decomposed.
(2) *Information expressed for instance as eco-exergy, it means in energy units, is not conserved.* Property 1 is included in this property, but in addition it should be stressed that living systems are able to multiply information by copying already achieved successful information, which implies that the information survives and

Table 6.3 β-values for different organisms

Early organisms	Plants	Animals	β-values
Detritus			1.00
Virus			1.01
Minimal cell			5.8
bacteria			8.5
Archaea			13.8
Protists	Algae		20
Yeast			17.8
		Mesozoa, Placozoa	33
		Protozoa, amoeba	39
		Phasmida (stick insects)	43
Fungi, moulds			61
		Nemertina	76
		Cnidaria (corals, sea anemones, jelly fish)	91
	Rhodophyta		92
		Gastroticha	97
Prolifera, sponges			98
		Brachiopoda	109
		Platyhelminthes (flatworms)	120
		Nematoda (round worms)	133
		Annelida (leeches)	133
		Gnathostomulida	143
	Mustard weed		143
		Kinorhyncha	165
	Seedless vascular plants		158
		Rotifera (wheel animals)	163
		Entoprocta	164
	Moss		174
		Insecta (beetles, flies, bees, wasps, bugs, ants)	167
		Coleodiea (sea squirt)	191
		Lipidoptera (buffer flies)	221
		Crustaceans	232
		Chordata	246
	Rice		275
	Gymnosperms (inl. pinus)		314

(*continued*)

Chapter 6: Ecosystems have complex dynamics (growth and development) 109

Table 6.3 (*Continued*)

Early organisms	Plants	Animals	β-values
		Mollusca, bivalvia, gastropoda	310
		Mosquito	322
	Flowering plants		393
		Fish	499
		Amphibia	688
		Reptilia	833
		Aves (birds)	980
		Mammalia	2127
		Monkeys	2138
		Anthropoid apes	2145
		Homo sapiens	2173

Note: β-values = exergy content relatively to the exergy of detritus (Jørgensen et al., 2005).

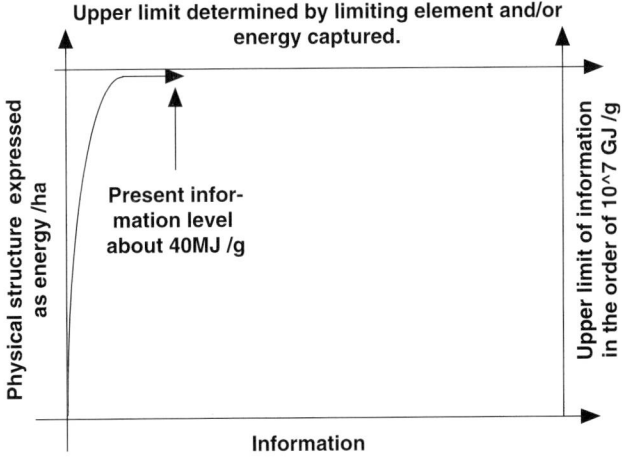

Figure 6.1 While further development of physical structure is limited either by a limiting element or by the amount of solar energy captured by the physical structure, the present most concentrated amount of information, the human body, is very far from its limit.

thereby gives the organisms additional possibilities to survive. The information is by autocatalysis (see Chapter 4) able to provide a pattern of biochemical processes that ensure survival of the organisms under the prevailing conditions determined by the physical–chemical conditions and the other organisms present in the ecosystem. By the growth and reproduction of organisms the information embodied in the genomes is copied. Growth and reproduction require input of food. If we calculate the eco-exergy of the food as just the about mentioned average of 18.7 kJ/g, the gain in eco-exergy may be more; but if we include in the energy content of the food the exergy content of the food, when it was a living organism or maybe even what the energy cost of the entire evolution has been, the gain in eco-exergy will be less than the eco-exergy of the food consumed. Another possibility is to apply emergy instead of energy. Emergy is defined later in this chapter (Box 6.2). The emergy of the food would be calculated as the amount of solar energy it takes to provide the food, which would require multiplication by a weighting factor $\gg 1$.

(3) *The disappearance and the copying of information, that are characteristic processes for living systems, are irreversible processes.* A made copy cannot be taken back and the death is an irreversible process. Although information can be expressed as eco-exergy in energy units it is not possible to recover chemical energy from information on the molecular level as know from the genomes. It would require a Maxwell's Demon that could sort out the molecules and it would, therefore, violate the second law of thermodynamics. There are, however, challenges to the second law (e.g., Capek and Sheehan, 2005) and this process of copying information could be considered one of them. Note that since the big bang enormous amounts of matter have been converted to energy ($E = mc^2$) in a form that makes it impossible *directly* to convert the energy again to mass. Similarly, the conversion of energy to information that is characteristic for many biological processes cannot be reversed directly in most cases. The transformation matter → energy → molecular information, which can be copied at low cost is possible on earth, but these transformation processes are irreversible.

(4) *Exchange of information is communication and it is this that brings about the self-organization of life.* Life is an immense communication process that happens in several hierarchical levels (Box 2.2). Exchange of information is possible with a very tiny consumption of energy, while storage of information requires that the information is linked to material, for instance are the genetic information stored in the genomes and is transferred to the amino-acid sequence.

A major design principle observed in natural systems is the feedback of energy from storages to stimulate the inflow pathways as a reward from receiver storage to the inflow source (H.T. Odum, 1971b). See also the "centripetality" in Chapter 4. By this feature the flow values developed reinforce the processes that are doing useful work. Feedback allows the circuit to learn. A wider use of the self-organization ability of ecosystems in environmental or rather ecological management has been proposed by H.T. Odum (1983, 1988).

E.P. Odum's idea of using attributes to describe the development and the conditions of an ecosystem has been modified and developed further during the past 15 years. Here we assess ecosystem development using ecological orientors to indicate that the development is not necessarily following in all details E.P. Odum's attributes because ecosystems are ontically open (Chapter 3). In addition, it is also rare that we can obtain data to demonstrate the validity of the attributes in complete detail. This recent development is presented in the next section.

Chapter 6: Ecosystems have complex dynamics (growth and development)

The concept of ecological indicators has been introduced ~15–20 years ago. These metrics indicate the ecosystem condition or the ecosystem health, and are widely used to understand ecosystem dynamics in an environmental management context. E.P. Odum's attributes could be used as ecological indicators; but also specific indicator species that show with their presence or absence that the ecosystem is either healthy or not, are used. Specific contaminants that indicate a specific disease are used as indicators. Finally, it should be mentioned that indicators such as biodiversity or thermodynamic variables are used to indicate a holistic image of the ecosystems' condition; further details see Chapter 10. The relationship between biodiversity and stability was previously widely discussed (e.g., May, 1973), who showed that there is not a simple relationship between biodiversity and stability of ecosystems. Tilman and his coworkers (Tilman and Downing, 1994) have shown that temperate grassland plots with more species have a greater resistance or buffer capacity to the effect of drought (a smaller change in biomass between a drought year and a normal year). However, there is a limit—each additional plant contributed less (see Figure 6.2). Previously, it has been shown that for models there is a strong correlation between eco-exergy (the definition; see Chapter 2) and the sum of many different buffer capacities. Many experiments (Tilman and Downing, 1994) have also shown that higher biodiversity increases the biomass and therefore the eco-exergy. There is in other words a relationship between biodiversity and eco-exergy and resistance or buffer capacity.

Box 6.1 gives the definitions for ecological orientors and ecological indicators. In ecological modeling, goal functions are used to develop structurally dynamic models. Also the definition of this third concept is included in the box.

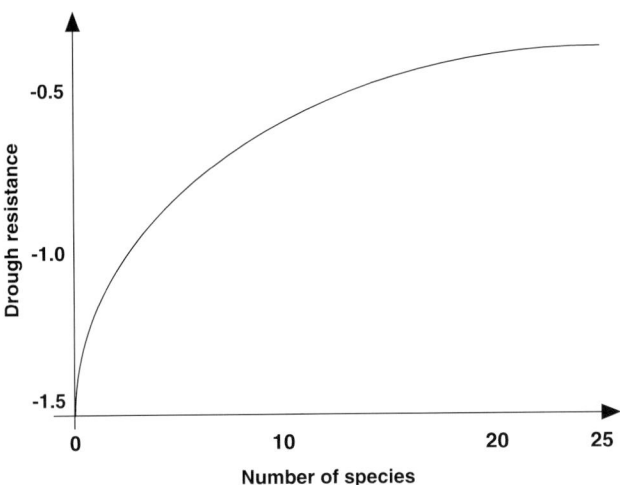

Figure 6.2 Results of the Tilman and Downing (1994) grassland experiments. The higher the number of species the higher the drought buffer capacity, although the gain per additional plant species decreasing with the number of species.

> **Box 6.1** Definitions of orientors, indicators, and goal functions
>
> *Ecological orientors*: Ecosystem variables that describe the range of directions in which ecosystems have a propensity to develop. The word orientors is used to indicate that we cannot give complete details about the development, only the direction.
> *Ecological indicators*: These indicate the present ecosystem condition or health. Many different indicators have been used such as specific species, specific contaminants, indices giving the composition of groups of organisms (for instance an algae index), E.P. Odum's attributes and holistic indicators included biodiversity and thermodynamic variables such as entropy or exergy.
> *Ecological goal functions*: Ecosystems do not have defined goals, but their propensity to move in a specific direction indicated by ecological orientors, can be described in ecological models by goal functions. Clearly, in a model, the description of the development of the state variables of the model has to be rigorously indicated, which implies that goals are made explicit. The concept should only be used in ecological modeling context.

It has been possible theoretically to divide most of E.P. Odum's attributes into three groups, defining three different growth and development forms for ecosystems (Jørgensen et al., 2000):

I. Biomass growth that is an attribute and also explains why P/B and R/B decreases with the development and the nutrients go from extrabiotic to intrabiotic pools.
II. Network growth that corresponds directly to increased complexity of the ecological network, more complex life and mineral cycles, a slower nutrient exchange rate and a more narrow niche specialization. It also implies a longer retention time in the system for energy and matter.
III. Information growth that explains the higher diversity, larger animals, longer life span, more symbiosis and feed back control and a shift from r-strategists to K-strategists.
IV. In addition, we may of course also have boundary growth—increased input, as we can observe for instance for energy during the spring. It is this initial boundary flow that is a prerequisite for maintaining ecosystems as open far-from-equilibrium systems.

6.3 ORIENTORS AND SUCCESSION THEORIES

The orientor approach that was briefly introduced above, describes ideal-typical trajectories of ecological properties on an integrated ecosystem level. Therefore, it follows the traditions of various concepts in ecological theory, which are related to environmental dynamics. A significant example is succession theory, describing "directional processes of colonization and extinction of species in a given site" (Dierssen, 2000). Although there are big intersections, these conceptual relationships have not become sufficiently obvious in the past, due to several reasons, which are mainly based on methodological problems and critical opinions which have been discussed eagerly after the release of

Odum's paper on the strategy of ecosystem development (1969). Which were the reasons for these controversies?

Traditional succession theory is basically oriented toward vegetation dynamics. The pioneers of succession research, Clements (1916) and Gleason (1917) were focusing mainly on vegetation. Consequently, also the succession definitions of Whittacker (1953), Egler (1954), Grime (1979), or Picket et al. (1987) are related to plant communities, while heterotrophic organisms often are neglected (e.g., Horn, 1974; Connell and Slayter, 1977). Therefore, also the conclusions of the respective investigations often have to be reduced to the development of vegetation components of ecosystems, while the orientor approach refers to the *whole ensemble* of organismic and abiotic subsystems and their interrelations. These conceptual distinctions for sure are preferable sources for misunderstandings.

A sufficient number of long-term data sets are not available. Therefore, as some authors state throughout the discussions of Odum's "strategy" paper (1969), the theoretical predictions of succession theory seem to be "based on untested assumptions or analogies" (e.g., Drury and Nisbet, 1973; Horn, 1974; Connell and Slayter, 1977), while there is only small empirical evidence. This situation becomes even more problematic if ecosystem data are necessary to test the theoretical hypotheses. Consequently, we will also in future have to cope with this lack of data, but we can use more and more empirical investigations, referring to the orientor principle, which have been reported in the literature (e.g., Marques et al., 2003; Müller et al., i.p.). We can hope for additional results from ecosystem analyses and Long Term Ecological Research Programs. Meanwhile validated models can be used as productive tools for the analysis of ecosystem dynamics.

The conceptual starting points differ enormously. Referring to the general objections against the maturity concept, Connell and Slayter (1977) funnel their heavy criticism about Odum's 24 ecosystem features into the questions of whether mature communities really are "internally controlled" and if "steady states really are maintained by internal feedback mechanisms". Having doubts in these facts, they state that, therefore, no characteristics can be deduced from this idea. Today, there is no doubt about the existence of self-organizing processes in all ecosystems (e.g., Jørgensen, 2002). Of course there are exterior constraints, but within the specific degrees of freedom, in fact the internal regulation processes are responsible for the development of ecosystems. Hence, the basic argument against the maturity concept has lost weight throughout the years.

Comparing successional dynamics, often different spatial and temporal scales are mixed. This point is related to the typical time scales of ecological investigations. They are most often carried out in a time span 2–4 years. Of course it is very difficult to draw conclusions over centuries from these short-term data sets. Also using paleo-ecological methods give rise to broad uncertainties, and when spatial differences are used to represent the steps of temporal developments, the questions of the site comparability introduces problems which might reduce the evidence of the findings enormously. Furthermore, there is the general problem of scale. If we transfer short-term results to long-term processes, then we cannot be sure to use the right algorithms and to take into account the correct, scale conform constraints and processes (O'Neill et al., 1986). And, looking at the spatial scale, the shifting mosaic hypothesis (Remmert, 1991) has shown that there will be huge differences if different spatial extents are taken into account, and that local instabilities can be leading to regional steady-state situations. What we can see is that there are many empirical traps we can fall into. Maybe the connection of empirical research and ecological modeling can be helpful as a "mechanism of self-control" in this context.

Due to the "ontic openness" of ecosystems, predictability in general is rather small but in many cases exceptions can be found. The resulting dilemma of a system's inherent uncertainty can be regarded as a consequence of the internal complexity of ecosystems, the non-linear character of the internal interactions and the often-unforeseeable dynamics of environmental constraints. Early on, succession researchers found the fundamentals of this argument, which are broadly accepted today. The non-deterministic potential of ecological developments has already been introduced in Tansley's (1953) polyclimax theory, which is based on the multiple environmental influences that function as constraints for the development of an ecosystem. Simberloff (1982) formulates that "the deterministic path of succession, in the strictest Clementsian mono-climax formulation, is as much an abstraction as the Newtonian particle trajectory" and Whittaker (1972) states, "the vegetation on the earth's surface is in incessant flux". Stochastic elements, complex interactions, and spatial heterogeneities take such important influences that the idea of Odum (1983) that "community changes...are predictable", must be considered in relative terms today, if detailed prognoses (e.g., on the species level) are desired. But this does not mean that general developmental tendencies can be avoided, i.e. this fact does not contradict the general sequence of growth forms as formulated in this volume. Quite the opposite: this concept realizes the fact that not all ecosystem features are optimized throughout the whole sequence, a fact that has been pointed out by Drury and Nisbet (1973) and others.

Disturbances are causes for separating theoretical prognoses from practical observations. One example for these non-deterministic events is disturbance, which plays a major role in ecosystem development (e.g., Drury and Nisbet, 1973; Sousa, 1984). Odum (1983) has postulated that succession "culminates in the establishment of as stable an ecosystem as its biologically possible on the site in question" and he notes that mature communities are able to buffer the physical environment to a greater extent than the young community. In his view stability and homoeostasis can be seen as the result (he even speaks about a purpose) of ecological succession from the evolutionary standpoint. But in between, the guiding paradigm has changed: today Holling's adaptive cycling model (1986) has become a prominent concept, and destruction is acknowledged as an important component of the continuous adaptation of ecosystems to changing environmental constraints. This idea also includes the feature of brittleness in mature states, which can support the role of disturbance as a setting of new starting points for an oriented development.

Terminology has inhibited the acceptance of acceptable ideas. The utilization of terms like "strategy", "purpose", or "goal" has led to the feeling that holistic attitudes toward ecological successions in general are loaded with a broad teleological bias. Critical colleagues argued that some of these theories are imputing ecosystems to be "intentionally" following a certain target or target state. This is not correct: the series of states is a consequence of internal feedback processes that are influenced by exterior constraints and impulses. The finally achieved attractor state thus is a result, not a cause.

Summarizing, many of the objections against the initial theoretical concepts of ecosystem development and especially against the stability paradigm have proven to be correct, and they have been modified in between. Analogies are not used anymore, and the number of empirical tests is increasing. On the other hand, the theory of self-organization has clarified many critical objections. Thus, a consensus can be reached if cooperation between theory and empiricism is enhanced in the future.

6.4 THE MAXIMUM POWER PRINCIPLE

Lotka (1925, 1956) formulated the maximum power principle. He suggested that systems prevail that develop designs that maximize the flow of *useful* (for maintenance and growth) energy, and Odum used this principle to explain much about the structure and processes of ecosystems (Odum and Pinkerton, 1955). Boltzmann (1905) said that the struggle for existence is a struggle for free energy available for work, which is a definition very close to the maximum exergy principle introduced in the next section. Similarly, Schrödinger (1946) pointed out that organization is maintained by extracting order from the environment. These two last principles may be interpreted as the systems that are able to gain the most free energy under the given conditions, i.e. to move most away from the thermodynamic equilibrium will prevail. Such systems will gain most biogeochemical energy available for doing work and therefore have most energy stored to use for maintenance and buffer against perturbations.

H.T. Odum (1983) defines the maximum power principle as a maximization of *useful* power. It is applied on the ecosystem level by summing up all the contributions to the *total* power that are useful. It means, that non-useful power is not included in the summation. Usually the maximum power is found as the sum of all flows expressed often in energy terms for instance kJ/24 h.

Brown et al. (1993) and Brown (1995) has restated the maximum power principle in more biological terms. According to the restatement it is the transformation of energy into work (consistent with the term useful power) that determines success and fitness. Many ecologists have incorrectly assumed that natural selection tends to increase efficiency. If this were true, then endothermy could never have evolved. Endothermic birds and mammals are extremely inefficient compared with reptiles and amphibians. They expend energy at high rates in order to maintain a high, constant body temperature, which, however, gives high levels of activities independent of environmental temperature (Turner, 1970). Brown (1995) defines fitness as reproductive power, dW/dt, the rate at which energy can be transformed into work to produce offspring. This interpretation of the maximum power principle is even more consistent with the maximum exergy principle that is introduced in the next section, than with Lotka's and Odum's original idea.

In the book *Maximum Power: The Ideas and Applications of H.T. Odum*, Hall (1995) has presented a clear interpretation of the maximum power principle, as it has been applied in ecology by H.T. Odum. The principle claims that power or output of useful work is maximized, not the efficiency and not the rate, but the tradeoff between a high rate and high efficiency yielding most useful energy or useful work. It is illustrated in Figure 6.3.

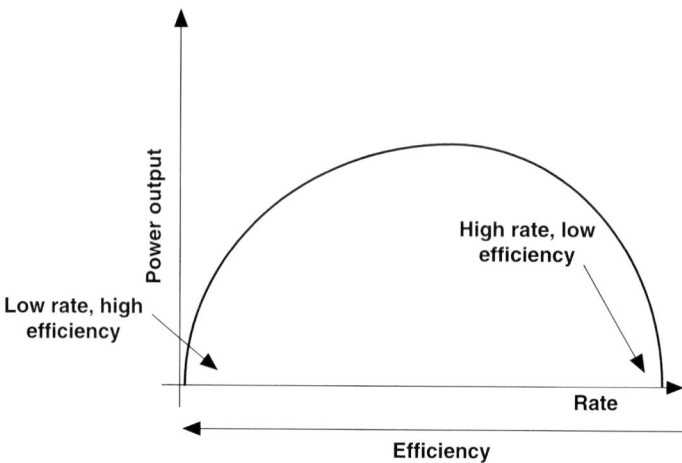

Figure 6.3 The maximum power principle claims that the development of an ecosystem is a tradeoff (a compromise) between the rate and the efficiency, i.e. the maximum power output per unit of time.

Hall is using an interesting semi-natural experiment to illustrate the application of the principle in ecology. Streams were stocked with different levels of predatory cutthroat trout. When predator density was low, there was considerable invertebrate food per predator, and the fish used relatively little maintenance energy searching for food per unit of food obtained. With a higher fish-stocking rate, food became less available per fish, and each fish had to use more energy searching for it. Maximum production occurred at intermediate fish-stocking rates, which means intermediate rates at which the fish utilized their food.

Hall (1995) mentions another example. Deciduous forests in moist and wet climates tend to have a leaf area index (LAI) of ~6 m^2/m^2. Such an index is predicted from the maximum power hypothesis applied to the net energy derived from photosynthesis. Higher LAI values produce more photosynthate, but do so less efficiently because of the metabolic demand of the additional leaf. Lower leaf area indices are more efficient per leaf, but draw less power than the observed intermediate values of roughly 6.

The same concept applies for regular fossil fuel power generation. The upper limit of efficiency for any thermal machine such as a turbine is determined by the Carnot efficiency. A steam turbine could run at 80 percent efficiency, but it would need to operate at a nearly infinitely slow rate. Obviously, we are not interested in a machine that generates electricity or revenues infinitely slowly, no matter how efficiently. Actual operating efficiencies for modern steam powered generator are, therefore, closer to 40 percent, roughly half the Carnot efficiency.

These examples show that the maximum power principle is embedded in the irreversibility of the world. The highest process efficiency can be obtained by endo-reversible conditions, meaning that all irreversibilities are located in the coupling of the system to its surroundings, there are no internal irreversibilities. Such systems will, however, operate

Chapter 6: Ecosystems have complex dynamics (growth and development)

too slowly. Power is zero for any endo-reversible system. If we want to increase the process rate, it will imply that we also increase the irreversibility and thereby decrease the efficiency. The maximum power is the compromise between endo-reversible processes and very fast completely irreversible processes.

The concept of emergy (embodied energy) was introduced by H.T. Odum (1983) and attempts to account for the energy required in formation of organisms in different trophic levels. The idea is to correct energy flows for their quality. Energies of different types are converted into equivalents of the same type by multiplying by the energy transformation ratio. For example fish, zooplankton, and phytoplankton can be compared by multiplying their actual energy content by their solar energy transformation ratios. The more transformation steps there are between two kinds of energy, the greater the quality and the greater the solar energy required to produce a unit of energy (J) of that type. When one calculates the energy of one type, that generates a flow of another, this is sometimes referred to as the embodied energy of that type. Figure 6.4 presents the concept of embodied energy in a hierarchical chain of energy transformation. One of the properties of high quality energies is their flexibility (which requires information). Whereas low quality products tend to be special, requiring special uses, the higher quality part of a web is of a form that can be fed back as an amplifier to many different web components.

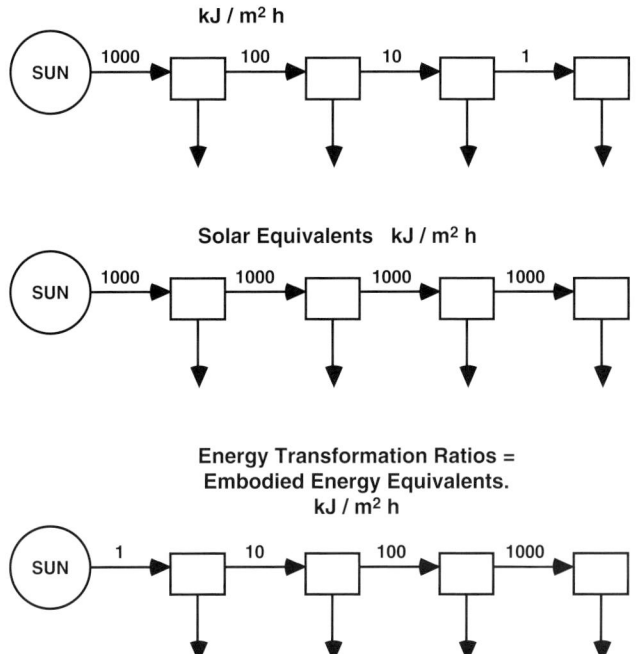

Figure 6.4 Energy flow, solar equivalents, and energy transformation ratios = embodied energy equivalents in a food chain (Jørgensen, 2002).

A good down to earth example of what emergy is, might be the following: in 1 year one human can survive on 500 fish each of the size of 500 g, that may have consumed 80,000 frogs with the size of 20 g. The frogs may have eaten 18×10^6 insects of the size of 1 g. The insects have got their food from 200,000 kg dry matter of plants. As the photosynthetic net production has an efficiency of 1 percent, the plants have required an input of $\sim 3.7 \times 10^9$ J, presuming an energy content of plant dry matter of 18.7 kJ/g. To keep one human alive costs, therefore 3.7×10^9 J, although the energy stock value of a human being is only in the order 3.7×10^5 J or 10,000 times less. The transformity is, therefore, 10,000.

H.T. Odum has revised the maximum power principle by replacing power with emergy–power (empower), meaning that all the contributions to power are multiplied by a solar equivalent factor that is named transformity to obtain solar equivalent joules (sej) (see Box 6.2). The difference between embodied energy flows and power, see Equation 6.1, simply seems to be a conversion to solar energy equivalents of the free energy.

Box 6.2 Emergy

"*Emergy* is the available energy of one kind previously used up directly and indirectly to make a service or product. Its unit is the emjoule [(ej)]" and its physical dimensions are those of energy (Odum, 1996). In general, since solar energy is the basis for all the energy flows in the biosphere, we use *solar emergy* (measured in *sej*, solar emjoules), the solar energy equivalents required (directly or indirectly) to make a product.

The total emergy flowing through a system over some unit time, referenced to its boundary source, is its empower, with units [sej/(time)] (Odum, 1988). If a system, and in particular an ecosystem, can be considered in a relatively steady state, the empower (or emergy flow) can be seen as nature's "labor" required for maintaining that state.

The emergy approach starts from Lotka's maximum power principle (1922, 1956) and corrects the function, which is maximized, since not all the energy types have the same ability of doing actual work. Thus power (flow of energy) is substituted by empower (flow of emergy), that is "in the competition among self-organizing processes, network designs that maximize empower will prevail" (Odum, 1996).

Transformity is the ratio of emergy necessary for a process to occur to the exergy output of the process. It is an intensive function and it is dimensionless, even though sej/J is used as unit.

Emergy can be written as a function of transformity and exergy as follows (*i* identifies the inputs):

$$Em = \sum_i \tau_i Ex_i$$

While transformity can be written as

$$\tau_k = Em_k / Ex_k$$

even though it is often calculated as

$$\tau_k = \frac{Em_k / \text{time}}{Ex_k / \text{time}}$$

By definition the transformity of sunlight is equal to 1 and this assumption avoids the circularity of these expressions. All the transformities (except that of solar energy) are, therefore, greater than 1.

Transformities are always measured relative to a planetary solar emergy baseline and care should be taken to ensure that the transformities used in any particular analysis are all expressed relative to the same baseline (Hall, 1995). However, all the past baselines can be easily related through multiplication by an appropriate factor and the results of an emergy analysis do not change by shifting the baseline (Odum, 1996).

Emergy and transformity are not state functions, i.e. they strongly depend on the process that is used to obtain a certain item. There are transformities that are calculated from global biosphere data (i.e. rain, wind, geothermal heat) and others that, being the result of more complex and variable processes have high variability: for example, electricity can be generated by many processes (using wood, water, coal, gas, tide, solar radiation, etc.) each with a different transformity (Odum, 1996).

In general transformity can be seen as a measure of "quality": while emergy, following "memorization" laws, can in general remain constant or grow along transformation chains, since as energy decreases, transformities increase. On the other hand, when comparing homologous products, the lower the transformity, the higher the efficiency in transforming solar emergy into a final product.

Emergy is a donor-referenced concept and a measure of convergence of energies, space and time, both from global environmental work and human services into a product. It is sometimes referred to as "energy memory" (Scienceman, 1987) and its logic (of "memorization" rather than "conservation") is different from other energy-based analyses as shown by the emergy "algebra". The rules of emergy analysis are:

- All source emergy to a process is assigned to the processes' output.
- By-products from a process have the total emergy assigned to each pathway.
- When a pathway splits, the emergy is assigned to each 'leg' of the split based on its percentage of the total energy flow on the pathway.
- Emergy cannot be counted twice within a system: (a) emergy in feedbacks cannot be double counted; (b) by-products, when reunited, cannot be added to equal a sum greater than the source emergy from which they were derived.

For in depth discussion of this issue and the differences between energy and emergy analysis see Odum (1996).

Embodied energy is, as seen from these definitions, determined by the biogeochemical energy *flow* into an ecosystem component, measured in solar energy equivalents. The stored emergy, *Em*, per unit of area or volume to be distinguished from the emergy flows can be found from:

$$Em = \sum_{i=1}^{n} \Omega_i c_i \quad (6.1)$$

where Ω_i is the quality factor which is the conversion to solar equivalents, as illustrated in Figure 6.4, and c_i is the concentration expressed per unit of area or volume.

The calculations reduce the difference between stored emergy (= embodied energy) and stored exergy (see next section), to the energy quality factor. The quality factor for exergy accounts for the information embodied in the various components in the system (detailed information is given in the next section), while the quality factor for emergy accounts for the solar energy cost to form the various components. Emergy calculates thereby how much solar energy (which is our ultimate energy resource) it has taken to obtain 1 unit of biomass of various organisms. Both concepts attempt to account for the quality of the energy. Emergy by looking into the energy flows in the ecological network to express the energy costs in solar equivalents. Exergy by considering the amount of biomass and information that has accumulated in that organism. One is measure of the path that was taken to get to a certain configuration, the other a measure of the organisms in that configuration.

6.5 EXERGY, ASCENDENCY, GRADIENTS, AND ECOSYSTEM DEVELOPMENT

Second law dissipation acts to tear down structure and eliminate gradients, but ecosystems have the ability to move away from thermodynamic equilibrium in spite of the second law dissipation due to an inflow of energy from solar radiation. Even a simple physical system as a Bernard cell is using an inflow of energy to move away from thermodynamic equilibrium. A Bernard cell consists of two plates, that are horizontally placed in water a few centimeter from each other. The lower plate has higher temperature than the upper plate. Consequently, energy is flowing from the lower to the upper plate. When the temperature difference is low the motion of the molecules is random. When the temperature exceeds a critical value the water molecules are organized in a convection pattern, series of rolls or hexagons. The energy flow increases due to the convection. The greater the flow of energy the steeper the temperature gradient (remember that work capacity = entropy times temperature gradient) and the more complex the resulting structure. Therefore, greater exergy flow moves the system further away from thermodynamic equilibrium—higher temperature gradient and more ordered structure containing information corresponding to the order. The origin of ordered structures is, therefore, openness and a flow of energy (see Chapter 2). Openness and a flow of energy are both necessary conditions (because it will always cost energy to maintain an ordered structure) and sufficient (as illustrated with the Bernard cell). Morowitz (1968, 1992) has shown that an inflow of energy always will create one cycle of matter, which is an ordered structure. Openness and a flow of energy is, however, not sufficient condition for ecosystems (see Chapter 2), as additional conditions are required to ensure that the ordered structure is an ecosystem.

Biological systems, especially, have many possibilities for moving away from thermodynamic equilibrium, and it is important to know along which pathways among the possible ones a system will develop. This leads to the following hypothesis (Jørgensen and Mejer, 1977, 1979; Jørgensen, 1982, 2001, 2002; Jørgensen et al., 2000): if a system receives an input of exergy, then it will utilize this exergy to perform work. The work performed is first applied to maintain the system (far) away from thermodynamic equilibrium whereby exergy is lost by transformation into heat at the temperature of the environment. If more exergy is available, then the system is moved further away from thermodynamic equilibrium, reflected in growth of gradients. If there is offered more than one pathway to depart from equilibrium, then the one yielding the highest eco-exergy storage (denoted Ex) will tend to be selected. Or expressed differently: among the many ways for ecosystems to move away from thermodynamic equilibrium, the one maximizing dEx/dt under the prevailing conditions will have a propensity to be selected.

Rutger de Wit (2005) has expressed preference for a formulation where the flow of exergy is replaced by a flow of free energy, which of course is fully acceptable and makes the formulation closer to classic thermodynamics. However, eco-exergy storage can hardly be replaced by free energy because it is a free-energy *difference* between the system and the same system at thermodynamic equilibrium. The reference state is therefore different from ecosystem to ecosystems, which is considered in the definition of eco-exergy. In addition, free energy is not a state function far from thermodynamic equilibrium—just consider the immediate loss of eco-exergy when an organism dies. Before the death the organism has high eco-exergy because it can utilize the enormous information that is embodied in the organism, but at death the organism loses immediately the ability to use this information that becomes, therefore, worthless. Moreover, the information part of the eco-exergy cannot be utilized directly as work; see the properties of information presented in Section 6.2.

Just as it is not possible to prove the three laws of thermodynamics by deductive methods, so can the above hypothesis only be "proved" inductively. A number of concrete cases which contribute generally to the support of the hypothesis will be presented below and in Chapters 8 and 9. Models are often used in this context to test the hypothesis. The exergy can be approximated by use of the calculation methods in Box 6.3. Strictly speaking exergy is a measure of the useful work which can be performed. Conceptually, this obviously includes the energetic content of the material, i.e. biomass, but also the state of organization of the material. One way to measure the organization is the information content of the material, which could be the complexity at the genetic or ecosystem levels. Currently, the organizational aspect of exergy is expressed as Kullbach's measure of information based on the genetic complexity of the organism:

$$Ex = B\,RT\,K \tag{6.2}$$

where B is the biomass, R the gas constant, T the Kelvin temperature, and K Kullbach's measure of information (further details see Box 6.3). The exergy of the organism is found on basis of the information that the organism carries:

$$Ex_i = \beta_i c_i \tag{6.3}$$

where Ex_i is the exergy of the ith species, β_i a weighting factor that considers the information the ith species is carrying in c_i (Table 6.2). Jørgensen et al. (2005) show how the β-values have been found for different organisms. A high uncertainty is, however, associated with the assessment of the β-values, which implies that the exergy calculations have a corresponding high uncertainty. In addition, the exergy is calculated based on models that are simplifications of the real ecosystems. The calculated exergy should, therefore, only be used relatively and considered an index and not a real absolute exergy value.

Box 6.3 Calculation of eco-exergy

It is possible to distinguish between the exergy of information and of biomass (Svirezhev, 1998). p_i defined as c_i/B, where

$$B = \sum_{i=1}^{n} c_i$$

is the total amount of matter in the system, is introduced as new variable in Equation 2.8:

$$Ex = B\,RT \sum_{i=1}^{n} p_i \ln \frac{p_i}{p_{i,o}} + B \ln \frac{B}{B_o}$$

As the biomass is the same for the system and the reference system, $B \approx B_o$ exergy becomes a product of the total biomass B (multiplied by RT) and Kullback measure:

$$K = \sum_{i=1}^{n} p_i \ln \frac{p_i}{p_{i,o}}$$

where p_i and $p_{i,o}$ are probability distributions, a posteriori and a priori to an observation of the molecular detail of the system. It means that K expresses the amount of information that is gained as a result of the observations. If we observe a system that consists of two connected chambers, then we expect the molecules to be equally distributed in the two chambers, i.e. $p_1 = p_2 = 1/2$. If we, on the other hand, observe that all the molecules are in one chamber, we get $p_1 = 1$ and $p_2 = 0$.

Specific exergy is exergy relatively to the biomass and for the ith component: Sp. ex.$_i = Ex_i/c_i$. It implies that the total specific exergy per unit of area or per unit of volume of the ecosystem is equal to RTK.

For the components of the ecosystem, 1 (covers detritus), 2, 3, 4.... N, the probability, $p_{1,o}$, consists at least of the probability of producing the organic matter (detritus), i.e. $p_{1,o}$, and the probability, $p_{i,a}$, to find the correct composition of the enzymes determining the biochemical processes in the organisms. Living organisms use 20 different amino acids and each gene determines on average a sequence of ~700 amino acids (Li and Grauer, 1991). $p_{i,a}$, can be found from the number of permutations among which the characteristic amino-acid sequence for the considered organism has been selected.

The total exergy can be found by summing up the contributions originating from all components. The contribution by inorganic matter can be neglected as the contributions by detritus and even to a higher extent from the biological components are much higher due to an extremely low probability of these components in the reference system. Roughly, the more complex (developed) the organism is the more enzymes with the right amino-acid sequence are needed to control the life processes, and therefore the lower is the probability $p_{i,a}$, The probability $p_{i,a}$, for various organisms has been found on basis of our knowledge about the genes that determine the amino-acid sequence. As the concentrations are multiplied by RT and $\ln(p_i/p_{i,o})$, denoted β_i; a table with the β-values for different organisms have been prepared (see Table 6.3). The contribution by detritus, dead organic matter, is in average 18.7 kJ/g times the concentration (in g/unit of volume). The exergy can now be calculated by the following equation:

$$\text{Exergy total} = \sum_{i=1}^{n} \beta_i c_i \quad \text{(as detritus equivalent)}$$

The β-values are found from Table 6.3 and the concentration from modeling or observations. By multiplication by 18.7, we get the exergy in kilojoules.
Notice that

$$Ex_{\text{bio}} = \sum_{i=1}^{n} c_i \quad \text{(as detritus equivalent)}$$

while

$$Ex_{\text{info}} = \sum_{i=1}^{n} (\beta_i - 1) c_i \quad \text{(as detritus equivalent)}$$

Consistency of the exergy-storage hypothesis, as we may call it, with other theories (goal functions, orientors; see Sections 6.2 and 6.3) describing ecosystem development will be demonstrated as a pattern in a later section of this chapter. It should, however, in this context be mentioned that exergy storage in the above-mentioned main hypothesis can be replaced by maximum power. Exergy focuses on the storage of biomass (energy) and information, while power considers the energy flows resulting from the storages.

Ascendency (Box 4.1) is a complex measure of the information and flows embodied the ecological network. The definition is given in Chapter 4. At the crux of ascendency lies the action of autocatalysis (Chapter 4). One of the chief attributes of autocatalysis is what Ulanowicz (1997) calls "centripetality" or the tendency to draw increasing amounts of matter and energy into the orbit of the participating members (Chapter 4). This tendency inflates ascendency both in the quantitative sense of increasing total system activity and qualitatively by accentuating the connections in the loop above and beyond pathways connecting non-participating members. At the same time, increasing storage of exergy is a particular manifestation of the centripetal tendency, and the dissipation of external exergy gradients to feed system autocatalysis describes centripetality in an almost tautological fashion.

In retrospect, the elucidation of the connections among ascendency, eco-exergy, and aggradation (Ulanowicz et al., 2006) has been effected by stages that are typical of theory-driven research. First, it was noted in phenomenological fashion how quantitative observations of the properties were strongly correlated; the correlation coefficient, r^2, was found for a number of models to be 0.99 (Jørgensen, 1995). Thereafter, formal definitions were used to forge theoretical ties among the separate measures. Finally, the perspective offered by these new theoretical connections facilitated a verbal description of the common unitary agency that gave rise to the independent trends that had been formalized as separate principles. Eco-exergy and ascendency represent two sides of the same coin or two different angles in the description of ecosystem development. A simple physical phenomenon as light requires both a description as waves and as particle to be fully understood. It is, therefore, understandable that ecosystem developments that are much more complicated than light require multiple description. Exergy covers the storage, maximum power the flows, and ascendency the ecological network and all three concepts contribute to the overall aggradation, moving away from thermodynamic equilibrium. All three concepts have well-structured roots in the theoretical soil. Their shortcomings are, however, that calculations of exergy, maximum power, and ascendency always will be incomplete due to the enormous complexity of ecosystems (see Section 6.1).

Ecosystems can also be understood as a (high) number of interacting gradients, which are formed by self-organizing processes (Mueller and Leupelt, 1998). Gradient maintenance costs exergy that is transformed by decomposition processes to heat at the temperature of the environment, i.e. the exergy is lost. The gradients can be classified in various ways, but we could also distinguish three types of gradients corresponding to the three growth forms (see Section 6.2): gradients due to organisms in the ecosystems (trees are good illustrations), gradients due to formation of a more complex network (for instance the spatial distribution of more or fewer niches), and gradients due to information (the level of information could be used directly as illustration). The first-mentioned class of gradients requires the most exergy for maintenance, while information gradients require very little or no exergy for maintenance. Gradients summation is captured in the exergy measure since work capacity is an extensive variable times a gradient (see Chapter 2).

Exergy storage is the simplest of the three concepts to calculate; but clearly the assessment of the β-values has some shortcomings. The latest list is more differentiated than

Chapter 6: Ecosystems have complex dynamics (growth and development)

the previous ones (Jørgensen et al., 1995; Fonseca et al., 2000) and is based on the latest results of the entire genome analyses for 11 species plus a series of complexity measures for a number of species, families, orders, or classes. The list will most probably be improved as genetics gains more information about the genomes and proteomes of more species. The total information of an ecosystem should furthermore include the information of the network. All ecological models that are used as basis for the exergy calculations are much simpler than the real network and the information contained in the network of the model become negligible compared to the exergy in the compartments. A calculation method to assess the information of the real ecological network is needed to account for the contribution to the total ecosystem exergy.

Power is very difficult to assess because the ecological observations are mostly based on concentrations and not on flows, which implies that it is hardly possible to validate the flow values resulting from ecological models. In addition, the number of flows in the real ecological network is magnitudes higher than the few flows that can be included in our primitive calculations.

Calculations of ascendency have the same shortcomings as calculations of power.

The three concepts may all have a solid theoretical basis but their applications in practice still have definite weaknesses that are rooted in the complexity of real ecosystems. Based on an integration of the three concepts, we are able to expand on the earlier hypothesis based on exergy alone and let it comprise acendency and power in addition to exergy.

6.6 SUPPORT FOR THE PRESENTED HYPOTHESES

Below are presented a few case studies from Jørgensen (2002) and Jørgensen et al. (2000) supporting the presented exergy storage hypothesis, but maximum power or ascendency could also have been applied as discussed in Section 6.5. More examples can be found in these references and in Chapter 8.

1. Size of genomes

In general, biological evolution has been toward organisms with an increasing number of genes and diversity of cell types (Futuyima, 1986). If a direct correspondence between free energy and genome size is assumed, then this can reasonably be taken to reflect increasing exergy storage accompanying the increased information content and processing of "higher" organisms.

2. Le Chatelier's principle

The exergy storage hypothesis might be taken as a generalized version of "Le Chatelier's Principle." Biomass synthesis can be expressed as a chemical reaction:

$$\text{energy} + \text{nutrients} = \text{molecules with more free energy (exergy) and organization} + \text{dissipated energy} \quad (6.4)$$

According to Le Chatelier's Principle, if energy is put into a reaction system at equilibrium, then the system will shift its equilibrium composition in a way to counteract the change. This means that more molecules with more free energy and organization will be formed. If more pathways are offered, then those giving the most relief from the disturbance (displacement from equilibrium) by using the most energy, and forming the most molecules with the most free energy, will be the ones followed in restoring equilibrium.

3. The sequence of organic matter oxidation

The sequence of biological organic matter oxidation (e.g., Schlesinger, 1997) takes place in the following order: by oxygen, by nitrate, by manganese dioxide, by iron (III), by sulphate, and by carbon dioxide. This means that oxygen, if present, will always out compete nitrate which will out compete manganese dioxide, and so on. The amount of exergy stored as a result of an oxidation process is measured by the available kJ/mole of electrons which determines the number of adenosine triphosphate molecules (ATPs) formed. ATP represents an exergy storage of 42 kJ/mole. Usable energy as exergy in ATPs decreases in the same sequence as indicated above. This is as expected if the exergy storage hypothesis were valid (Table 6.4). If more oxidizing agents are offered to a system, the one giving the highest storage of free energy will be selected.

In Table 6.3, the first (aerobic) reaction will always out compete the others because it gives the highest yield of stored exergy. The last (anaerobic) reaction produces methane; this is a less complete oxidation than the first because methane has a greater exergy content than water.

4. Formation of organic matter in the primeval atmosphere

Numerous experiments have been performed to imitate the formation of organic matter in the primeval atmosphere on earth 4 billion years ago (Morowitz, 1968). Energy from various sources were sent through a gas mixture of carbon dioxide, ammonia, and methane. There are obviously many pathways to utilize the energy sent through simple gas mixtures, but mainly those forming compounds with rather large free energies (amino acids and RNA-like molecules with high exergy storage, decomposed when the compounds are oxidized again to carbon dioxide, ammonia and methane) will form an appreciable part of the mixture (according to Morowitz, 1968).

Table 6.4 Yields of kJ and ATP's per mole of electrons, corresponding to 0.25 moles of CH_2O oxidized (carbohydrates)[1]

Reaction	kJ/mole e^-	ATP's/mole e^-
$CH_2O + O_2 = CO_2 + H_2O$	125	2.98
$CH_2O + 0.8\ NO_3^- + 0.8\ H^+ = CO_2 + 0.4\ N_2 + 1.4\ H_2$	119	2.83
$CH_2O + 2\ MnO_2 + H^+ = CO_2 + 2\ Mn^{2+} + 3\ H_2O$	85	2.02
$CH_2O + 4\ FeOOH + 8\ H^+ = CO_2 + 7\ H_2O + Fe^{2+}$	27	0.64
$CH_2O + 0.5\ SO_4^{2-} + 0.5\ H^+ = CO_2 + 0.5\ HS^- + H_2O$	26	0.62
$CH_2O + 0.5\ CO_2 = CO_2 + 0.5\ CH_4$	23	0.55

[1]The released energy is available to build ATP for various oxidation processes of organic matter at pH 7.0 and 25° C.

5. Photosynthesis

There are three biochemical pathways for photosynthesis: (1) the C3 or Calvin–Benson cycle, (2) the C4 pathway, and (3) the Crassulacean acid metabolism (CAM) pathway. The latter is least efficient in terms of the amount of plant biomass formed per unit of energy received. Plants using the CAM pathway are, however, able to survive in harsh, arid environments that would be inhospitable to C3 and C4 plants. CAM photosynthesis will generally switch to C3 as soon as sufficient water becomes available (Shugart, 1998). The CAM pathways yield the highest biomass production, reflecting exergy storage, under arid conditions, while the other two give highest net production (exergy storage) under other conditions. While it is true that a gram of plant biomass produced by the three pathways has different free energies in each case, in a general way improved biomass production by any of the pathways can be taken to be in a direction that is consistent, under the conditions, with the exergy storage hypothesis.

6. Leaf size

Givnish and Vermelj (1976) observed that leaves optimize their size (thus mass) for the conditions. This may be interpreted as meaning that they maximize their free-energy content. The larger the leaves the higher their respiration and evapotranspiration, but the more solar radiation they can capture. Deciduous forests in moist climates have a LAI of ~6 m^2/m^2 (see also Section 2.4). Such an index can be predicted from the hypothesis of highest possible leaf size, resulting from the tradeoff between having leaves of a given size versus maintaining leaves of a given size (Givnish and Vermelj, 1976). Size of leaves in a given environment depends on the solar radiation and humidity regime, and while, for example, sun and shade leaves on the same plant would not have equal exergy contents, in a general way leaf size and LAI relationships are consistent with the hypothesis of maximum exergy storage.

7. Biomass packing

The general relationship between animal body weight, W, and population density, D, is $D = A/W$, where A is a constant (Peters, 1983). Highest packing of biomass depends only on the aggregate mass, not the size of individual organisms. This means that it is biomass rather than population size that is maximized in an ecosystem, as density (number per unit area) is inversely proportional to the weight of the organisms. Of course the relationship is complex. A given mass of mice would not contain the same exergy or number of individuals as an equivalent weight of elephants. Also, genome differences (Example 1) and other factors would figure in. Later we will discuss exergy dissipation as an alternative objective function proposed for thermodynamic systems. If this were maximized rather than storage, then biomass packing would follow the relationship $D = A/W^{0.65-0.75}$ (Peters, 1983). As this is not the case, biomass packing and the free energy associated with this lend general support for the exergy storage hypothesis.

8. Cycling

If a resource (for instance, a limiting nutrient for plant growth) is abundant, then it will typically recycle faster. This is a little strange because recycling is not needed when a

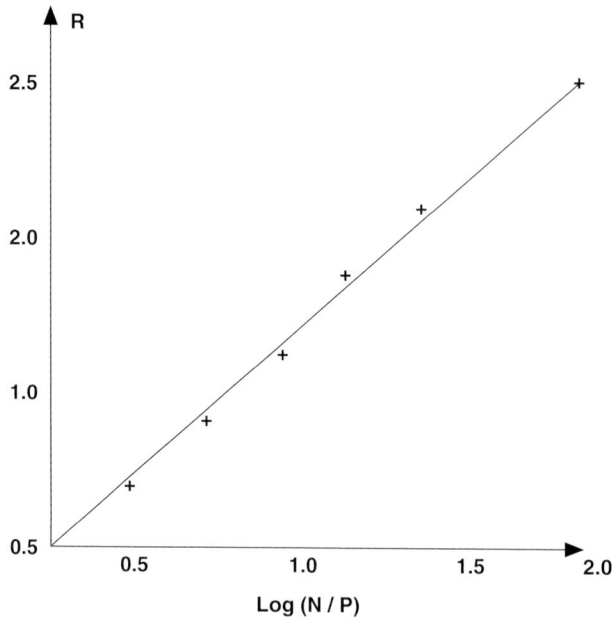

Figure 6.5 Log–log plot of the ratio of nitrogen to phosphorus turnover rates, R, at maximum exergy versus the logarithm of the nitrogen/phosphorus ratio, log N/P. The plot is consistent with Vollenweider (1975).

resource is non-limiting. A modeling study (Jørgensen, 2002) indicated that free-energy storage increases when an abundant resource recycles faster. Figure 6.5 shows such results for a lake eutrophication model. The ratio, R, of nitrogen (N) to phosphorus (P) cycling which gives the highest exergy is plotted in a logarithmic scale versus log (N/P). The plot in Figure 6.5 is also consistent with empirical results (Vollenweider, 1975). Of course, one cannot "inductively test" anything with a model, but the indications and correspondence with data do tend to support in a general way the exergy storage hypothesis. The cycling ratio giving the highest ascendency is also correlated similarly to the N/P ratio (Ulanowicz, personal communication). In the light of the close relationship between exergy and ascendency this result is not surprising (see above, Jørgensen, 1995; Ulanowicz et al., 2006).

9. Structurally dynamic modeling

Dynamic models whose structure changes over time are based on non-stationary or time-varying differential or difference equations. We will refer to these as *structurally dynamic models*. A number of such models, mainly of aquatic systems (Jørgensen, 1986, 1988, 1990; Nielsen, 1992a,b; Jørgensen and Padisak, 1996; Coffaro et al., 1997; Jørgensen and de Bernardi, 1997, 1998), but also as population dynamic models (Jørgensen, 2002) and terrestrial systems (Jørgensen and Fath, 2004) have been investigated to see how structural changes are reflected in exergy changes. The technicalities of parameter fitting aside, this overall result means that system structure must change if its eco-exergy storage is to be continually maximized. Changes in parameters, and thus system structure, not only reflect changes in external boundary conditions, but also mean that such changes are necessary for the ongoing maximization of exergy. For all models investigated along these lines, the changes obtained were in accordance with actual observations (see references). These studies therefore affirm, in a general way, that systems adapt structurally to maximize their content of eco-exergy. The shortcomings of assessing the exergy content of an ecosystem have been discussed above. At least in models the applicability of the exergy calculations have shown their more practical use, which can be explained by a robustness in the model calculations.

It is noteworthy that Coffaro et al. (1997), in his structural-dynamic model of the Lagoon of Venice, did not calibrate the model describing the spatial pattern of various macrophyte species such as *Ulva* and *Zostera*, but used exergy-index optimization to estimate parameters determining the spatial distribution of these species. He found good accordance between observations and model, as was able by this method *without* calibration, to explain more than 90 percent of the observed spatial distribution of various species of *Zostera* and *Ulva*. Box 6.4 gives an illustration of a structurally dynamic model (SDM).

10. Seasonal changes

In natural history it is often observed, particularly at latitudes where there are winters, that taxonomically more primitive forms tend to pass through their non-dormant phenological states earlier in growing seasons and more advanced forms later. It is as though ecosystems must be rebuilt after the "creative destruction" of winter, and until they are reconstituted the active life-history stages of more complex forms of life cannot be supported. Does the maximum exergy storage principle complies with the annual activity cycles of species and communities? Phenological fluctuations of biota, in fact the growth of individual organisms themselves are generally parallel to the four stages of succession, and also the three growth forms (Jørgensen et al., 2000). This is true for the progression of individual species and their assemblages and is best seen at mid to high latitudes. Toward the tropics a great variety of the life history stages of the rich assortment of species is expressed at any given time. At higher latitudes phenological cycles are more obviously entrained to seasonal fluctuations. Focusing at mid-latitudes, and letting "time" be relative to the unit in question (i.e. biological time, whether for a species or whole ecosystem), "winter" represents the initial condition (Stage 0). During "spring", the growth forms unfold in quick succession. Form I dominates early (Stage I), Form II later (Stage II), and Form III in "summer", which advances toward seasonal maturity (Stage III). Ephemeral species pass quickly through their own Stage III to seed set, dispersal, senescence (Stage IV), and often, disappearance. Permanent species remain more or less in Stage III until near the end of the growing season when they or their parts pass into quasi-senescent states (Stage IV), as in leaf fall and hibernation.

Exergy storage and utilization patterns may be intuited from the principles laid down for succession (Figures 6.7 and 6.8 related text) to follow these seasonal trends

Box 6.4 Illustration of structurally dynamic modeling

Structurally dynamic model of Darwin's finches (Jørgensen and Fath, 2004). The models reflect therefore—as all models—the available knowledge which in this case is comprehensive and sufficient to validate even the ability of the model to describe the changes in the beak size as a result of climatic changes, causing changes in the amount, availability, and quality of the seeds that make up the main food item for the finches. The medium ground finches, *Geospiza fortis*, on the island Daphne Major were selected for these modeling case due to very detailed case specific information found in Grant (1986). The model has three state variables: seed, Darwin's Finches adult, and Darwin's finches juvenile. The juvenile finches are promoted to adult finches 120 days after birth. The mortality of the adult finches is expressed as a normal mortality rate (Grant, 1986) plus an additional mortality rate due to food shortage and an additional mortality rate caused by a disagreement between bill depth and the size and hardness of seeds.

The beak depth can vary between 3.5 and 10.3 cm (Grant, 1986) the beak size = $\sqrt{DH}$, where D is the seed size and H the seed hardness which are both dependent on the precipitation, particularly in the months January–April (Grant, 1986). It is possible to determine a handling time for the finches for a given $\sqrt{DH}$ as function of the bill depth (Grant, 1986) which explains that the accordance between $\sqrt{DH}$ and the beak depth becomes an important survival factor. The relationship is used in the model to find a function called "diet" which is compared with $\sqrt{DH}$ to find how well the bill depth fits to the $\sqrt{DH}$ of the seed. This fitness function is based on information given by Grant (1986) about the handling time. It influences as mentioned above the mortality of adult finches, but has also impact on the number of eggs laid and the mortality of the juvenile finches. The growth rate and mortality of seeds is dependent on the precipitation which is a forcing function know as function of time (Grant, 1986). A function called shortage of food is calculated from the food required of the finches which is known (Grant, 1986), and from the food available (the seed state variable). How the food shortage influences the mortality of juvenile finches and adult finches can be found in Grant (1986). The seed biomass and the number of *G. fortis* as function of time from 1975 to 1982 are known (Grant, 1986). These numbers from 1975 to 1976 have been used to calibrate the following parameters:

(i) the influence of the fitness function on (a) the mortality of adult finches, (b) the mortality of juvenile finches, and (c) the number of eggs laid;
(ii) the influence of food shortage on the mortality of adult and juvenile finches is known (Grant, 1986). The influence is therefore calibrated within a narrow range of values;
(iii) the influence of precipitation on the seed biomass (growth and mortality).

All other parameters are known from the literature.

Chapter 6: Ecosystems have complex dynamics (growth and development)

The exergy density is calculated (estimated) as 275 × the concentration of seed + 980 × the concentration of Darwin's finches (see Table 6.2). Every 15 days it is found if a feasible change in the beak size taken the generation time and the variations in the beak size into consideration will give a higher exergy. If it is the case, then the beak size is changed accordingly. The modeled changes in the beak size were confirmed by the observations. The model results of the number of Darwin's finches are compared with the observations (Grant, 1986) in Figure 6.6. The standard deviation between modeled and observed values was 11.6 percent and the correlation coefficient, r^2, for modeled versus observed values is 0.977. The results of a non structural dynamic model would not be able to predict the changes in the beak size and would, therefore, give too low values for the number of Darwin's finches because their beak would not adapt to the lower precipitation yielding harder and bigger seeds.

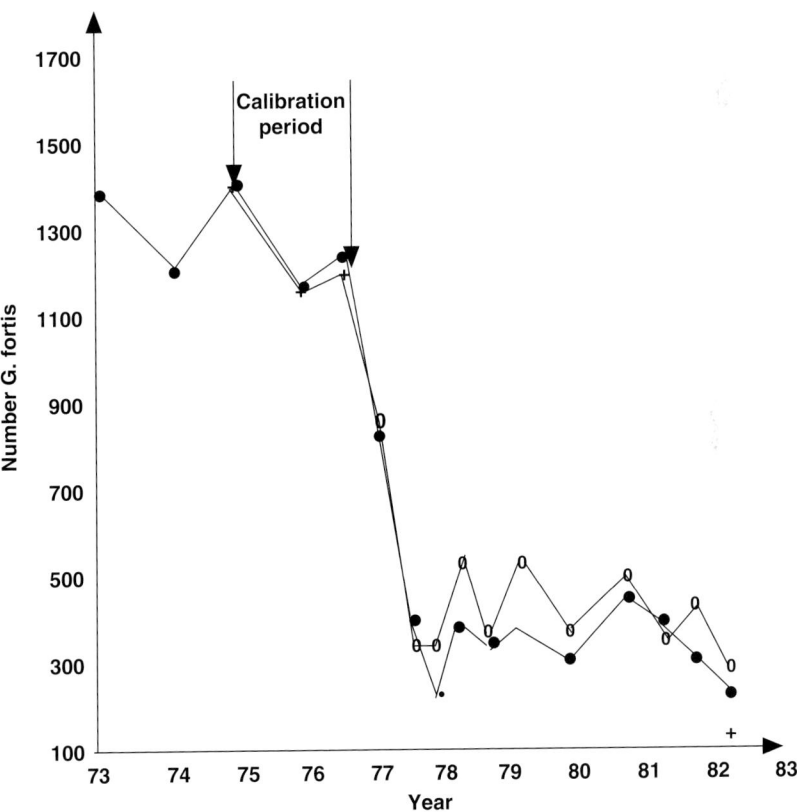

Figure 6.6 The observed number of finches (•) from 1973 to 1983, compared with the simulated result (O); 75 and 76 were used for calibration and 77/78 for the validation referred to in Box 6.5.

132 A New Ecology: Systems Perspective

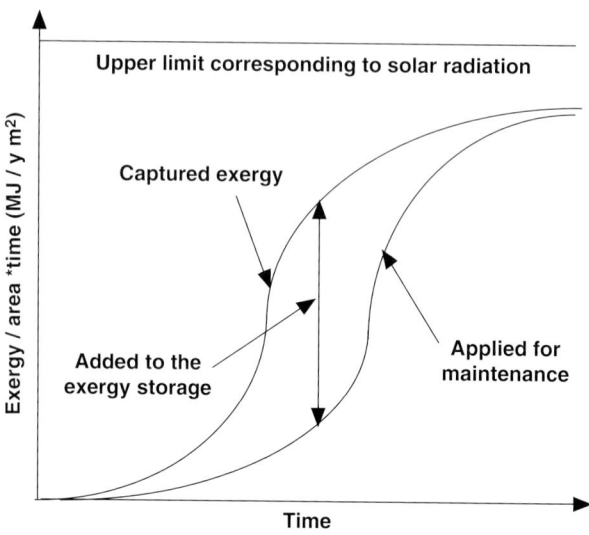

Figure 6.7 Exergy utilization of an ecosystem under development is shown versus time. Notice that the consequence of the growth in exergy is increased utilization of exergy for maintenance.

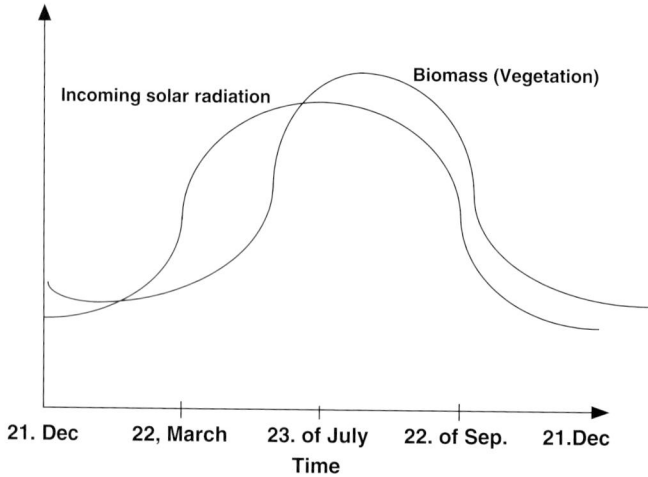

Figure 6.8 The seasonal changes in incoming solar radiation and biomass (vegetation) are shown for a typical temperate ecosystem. The slope of the curve for biomass indicates the increase in exergy due to Growth Form I. The Growth Form I can continue as long as the captured solar radiation is larger than the exergy applied for maintenance. Therefore the biomass has its maximum around August 1st. The biomass is at minimum around February 1st because at that time is the captured exergy and the exergy applied for maintenance in balance.

also, in mass, throughflow, and informational characteristics. In winter, biomass and information content are at seasonal lows. The observations of the seasonal changes may be considered an indirect support for the hypothesis. In spring, the flush of new growth (dominantly Form I) produces rather quickly a significant biomass component of exergy (Figure 6.8), but the information component remains low due to the fact that most active flora, fauna, and microbiota of this nascent period tend to be lower phylogenetic forms.

These lower forms rapidly develop biomass but make relatively low informational contributions to the stored exergy. As the growing season advances, in summer, Growth Forms II and III become successively dominant. Following the expansion of system organization that this represents, involving proliferation of food webs and interactive networks of all kinds, and all that this implies, waves of progressively more advanced taxonomic forms can now be supported to pass through their phenological and life cycles. Albedo and reflection are reduced, dissipation increases to seasonal maxima following developing biomass, and as seasonal maxima are reached further increments taper to negligible amounts (Figure 6.8). The biotic production of advancing summer reflects more and more advanced systemic organization, manifested as increasing accumulations of both biomass and information to the exergy stores. In autumn, the whole system begins to unravel and shut down in pre-adaptation to winter, the phenological equivalent of senescence. Networks shrink, and with this all attributes of exergy storage, throughflow, and information transfer decline as the system slowly degrades to its winter condition. Biological activity is returned mainly to the more primitive life forms as the ecosystem itself returns to more "primitive" states of exergy organization required for adaptation to winter. The suggestion from phenology is that the exergetic principles of organization apply also to the seasonal dynamics of ecosystems.

6.7 TOWARD A CONSISTENT ECOSYSTEM THEORY

Ecosystem properties can only be revealed by a plurality of views. It is, therefore, not surprising that there are many different ecosystem theories published in the scientific literature. It is also important to try to understand the theories in relation to each other and examine if they are contradictory or form a pattern that can be used to give a better understanding of the nature of ecosystems and to solve the global environmental problems. The goal is to give a common framework of reference for *further* development of a more profound and comprehensive ecosystem theory than the one we are able to present today. The pattern should serve as a "conceptual diagram", which can be used as a basis for further discussion of ecosystems. We are still in an early stage of an ecosystem-theoretical development and it may be argued that this attempt is premature, but the experience from modeling has taught us that it is better to conclude one's thoughts in a conceptual diagram at an early stage and then be ready to make changes than to let all modeling efforts wait until all details are known, as this will never be the case due to the immense complexity of nature (Jørgensen, 2002). Moreover, recent development in ecosystem theory has made it possible to conclude that the theories presented here are indeed consistent and complimentary (Fath et al., 2001). The special issue in Ecological Modelling 158.3 (2002) has demonstrated that the theory can be applied to explain ecological observations, although the ecosystem theory presented here does not contain laws in the classical physical sense that we can make exact predictions. The theory is rather closer to quantum mechanics (we have to accept an uncertainty), chaos theory (sometimes predictions of complex systems are impossible), and the Prigogine thermodynamics (all processes are irreversible). Given the limitations in our theory, that ecosystems are enormously complex and we can, therefore, not know all details and that ecosystems have an ontic openness (see Chapter 3), it is still possible to apply the theory in ecology and environmental management.

The core pattern concerns the systemness of life and how these interactions lead to complex organization and dynamics. Understanding, measuring, and tracking these patterns is of paramount importance and the various holistic indicators have been developed to do so. Taken together, we can use this systems-oriented thermodynamic approach to formulate a limited number of propositions or hypotheses to explain a very large number of ecological observations. These recent developments in systems ecology represent a profound paradigm shift. The paradigm that is now receding has dominated our culture for several hundred years. It views the universe as a mechanical system composed of elementary building blocks. The new paradigm is based on a holistic worldview. The world is seen as an integrated whole and recognizes the fundamental interdependence of all phenomena.

In the paper by Jørgensen et al. (2000), Figure 6.9 illustrated the concomitant development of ecosystems, exergy captured (most of that being degraded) and exergy stored (biomass, structure, information). Data points correspond to different ecosystems (see Table 6.5, which shows the values). Debaljak (2001) showed that he gets the same shape of the curve when he determines exergy captured and exergy stored in managed forest and virgin forest on different stages of development (see Figure 6.10). The exergy captured was determined as in Table 6.5 by measurement of the temperature of the infrared radiation, while the exergy storage was determined by a randomized measurement of the size of all trees and plants. The stages are indicated on the figure, where also pasture is included for comparison. Catastrophic events as storm or fire may cause destructive regeneration, which is described below.

Holling (1986) (see Figure 6.11) has suggested how ecosystems progress through the sequential phases of renewal (mainly Growth Form I), exploitation (mainly Growth Form II), conservation (dominant Growth Form III), and creative destruction. The latter phase fits also into the three growth forms but will require a further explanation. The creative destruction phase is either a result of external or internal factors. In the first case (for instance hurricanes and volcanic activity), further explanation is not needed as an ecosystem has to use the growth forms under the prevailing conditions, which are determined by the external factors. If the destructive phase is a result of internal factors, then the question is "why would a system be self-destructive?"

A possible explanation is that a result of the conservation phase is that almost all nutrients will be contained in organisms which implies that there are no nutrients available to test new and possibly better solutions to move further away from thermodynamic equilibrium or expressed in Darwinian terms to increase the probability of survival. Holling also implicitly indicates this by calling this phase creative destruction.

Chapter 6: Ecosystems have complex dynamics (growth and development)

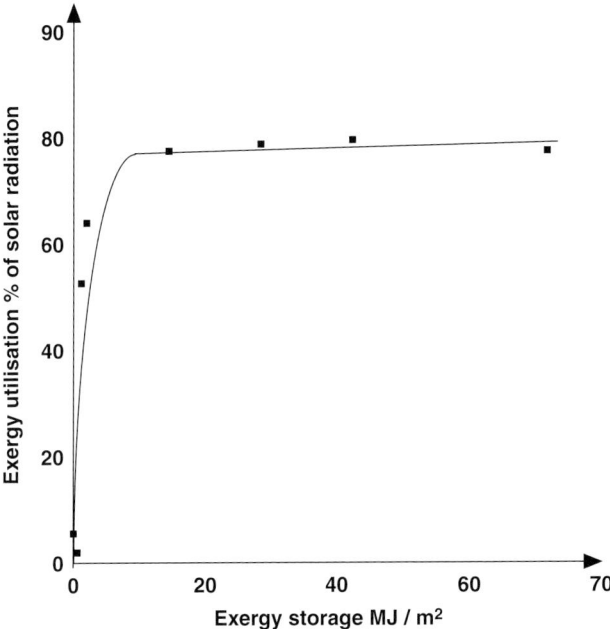

Figure 6.9 The exergy captured (percent of solar radiation; Kay and Schneider, 1992) is plotted versus the exergy stored (MJ/m^2), calculated from characteristic compositions of the focal eight ecosystems. The numbers from Table 6.5 are applied to construct this plot. Notice that exergy utilization is parallel (proportional) to energy absorbed.

Table 6.5 Exergy utilization and storage in a comparative set of ecosystems

Ecosystem	Exergy utilization (percent)	Exergy storage (MJ/m^2)
Quarry	6	0
Desert	2	0.073
Clear-cut forest	49	0.594
Grassland	59	0.940
Fir plantation	70	12.70
Natural forest	71	26.00
Old-growth deciduous forest	72	38.00
Tropical rain forest	70	64.00

Therefore, when new solutions are available, it would in the long run be beneficial for the ecosystem to decompose the organic nutrients into inorganic components that can be utilized to test the new solutions. The creative destruction phase can be considered a method to utilize the three other phases and the three growth forms more effectively in the long run (Fath et al., 2004). This is indicated on the figure as "trend of each further

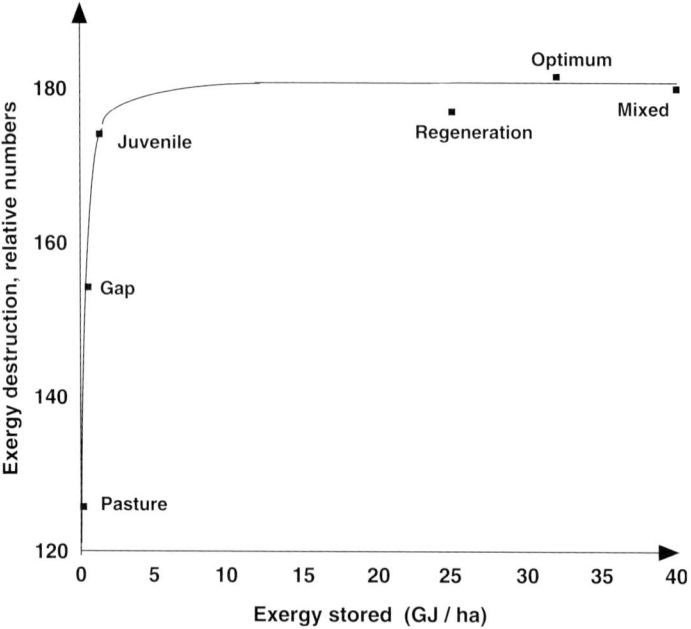

Figure 6.10 The plot shows the result by Debeljak (2001). He examined managed a virgin forest in different stages. Gap has no trees, while the virgin forest changes from optimum to mix to regeneration and back to optimum, although the virgin forest can be destroyed by catastrophic events as fire or storms. The juvenile stage is a development between the gap and the optimum. Pasture is included for comparison.

cycle" and it is shown that the ecosystem is moving toward a higher specific exergy (and biomass if possible, as growth of biomass is dependent on the available amount of the limiting element), if the inorganic components are available to form more biomass for each cycle.

Five of the presented hypotheses to describe ecosystem growth and development, are examined with respect to three growth forms, excluding the boundary growth:

A. Entropy production tends to be minimum (proposed by Prigogine (1947, 1955, 1980) for linear systems at steady non-equilibrium state, not very far from thermodynamic equilibrium systems). Mauersberger (1981, 1983, 1995) applied this to derive expressions for bioprocesses at a stable stationary state (see also Chapter 2). Reduction of the entropy production means that the energy utilization is increased, which is obtained by an increased cycling of the energy and reduced loss of energy to the environment, or expressed differently: the retention time of a given portion of energy in the system is increased.

B. Natural selection tends to maximize the energy flux through the system, so far as compatible with the constraints to which the system is subject (H.T. Odum, 1983). This is the maximum power principle (see Section 6.4).

Chapter 6: Ecosystems have complex dynamics (growth and development) 137

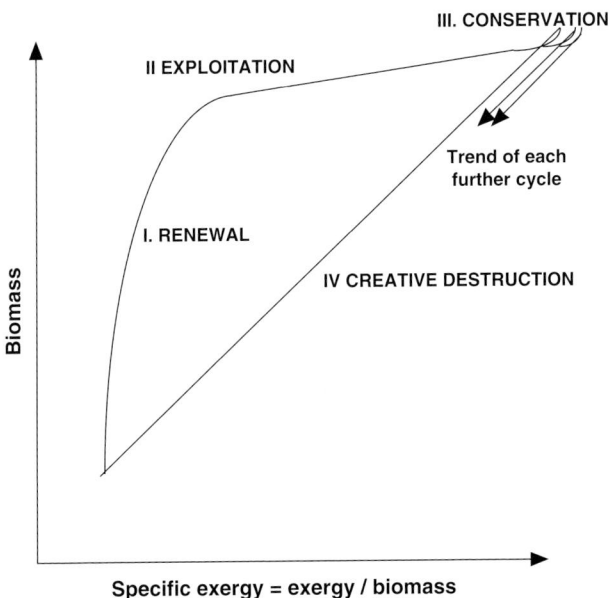

Figure 6.11 Holling's four phases of ecosystems, described terms of biomass versus specific exergy. The presentation is inspired by Ulanowicz (1997).

C. Ecosystems organize to maximize exergy degradation (Kay, 1984). Living systems transform more exergy to heat at the temperature of the environment, or said differently, produce more entropy, than their non-living complements. Living systems, therefore, increases the destruction of exergy but at the same time living systems increase the order and organization.
D. A system that receives a throughflow of exergy will have a propensity to move away from thermodynamic equilibrium, and if more combinations of components and processes are offered to utilize the exergy flow, the system has the propensity to select the organization that gives the system as much stored exergy as possible (see Section 6.5, Jørgensen and Mejer (1977, 1979), Jørgensen (1982, 2002), Mejer and Jørgensen (1979)).
E. Ecosystems have a propensity to develop toward a maximization of the ascendency (Ulanowicz, 1986; see also Section 6.5).

The usual description of ecosystem development illustrated for instance by the recovery of Yellowstone Park after fire, an island formed after a volcanic eruption, reclaimed land, etc. is well covered by E.P. Odum (1969): at first the biomass increases rapidly which implies that the percentage of captured incoming solar radiation increases but also the energy needed for the maintenance. Growth Form I is dominant in this first phase, where exergy stored increases (more biomass, more physical structure to capture more solar radiation), but also the throughflow (of useful energy), exergy dissipation and the entropy production increases due to increased need of energy for maintenance. Living systems are effectively capturing (lower the albedo) and utilizing energy (exergy) (both the exergy decomposed due to respiration and the exergy stored in the living matter are increased) from the ambient physical systems.

Growth Forms II and III become, in most cases, dominant later, although an overlap of the three growth forms takes place. The smaller the ecosystem is, which implies that it has a high relative openness (see Chapter 2), the faster will Growth Forms II and III, particularly Growth Form III, contribute to the development of the ecosystem (Patricio et al., 2006). The recovery of small inter-tidal rocky communities has been examined in this paper. It was found that biodiversity and specific eco-exergy (= eco-exergy/biomass) recover much faster than biomass and eco-exergy, i.e. the Growth Forms II and III are dominant in the initial phase of recovery.

When the percentage of solar radiation captured reaches ~80 percent it is not possible to increase the amount of captured solar radiation further (due in principle to the second law of thermodynamics). Further growth of the physical structure (biomass) does, therefore, not improve the exergy balance of the ecosystem. In addition, all or almost all the essential elements are in the form of dead or living organic matter and not as inorganic compounds ready to be used for growth. The Growth Form I will and can therefore not proceed, but Growth Forms II and III can still operate. The ecosystem can still improve the ecological network and can still evolve novel, more complex organisms and environments. One tendency is to increase the occurrence of larger, long-lived organisms (Cope's law: the later descendent may be increasingly larger than their ancestors; for instance, the horse today is much bigger than the horse fossils from 20 to 30 million years ago) and less developed with more developed and more non-nonsense genes. Growth Forms II and III do not require, however, more exergy for maintenance. Exergy degradation is, therefore, not increasing but is maintained at a constant level (see Figures 6.9 and 6.10). The accordance with the five descriptors plus specific entropy production and the three growth forms based on this description of ecosystem development is shown in Table 6.6

Table 6.6 Accordance between growth forms and the proposed descriptors

Hypothesis	Growth Form I	Growth Form II	Growth Form III
Exergy storage	Up	Up	Up
Power/throughflow	Up	Up	Up
Ascendency	Up	Up	Up
Exergy dissipation	Up	Equal	Equal
Retention time	Equal	Up	Up
Entropy production	Up	Equal	Equal
Exergy/biomass = specific exergy	Equal	Up	Up
Entropy/biomass = specific entropy production	Equal	Down	Down
Ratio indirect/direct effects	Equal	Up	Up

Ecosystem development is accomplished by three growth forms (in addition to boundary growth), which support the results in Table 6.6:

I. Biomass growth is the increase in physical structure of the ecosystem, which occurs primarily by the capture and conversion of incoming solar radiation into organic compounds. This first stage implies that more exergy is degraded due to an increased demand constructing and maintaining biomass. The most dynamic indicator of this growth form is the eco-exergy degradation, although the eco-exergy storage, throughflow, and ascendency also will increase. While there is an upper limit to the amount of solar radiation available, and to the amount that can be harvested by ecosystems, system development continues through Growth Forms II and III.
II. Network growth entails a richer, more complex interaction structure, which through increased cycling offers better utilization of the available energy. In turn, both throughflow and exergy storage increase without an increase in exergy degradation. It means that specific exergy degradation and specific entropy production are decreasing during this stage. Throughflow, eco-exergy, specific eco-exergy, and ascendency can all be used as dynamic indicators for this growth form.
III. Information growth represents an increase in the genetic and organizational diversity in the ecosystem. One way this occurs is through an increase in number of species with longer and more complex life histories, larger body size, and complex physiologies. This implies that, similar to Growth Form II, throughflow and exergy storage increase while specific exergy degradation and entropy production decrease. Eco-exergy, specific eco-exergy, and ascendency can all be used as indicators for this growth form, with specific eco-exergy capturing the genetic information and ascendency the organizational information.

6.8 EXERGY BALANCES FOR THE UTILIZATION OF SOLAR RADIATION

In Jørgensen and Svirezhev (2004) the following expression has been shown for eco-exergy gained, Ex, as a result of the energy of the incoming solar radiation, E_{in}:

$$Ex = (E_{in} - R)\left[K - \ln\left(\frac{(E_{in} - R)}{E_{in}}\right)\right] + R \qquad (6.5)$$

Where R is the difference between the total incoming and outgoing radiation and K Kullback's measure of information. If we introduce the radiation efficiency $\eta_{rad} = R/E_{in}$ and the exergy efficiency, $\eta_{Ex} = Ex/E_{in}$, (Equation 6.5) can be reformulated as:

$$\eta_{Ex} = (1 - \eta_{rad})K + (1 - \eta_{rad})\ln(1 - \eta_{rad}) + \eta_{rad}$$

η_{Ex} is, therefore, a function of two independent variables, η_{rad} and K, but is independent of any parameter.

Figure 6.12 shows the relationship between η_{Ex} and η_{rad} for three different values of K. The active surface of an ecosystem will as seen in the figure operate as a "classical thermodynamic engine" performing mechanical and chemical work below the straight line, while the active surface will generate information when K is high and η_{rad} is not too high.

The plots in Figures 6.9 and 6.10 can be understood from these results: when an ecosystem has attained a certain level of eco-exergy—information it will be able to continue to generate information—increasing the eco-exergy without necessarily to increase the physical structure. In most examples of ecosystem restoration, the disturbed ecosystem is embedded in a larger ecosystem, which acts as a species reserve. Since it is an open system these organisms re-populate the disturbed area rapidly. Growth Form III may be significant at an early state of redevelopment (see Patricio et al., 2006) through information importation from the reservoir. Available diversity is sufficient to reconstruct the network and information level quickly.

"Intelligent Design" is therefore not needed to explain the evolution, as an ecosystem is designed to generate information.

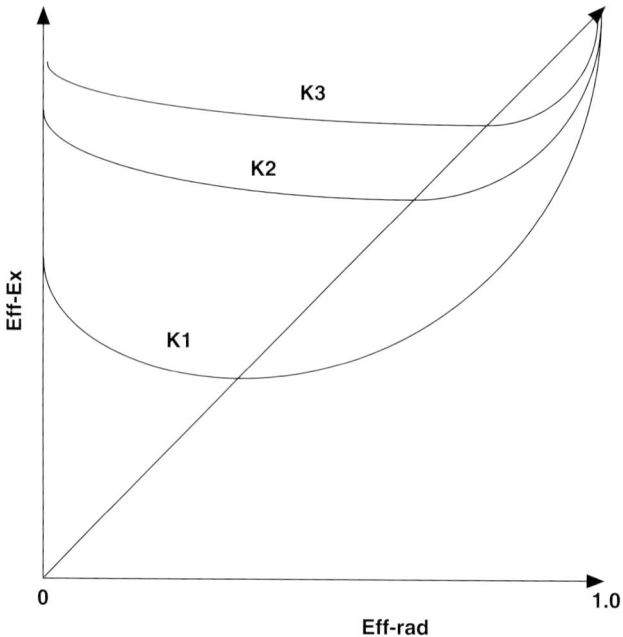

Figure 6.12 η_{Ex} is plotted versus η_{rad} for three information level $K_3 > K_2 > K_1$. Increasing Kullback's measure of information implies that the ecosystem will generate more information up to a higher η_{rad}. It shows the relationship between η_{Ex} and η_{rad} for three different values of K. The active surface of an ecosystem will as seen in the figure operate as a "classical thermodynamic engine" performing mechanical and chemical work below the straight line, while the active surface will generate information when K is high and η_{rad} is not too high.

6.9 SUMMARY AND CONCLUSIONS

Ecosystems have a very complex dynamics. It is rooted in the enormous number of different life forms, that is the result of the evolution and the enormous variability in the life conditions. Ecosystem development can be described by a wide spectrum of ecological indicators and orientors from single species and concentrations of specific chemical pollutants to holistic indicators such as biodiversity and thermodynamic functions. Physical–chemical systems can usually be described by matter and energy relations, while biological systems require a description that encompasses matter, energy, and information. Information has four properties that significantly different from the properties of matter and energy:

(1) *Information can, unlike matter and energy, disappear without trace.*
(2) *Information expressed for instance as eco-exergy is not conserved.*
(3) *The disappearance and the copying of information that are characteristic processes for living systems, are irreversible processes.*
(4) *Exchange of information is communication and it is this that brings about the self-organization of life.*

The complex dynamic of ecosystems can be determined by the following hypothesis: *If a system receives an input of exergy (energy that can do work), it will perform work. The work performed is first applied to maintain the system far away from thermodynamic equilibrium whereby work capacity is lost by transformation into heat at the temperature of the environment. If more exergy is available, then the system is moved further away from thermodynamic equilibrium, reflected in growth of gradients. If there is offered more than one pathway to depart from equilibrium, then the one yielding the most stored exergy, power, and ascendency will tend to be selected. Or expressed differently: among the many ways for ecosystems to move away from thermodynamic equilibrium, the one optimizing dEx/dt, $d(power)/dt$, $d(ascendency)/dt$, and $d(gradients)/dt$ under the prevailing conditions will have a propensity to be selected.* The eco-exergy stored, the ascendency, the power, and the gradients are just to be considered four different views (biomass—information accumulation, the ecological network, the flows, and the differences of intensive variables) of the same reaction of ecosystems by moving away from thermodynamic equilibrium. This chapter, several references, and Chapter 9 present several ecological observations and rules that support this hypothesis.

Light requires description as particles (photons) and waves. It is, therefore, not surprising that ecosystem dynamics can be described in several different ways. Due to the high complexity of ecosystems, it is however, not possible to apply these different descriptors of ecosystem dynamics for the entire ecosystems with all its detailed information, but only for models of ecosystems. Dependent on the knowledge about the ecosystem, it can be advantageous to apply eco-exergy (when the concentrations or biomasses of the focal species are known), ascendency (when a good model of the ecological network is known) or power (when the flows are known). Eco-exergy can be found as the sum of the multiple of concentrations and weighting factors considering the information that the various organisms are carrying. Eco-exergy indicates the distance from thermodynamic equilibrium according to the definition; see also Chapter 2, while emergy indicates the cost in the ultimate energy source on earth, solar radiation. The ratio between the two thermodynamic concepts indicates the efficiency of the system—more efficiency if more exergy is obtained relatively to the solar radiation. In Chapter 9 it is shown how this ratio can be applied as a powerful indicator for an ecosystem.

Ecosystems have three growth forms or methods to move away from thermodynamic equilibrium: biomass growth, network growth, or information growth. When an ecosystem has attained a certain level of energy capture, it will be able to continue to generate organismal and structural information such that eco-exergy, throughflow, and ascendency increase throughout development without n-ecessarily an increase in the physical biomass. The above-mentioned three descriptors, eco-exergy, power, and ascendency will all increase with the three growth forms, while exergy destruction or entropy production only increases with the first growth form and specific entropy only decreases with the second and third growth form.

7

Ecosystems have complex dynamics – disturbance and decay

Du siehst, wohin du siehst nur Eitelkeit auf Erden.
Was dieser heute baut, reißt jener morgen ein:
Wo itzund Städte stehn, wird eine Wiese sein
Auf der ein Schäferskind wird spielen mit den Herden:

Was itzund prächtig blüht, soll bald zertreten werden.
Was itzt so pocht und trotzt ist morgen Asch und Bein
Nichts ist, das ewig sei, kein Erz, kein Marmorstein.
Itzt lacht das Glück uns an, bald donnern die Beschwerden.

Der hohen Taten Ruhm muß wie ein Traum vergehn.
Soll denn das Spiel der Zeit, der leichte Mensch bestehn?
Ach! was ist alles dies, was wir für köstlich achten,

Als schlechte Nichtigkeit, als Schatten, Staub und Wind;
Als eine Wiesenblum, die man nicht wiederfind't.
Noch will, was ewig ist, kein einig Mensch betrachten!

(Andreas Gryphius, 1616–1664: Es ist alles eitel)

7.1 THE NORMALITY OF DISTURBANCE

Up to this point, the focus of this book has been on growth and development processes in ecosystems. In fact, these are most important features of ecosystem dynamics and they provide the origins of various emergent ecosystem properties. But the picture remains incomplete if disturbance and decay are not taken into account. On the following pages we will try to include those "destructive" processes into the "new" ecosystem theory as elaborated in this book. As a starting point for these discussions we can refer to common knowledge and emotion, as it is described in the poem of Andreas Gryphius (see above) who outlines the transience of human and environmental structures: Nothing lasts forever, towns will turn into meadows, flourishing nature can easily be destroyed, our luck can turn into misfortune, and in the end, what remains is emptiness, shadow, dust and wind. Although the poet seems to be comprehensible concerning the significance of decay, we cannot agree with his pessimistic ultimate: In the end, the death of organisms and disturbance of ecosystems can be useful elements of the growth, development and survival of the whole structure, i.e. if they expire within suitable thresholds and if we observe their outcomes over multiple scales.

On a small scale, we can notice that the individual living components of ecosystems have limited life spans that range from minutes to millennia (see Table 7.1). Death and decay of *organisms* and their subsystems are integral elements of natural dynamics. From a functional viewpoint, these processes are advantageous, to replace highly loaded or exhausted components (e.g., short life expectancies of some animal cells), or to adjust physiologies to changing environmental conditions (e.g., leaf litter fall in autumn). As a consequence of these processes, energy and nutrients are provided for the saprophagous branches of food webs, which in many cases show higher turnover rates than the phytophagous branches of the energy and nutrient flow networks. In those situations of death self-organized units give up their autonomy and their ability to capture and actively transform exergy, their structures are subject to dissipation. Reactivity, self-regulation, and the ability for replication are desist, releasing the internal order and constituents which thus potentially become ingredients of the higher system-level self-organization (see Chapter 3 "Ecosystems have Ontic Openness").

Also *populations* have limited durations at certain places on earth. Operating in a hierarchy of constraints, populations break down, e.g., if the exterior conditions are modified, if imperative resources are depleted, if the living conditions are modified by human actions, or if competition processes result in a change of the community assemblage. Following the thermodynamic argumentation of this book (see Chapters 2 and 6), in these situations a modified collection of organisms will take over, being able to increase the internal flows

Table 7.1 Some data about life expectancies of cells and organisms

Example	Average life span
Generation time of *E. coli*	20 min
Life spans of some human cells	
Small intestine	1–2 days
White blood cells	1–3 days
Stomach	2–9 days
Liver	10–20 days
Life span of some animals	
Water flea	0.2 years
Mouse	3–4 years
Nightingale	4 years
Dog	12–20 years
Horse	20–40 years
Giant tortoise	177 years
Life span of some plants	
Sun flower	1 year
Corylus avellana	4–10 years
Fagus sylvatica	200–300 years
Pinus aristata	4900 years

and to reduce the energetic, material, and structural losses into the environment in a greater quantity than the predecessors. During such processes, of course, only the very immediate conditions can be influential: The developmental direction is defined due to a short-term reaction, which increases orientor values at the moment the decision is made, on the basis of the disposable elements and the prevailing conditions. Thereafter, the structural fate of the system is predefined by new constraints; an irreversible reaction has taken place, and the sustainability of this pathway will be an object of the following successional processes.

Of course, such community dynamics have consequences for the abiotic processes and structures. Therefore, also *ecosystems* themselves exist for a limited period of time only. Their typical structural and organizational features are modified, not only if the external conditions change significantly, but also if due to internal competition processes certain elements attain dominance displacing other species. These processes can be observed on many different scales with distinct temporal characteristics—slow processes can occur as results of climatic changes (e.g., postglacial successions throughout the Holocene), shifts of biomes (e.g., Pleistocene dynamics of rain forests), or continuous invasions of new species. On the contrary, abrupt processes often modify ecosystems very efficiently within rather short periods of time.

The most commonly known extreme event has taken place at the end of the Cretaceous age, 65 million years ago, when—purportedly due to an asteroid impact—enormous changes of the global community structures took place, no organism bigger than 25 kg survived on land: planktonic foraminifera went extinct by 83%, the extinctions of ammonites reached 100%, marine reptiles were affected by 93%, and the nonavian dinosaurs were driven totally extinct. No doubt, this was a big loss of biodiversity, and many potential evolutionary pathways disappeared; but, as we know 65 million years later, this event was also a starting shot for new evolutionary traits and for the occupation of the niches by new species, e.g., for the rapid development of mammals or organisms which are able to read or write books (see Box 7.1).

Box 7.1 Creativity needs disturbance

Necessity is the mother of invention.
Constraints mean problems in the first hand, but problems require solutions, and (new) solutions require creativity. Let us exemplify this by evolutionary processes, the genetic code and language. The constraints in the chemical beginning of the evolution were that whenever a primitive but relatively well-functioning assemblage of organic molecules was formed, the composition that made the entity successful was forgotten with its breakdown. The next entity would have to start from scratch again. If at least the major part of the well-functioning composition could be remembered, then the entities would be able to improve their composition and processes generation by generation.

For organisms the problem is to survive. When new living conditions are emerging the accompanying problems for the phenotypes are solved by new properties of the genotypes or their interactions in the ecological networks. The survival based on the two

(continued)

growth forms "biomass growth" and "network growth" are ensured by adaptation to the currently changed prevailing conditions for life. But information growth is needed, too, because survival under new emergent conditions requires a system to transfer information to make sure that solutions are not lost. These problems on the need for information transfer have been solved by development of a genetic system that again put new constraints on survival. It is only possible to ensure survival in the light of the competition by use of the adopted genetic system. But the genes have also created new possibilities because mutations and later in the evolution sexual recombinations create new possible solutions. Therefore, as shown in Figure 7.1 what starts with constraints and new and better properties of the organisms or their ecological networks ends up as new possibilities through a coding system that also may be considered initially as constraints.

An organism's biochemistry is determined by the composition of a series of enzymes that again are determined by the genes. Successful organisms will be able to get more offspring than less successful organisms and as the gene composition is inherited, the successful properties will be more and more represented generation after generation. This explains that the evolution has been directed toward more and more complex organisms that have new and emerging properties.

The genetic code is a language or an alphabet. It is a constraint on the living organisms that have to follow the biochemical code embodied in the genes. The sequence of three amino bases with four possibilities determines the sequence of amino acids in

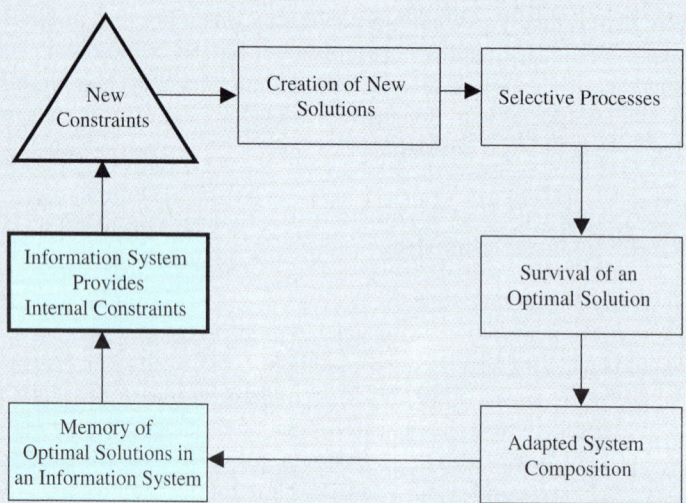

Figure 7.1 Life conditions are currently changed and have a high variability in time and space. This creates new challenges (problems) to survival. Organisms adapt or a shift to other species takes place. This requires an information system that is able to transfer the information about good solutions to the coming generations of organisms. Consequently, an information system is very beneficial, but it has to be considered as a new source of constraints that however can open up for new possibilities.

the proteins. There are, in other words, $4 \times 4 \times 4 = 64$ different codings of the three amino bases; but as there are only 20 amino acids to select from, it contains amino base coding redundant amino base coding combinations in the sense that for some amino acids two or more combinations of amino bases are valid. As an alphabet is a constraint for an author (he has to learn it and he is forced to use it if he wants to express his thoughts), the genetic code is a constraint for the living organism. But as the alphabet gives a writer almost unlimited opportunities to express thoughts and feelings, so the genetic code has given the living organisms opportunity to evolve, becoming more and more complex, more and more creative, having more and more connectivity among the components and becoming more and more adaptive to the constraints that are steadily varying in time and space. The genetic code, however, has not only solved the problem associated with these constraints, but it has also been able to give the living organisms new emergent properties and has enhanced the evolution.

When the human language was created a couple of millions years ago, it first provided new constraints for humans. They had to learn the language and use it, but once they have mastered the language it also gave new opportunities because it made it possible to discuss cooperation and a detailed better hunting strategy, e.g., which would increase the possibility for survival. The written language was developed to solve the problem of making the message transfer more independent of time and space. To learn to write and read were new constraints to humans that also open up many new possibilities of expressing new ideas and thoughts and thereby move further away from thermodynamic equilibrium.

Animals also communicate through sounds or chemicals for warnings, for instance by marking of hunting territories by urine. The use of these signals has most likely been a factor that has reduced the mortality and increased the change of survival.

We will use a numeric example to illustrate the enormous evolutionary power of the genes to transfer information from generation to generation. If a chimpanzee would try to write this book by randomly using a computer key board, the chimpanzee would not have been able to write the book even if he started at the big bang 15 billion years ago, but if we could save the signs that were correct for the second round and so on, then 1/40 of the volume would be correct in the first round (assuming 40 different signs), $(39 \times 39)/(40 \times 40)$ would still be incorrect after the second round, $(39 \times 39 \times 39)/(40 \times 40 \times 40)$ after the third round and so on. After 500 rounds, which may take a few years, there would only be 5 "printed" errors left, if we presume that this book contains 500,000 signs. To write one round of the volume would probably require 500,000 s or about a week. To make 500 rounds would there take about 500 weeks or about 9 years.

The variation in time and space of the conditions for living organism has been an enormous challenge to life because it has required the development of a wide range of organisms. The living nature has met the challenge by creation of an enormous differentiation. There are five million known species on earth and we are currently finding new species. It is estimated that the earth has about 10^7 species. We see the same pattern as we have seen for the genetic constraints: The constraints are a challenge for the living nature, but the solution gives new emergent possibilities with an unexpected creative power.

Table 7.2 shows that there have been several extinction events during the history of the Earth. An interesting hypothesis concerning global extinction rates was published by Raup and Sepkoski (1986). The authors have observed the development of families of marine animals during the last 250 million years. The result, which is still discussed very critically in paleontology, was that mass extinction events seem to have occurred at rather regular temporal intervals of approximately 26 million years. Explanations were discussed as astronomic forces that might operate with rather precise schedules, as well as terrestrial events (e.g., volcanism, glaciation, sea level change). We will have to wait for further investigations to see whether this hypothesis has been too daring.

Today we can use these ideas to rank the *risk of perturbations* in relation to their temporal characteristics. While mass extinctions seem to be rather rare (Table 7.2), smaller perturbations can appear more frequently (Figure 7.2). In hydrology, floods are distinguished due the temporal probability of their occurrence: 10-, 100-, and 1000-year events are not only characterized by their typical probabilities (translated into typical frequencies), but also by their extents. The rarer the event is, the higher is the risk of the provoked damages. A 100-year flood will result in bigger disturbances than a 10-year event. Also the effects of other disturbance types can be ordered due to their "typical frequencies" (Table 7.3). An often discussed example is fire. The longer the period between two

Table 7.2 Five significant mass extinctions

Geological period	Million years bp	Families lost (%)	Potential reason
Ordovician	440	25	Sudden global cooling
Devonian	370	19	Global climate change
Permian	245	54	Global climate change induced by a bolide
Triassic	210	23	?
Cretaceous	65	17	Asteroid strike

Source: Eldredge (1998).

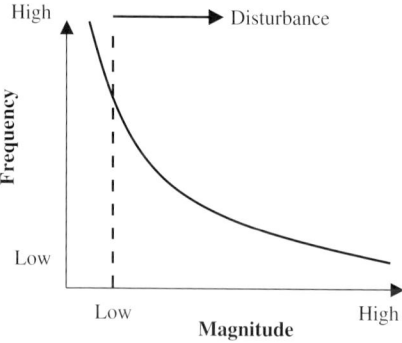

Figure 7.2 Interrelationship between frequencies and magnitudes of perturbations and disturbances, after White and Jentsch (2001).

events, the higher is the probability that the amount of fuel (accumulated burnable organic material) has also increased, and therefore the consequences will be higher if the fire interval has been longer. Similar interrelations can be found concerning the other significant sources of "natural" disturbances, such as volcanoes, droughts, soil erosion events, avalanches, landslides, windstorms, pests, or pathogen outbreaks. The consequences of such rare events can be enormous, and they can be compounded due to human interventions and management regimes. Further information about the hierarchical distinction of rare events included the required time for recovery (Box 7.2).

Table 7.3 Temporal characteristics of some disturbances

Example	Typical temporal scale (orders of magnitude)
Plate tectonics	$\sim 10^5$ years
Climatic cycles	$\sim 10^4$ years
Killing frost	$\sim 10^2$ years
Drought cycles	~ 10 years
El Nino	~ 10 years
Seasonal change	1 year

Source: Di Castri and Hadley (1988), Müller (1992) and Gundersson and Holling (2002).

Box 7.2 Hierarchical distinction of rare events

In Section 2.6, hierarchy theory has been introduced briefly. A key message of this concept is that under steady state conditions the slow processes with broad spatial extents provide constraints for the small-scale processes, which operate with high frequencies. When disturbances occur these hierarchies can be broken and as a consequence (as demonstrated in Section 7.5) small-scale processes can determine the developmental directions of the whole ensemble.

In Figure 7.3 disturbance events are arranged hierarchically, based on quantifications and literature reviews from Vitousek (1994) and Di Castri and Hadley (1988). Here we can also find direct interrelations between spatial and temporal characteristics, i.e., concerning the processes of natural disasters: The broader the spatial scale of a disturbance, the longer time is necessary for the recovery of the system. Furthermore, as shown in Section 7.1, we can assume that events that provoke long recovery times occur with smaller frequencies than disturbances with smaller effects.

Gigon and Grimm (1997) argue that the chain of disturbance effects can also be comprehended from a hierarchical viewpoint. The disturbing event occurs with typical spatio-temporal characteristics, and initially it mainly hits those ecosystem structures that operate on the same scales. Thereafter, an indirect effect chain starts because the internal constraints have changed abruptly. Thus, in the next step, potentially those components should be effected that operate on a lower scale than the initially changed

(continued)

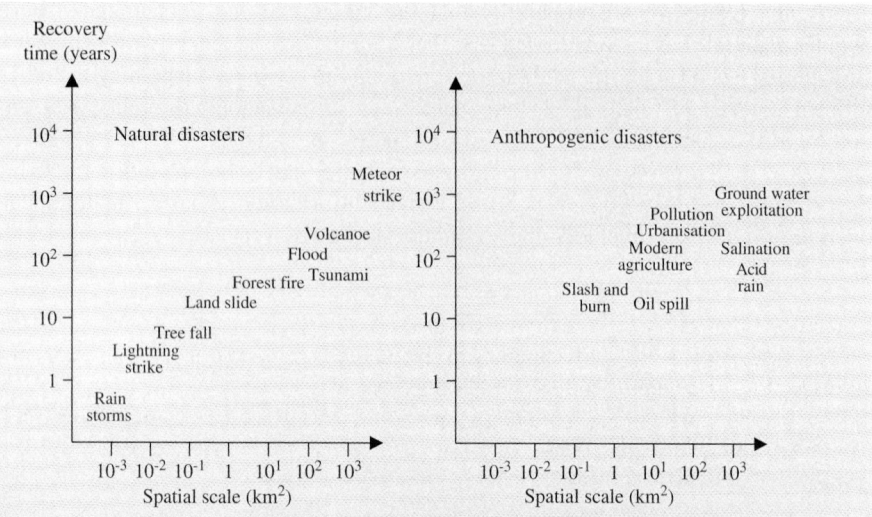

Figure 7.3 Spatial and temporal characteristics of some natural and anthropogenic disasters, after Vitousek (1994) and Di Castri and Hadley (1988). The temporal dimension is being depicted by the specific recovery times after the disturbances have taken place.

holon. Consequently, the biological potential is modified and then also higher levels of the hierarchy can be affected.

In the 1900s, another important feature of disturbance has been discussed: There are certain disasters, which provoke disturbances that are necessary for the long-term development and stability of the affected system. For instance, forest fires are events that necessarily belong into the developmental history of forests. Therefore, the concepts of stratified stability or incorporated disturbances have been set up (e.g., Urban et al., 1987; van der Maarel, 1993). They can today be used as illustrative examples for the natural functioning of the adaptive cycle concept.

This cannot be assigned to the anthropogenic disturbances. Although in the figure only a small selection of such processes can be found, it is obvious that the balance of the natural disasters is not reached by these processes. The influences seem to be so manifold and complex, that only a minor scale dependency can be found. Furthermore, the recovery potential may be based on internal processes and is therefore not dependent on the quantification of openness.

The figure can also be used to illustrate the quantification of openness as introduced in Section 2.6 (Table 2.3). The recovery time is approximately proportional to the periphery of the affected area and can be represented by the square root of the area. As seen in the figure for natural disasters, a meteor strike is affecting an area of approximately 6 orders of magnitude higher than rainstorms. The recovery time after the strike should therefore require 3 orders of magnitude longer time than after the rainstorm. This is approximate due to the relationships of the peripheries, which expresses the exposure of an area to the environment.

7.2 THE RISK OF ORIENTOR OPTIMIZATION

Translating these general points into our ecosystem theory, it is obvious that *two general processes* are governing the dynamics of ecosystems. Besides growth and development processes, living systems are also susceptible to influences that move them back toward thermodynamic equilibrium. On the one hand, there are long phases of *complexification*. Starting with a pioneer stage, orientor dynamics bring about slow mutual adaptation processes with long durations, if there is a dominance of biological processes (see Ulanowicz, 1986a; Müller and Fath, 1998).

A system of interacting structural gradients is created that provokes very intensive internal flows and regulated exchanges with the environment (Müller, 1998). The processes are linked hierarchically, and the domain of the governing attractor (Figure 7.4) remains rather constant, whereupon optimization reactions provoke a long-term increase of orientors, efficiencies, and information dynamics.

The highest state of internal mutual adaptation is attained at the *maturity* domain (Odum, 1969). But the further the system has been moved away from thermodynamic equilibrium, the higher seems to be the risk of getting moved back (Schneider and Kay, 1994) because the forces are proportional to the gradients. The more the time has been used for

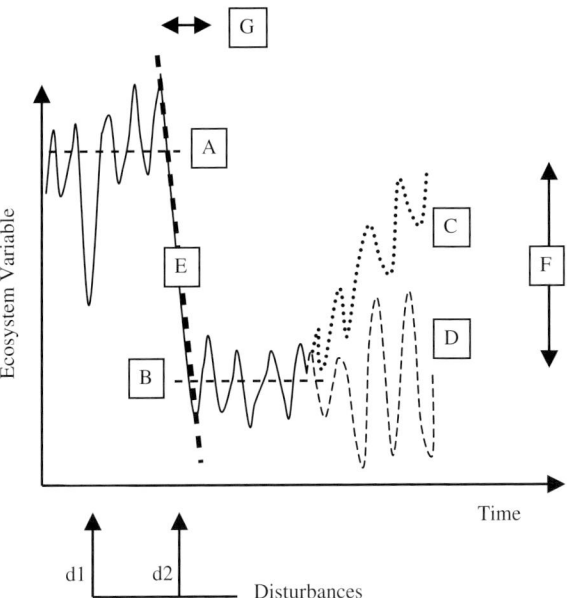

Figure 7.4 Some characteristics of disturbances, after White and Jentsch (2001). The state of the ecosystem is indicated by one ecosystem variable. Due to the disturbance d2 the system is shifted from state A to B, the indicator value thus decreases significantly. The effective disturbance d2 has a higher abruptness (E), a longer duration (G), and a higher magnitude (F) than d1 which does not affect the system. Throughout the following development a high impact affects the trajectory D, which provides a long-term decrease of the ecosystem variable, while a more resilient ecosystem turns back to orientor dynamics (C).

Table 7.4 Some characteristics of mature ecosystems and their potential consequences for the system's adaptability[1]

Orientor function	Risk related consequences
High exergy capture	The system operates on the basis of high energetic inputs → high vulnerability if the input pathways are reduced
High exergy flow density	Many elements of the flow webs have lost parts of their autonomy as they are dependent on inputs which can be provided only if the functionality of the whole system is guaranteed → high risk of losing mutually adapted components
High exergy storage and residence times	Exergy has been converted into biomass and information → high amount of potential fuel and risk of internal eutrophication
High entropy production	Most of the captured exergy is used for the maintenance of the mature system → minor energetic reserves for structural adaptations
High information	High biotic and abiotic diversity → risk of accelerated structural breakdown if the elements are correlated
High degree of indirect effects	Many interactions between the components → increase of mutual dependency and risk of cascading chain effects
High complexity	Many components are interacting hierarchically → reduced flexibility
High ascendancy and trophic efficiency	Intensive flows and high flow diversities have resulted in a loss reduction referring to all single energetic transfers → changing one focal element can bring about high losses
High degree of symbiosis	Symbiosis is linked with dependencies, i.e., if it is inevitable for one or both partners → risk of cascading chain effects
High intra-organismic storages	Energy and nutrients are processed and stored in the organismic phase → no short term availability for flexible reactions
Long life spans	Focal organisms have long-life expectancies → no flexible reactivity
High niche specialization and K selection	Organisms are specialized to occupy very specific niche systems and often have a reduced fecundity → reduced flexibility

[1]Maturity is attained due to a long-term mutual adaptation process. In the end of the development the interrelations between the components are extremely strong, sometimes rigid. Reactivity is reduced. If the constraints change this high efficient state runs the risk of being seriously disturbed.

complexification, the higher is the risk of being seriously hit by disturbance (Table 7.3), and the longer the elements of the system have increased their mutual connectedness, the stronger is the mutual interdependency (Chapter 5) and the total system's brittleness (Holling, 1986). Table 7.4 combines some features of mature ecosystems and lists some risk-related consequences of the orientor dynamics. In general, it can be concluded that the adaptability after changes of the constraints may be decreased when a high degree of maturity is attained.

7.3 THE CHARACTERISTICS OF DISTURBANCE

In such mature states, if certain thresholds are exceeded, fast dynamics can easily become *destructive*. If there is a change of the exterior conditions, or if strong physical processes become predominant, then the inherent brittleness (Holling, 1986) enhances the risk of gradient degradation, thus the flow schemes are interrupted, and energy, information, and nutrients are lost. Hierarchies break down, the attractors are modified, and the system experiences a reset to a new starting point.

Ecologists have studied these events with emphasis on the processes of disturbances. Picket and White (1985) have used a structural approach to define these events: "any relatively discrete event in space and time that disrupts ecosystem, community, or population structure and changes resources, substrates, or the physical environment is called disturbance." Certainly, functional features are also exposed to respective changes, ecosystem processes, and interactions are also disrupted. Chronic stress or background environmental variabilities are not included within this definition, although these relations can also cause significant ecosystem changes. If a disturbance exceeds certain threshold values, then flips and *bifurcations* can occur, which provoke irreversible changes of the system's trajectory. Therefore, understanding ecosystems requires an understanding of their disturbance history.

A focal problem of any disturbance definition is how to indicate the "normal state" of an ecosystem (White and Jentsch 2001) because most biological communities "are always recovering from the last disturbance" (Reice, 1994). For our orientor-based viewpoint it might be appropriate to distinguish the temporal phases during which orientor dynamics are executed from phases of decreasing complexifications caused by exceeding threshold values.

Some basic terms from disturbance ecology are introduced in Figure 7.4. Disturbances exhibit certain *magnitudes* (sizes, forces, and intensities of the events, as variables of the source components), *specificities* (spectrum of disturbed elements), and *severities* (the impacts of the events on system properties). They can be characterized by various *temporal indicators*, such as their spatio-temporal scales, their duration, abruptness, recurrence interval, frequency, or return times. In the literature, exogeneous disturbances resulting from processes outside the system are distinguished from endogeneous disturbances. The latter result from internal ecosystem processes, e.g., as a product of successional development.

Disturbance can have various effects on structural *biodiversity*. It is clear that high magnitudes can easily reduce diversity enormously, while minor inputs might have no effects at all. Connell and Slayter (1977) have found that the highest species numbers are produced by *intermediate disturbances*, because such situations provide suitable living conditions for the highest number of species with relation to their tolerance versus the prevailing disturbances (Sousa, 1984). Furthermore, disturbance is a primary cause of *spatial heterogeneity* in ecosystems, thus it also determines the potential for biodiversity (Jentsch et al., 2002). This concept has been widely discussed within the pattern process hypotheses of patch dynamics (Remmert, 1991). Other ideas concerning the crucial role of disturbance have been formulated, e.g., by Drury and Nisbet (1973) and Sousa (1984). Natural disturbances are an inherent part of the internal dynamics of ecosystems (O'Neill et al., 1986) and can set the timing of successional cycles. Natural disturbances thus seem to be crucial for the long-term ecosystem resilience and integrity.

Taking into account these high dynamic disturbance features, correlating them with the orientor principles (which also are based on changes), focusing on long-term dynamics, and adopting Heraclitus' knowledge from 500 BC ("nothing is permanent but change!"), it becomes rather difficult to find good arguments for an introduction of the *stability principle*. This conception has been the dominant target of environmental management in the last decades (Svirezhev, 2000), and it was strongly interrelated with the idea of a "balance of nature" or a "natural equilibrium" (Barkmann et al., 2001).

Stability has been described by several measures and concepts, such as resistance (the system is not affected by a disturbance), resilience (the systems is able to return to a referential state), or buffer capacity, which measures the overall sensitivities of system variables related to a certain environmental input. Indicators for the stability of ecosystems are for instance the structural effects of the input (recoverability to what extent—e.g., represented by the percentage of quantified structural elements—do the state variables of a system recover after an input?), the return times of certain variables (how long does it take until the referential state is reached again?), or the variance of their time series values after a disturbance (how big are the amplitudes of the indicator variable and how does that size develop?). All of these measures have to be understood in a multivariate manner; due to indirect effects, disturbances always affect many different state variables.

Our foregoing theoretical conceptions show both, that (a) the basic feature of natural systems is a thermodynamic *disequilibrium* and that (b) ecosystems are following *dynamic orientor trajectories* for most time of their existence. Steady state thus is only a short-term interval where the developmental dynamics are artificially frozen into a small-scale average value. Therefore, more progressive indicators of ecosystem dynamics should not be reduced to small temporal resolutions that exclude the long-term development of the system. They should much more be oriented toward the long-term orientor dynamics of ecosystem variables and try to represent the respective potential to continue

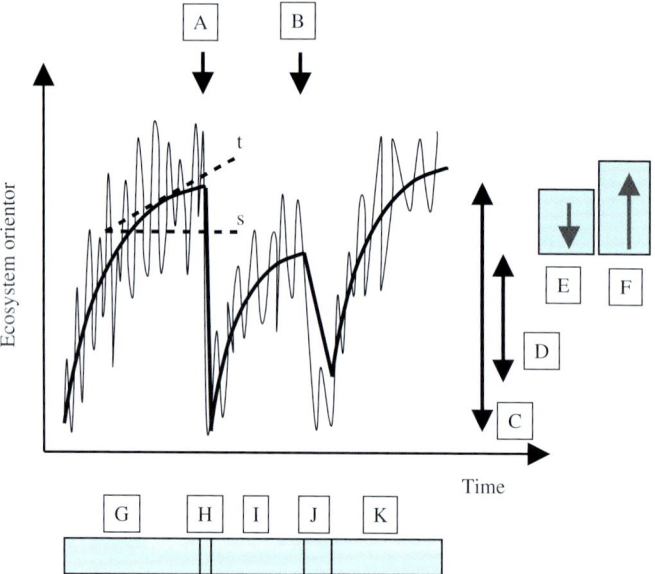

Figure 7.5 Sketch of the dynamics of ecosystem variables on two scales, both variables are influenced by the disturbances (A and B) with different magnitudes (C and D) and durations (H and J), and both variables are due to orientor dynamics during the phases G, I, and K. The development of the fast variable shows a high variance, which can be averaged to the slow dynamics. The long-term effects of the disturbances A and B can be distinguished on the basis of the orientor differences E (reduced resilience and recovery potential) and F (enhanced potential for resilience and recovery).

to change instead of evaluating a system due to its potential to return to one defined (non-developmental and perhaps extremely brittle) state. A good potential seems to lie in the concept of *resilience*, if we define it as the capacity of a disturbed system to return to its former complexifying *trajectory* (not to a certain referential state). Therefore, the reference situation (or the aspired dynamics of ecosystem management) would not be the static lines in Figure 7.5, but the orientor trajectory *t*. Similar ideas and a distinction of stability features with reference to the systems' stability are discussed in Box 7.3.

Box 7.3 Stability is related to uncorrelated complexity: After Ulanowicz (2002a,b)

Summary: The complexity of the pattern of ecosystem transfers can be gauged by the Shannon–Weaver diversity measure applied to the various flows. This index, in turn, can be decomposed into a component that refers to how the flows are constrained by (correlated with) each other and another that represents the remaining degrees of freedom, which the system can reconfigure into responses to novel perturbations. It is the latter (uncorrelated) complexity that supports system stability.

Development: In order to see how system stability is related only to part of the overall system complexity, it helps to resolve the complexity of a flow network into two components, one of which represents coherent complexity and the other, its incoherent counterpart (Rutledge et al., 1976.)

Prior to Rutledge et al., complexity in ecosystems had been reckoned in terms of a single distribution, call it $p(a_i)$. The most common measure used was the Shannon (1948) "entropy,"

$$H = -\sum_i p(a_i) \log[p(a_i)]$$

Rutledge et al. (1976) showed how information theory allows for the comparison of two different distributions. Suppose one wishes to choose a "reference" distribution with which to compare $p(a_i)$. Call the reference distribution $p(b_j)$. Now Bayesian probability theory allows one to define the joint probability, $p(a_i,b_j)$, of a_i occurring jointly with b_j. Ulanowicz and Norden (1990) suggested applying the Shannon formula to the joint probability to measure the full "complexity" of a flow network as,

$$H = -\sum_{i,j} p(a_i,b_j) \log[p(a_i,b_j)]$$

Then, using Rutledge's formulation, this "capacity" could be decomposed into two complementary terms as,

$$H = \sum_{i,j} p(a_i,b_j) \log\left[\frac{p(a_i,b_j)}{p(a_i)p(b_j)}\right] - \sum_{i,j} p(a_i,b_j) \log\left[\frac{p(a_i,b_j)^2}{p(a_i)p(b_j)}\right]$$

where the first summation represents the coherence between the a_i and the b_j, and the second on the remaining dissonance between the distributions.

(continued)

The genius of Rutledge et al. (1976) was to identify $p(a_i)$ and $p(b_j)$ with the compartmental distributions of inputs and outputs, respectively. That is, if T_{ij} represents the quantity of flow from compartment i to j, and $T_{..}$ represents the sum of all the flows (a dot in place of a subscript means summation over that index), then

$$p(a_i, b_j) \sim \frac{T_{ij}}{T_{..}}, \quad p(a_i) \sim \sum_j \frac{T_{ij}}{T_{..}}, \text{ and } p(b_j) \sim \sum_i \frac{T_{ij}}{T_{..}}$$

Substituting these estimates into the decomposition equation yields,

$$H = \sum_{i,j} \frac{T_{ij}}{T_{..}} \log\left[\frac{T_{ij}}{T_{..}}\right] = \sum_{i,j} \frac{T_{ij}}{T_{..}} \log\left[\frac{T_{ij} T_{..}}{T_{i.} T_{.j}}\right] - \sum_{i,j} \frac{T_{ij}}{T_{..}} \log\left[\frac{T_{ij}^2}{T_{i.} T_{.j}}\right]$$

or

$$H = I + D$$

where I is known as the "average mutual information" inherent in the flow structure and D the residual disorder. In other words, the complexity has been decomposed into a term that measures how well the flows are constrained (coordinated) and how much they remain independent (free.)

Rutledge et al. (1976) suggested that the ability of the ecosystem network to respond in new ways to novel disturbance is related to D, while Ulanowicz (1980) argued that I quantifies the organization inherent in the flow network. It is important to notice that I and D are complementary, which is to say that, other things being equal, any change in I will be accompanied by a complementary change in D. The system cannot "have its cake and eat it, too." Coherent performance, I, comes at the expense of reliability, D, and vice-versa.

In other words, one should expect system stability to be more related to the value of the disordered complexity, D, and less correlated to the overall complexity, H, as the latter also encompasses the complexity encumbered by system constraints.

7.4 ADAPTABILITY AS A KEY FUNCTION OF ECOSYSTEM DYNAMICS

Having introduced general aspects of disturbance ecology, we can now start to integrate the complexification and the disturbance-induced dynamics of ecosystems. The respective approach is based upon the concepts of the "Resilience Alliance" (see e.g., Holling, 1986, 2004; Gundersson and Holling, 2002; Elmquist et al., 2003; Walker and Meyers 2004; Walker et al., 2004), but they have been restricted to ecosystem dynamics and combined with the sequence of growth forms after Jørgensen et al. (2000) (see also Ulanowicz, 1986a,b, 1997; Fath et al., 2004). Under these prerequisites, we can distinguish the following principle steps of ecosystem development:

– *Start* of the succession (pioneer stage, boundary growth after Jørgensen et al., exploitation function after Holling, 1986): In this initial state, an input of low-entropy material into the system starts the sere. The developmental potential depends on the genetic information that is available in the seed bank or by lateral inputs. Due to a minor connectivity between the elements, self-regulation is low, leakyness is high, and the sum of potential developmental opportunities (developmental uncertainty) is high. The system provides a very high adaptability and flexibility.

- *Fast growth* (pioneer stage, structural growth after Jørgensen et al., exploitation function after Holling): Pioneer stages can also be characterized by a high and rapid increase of biomass, correlated with an increase of the numbers and sizes of the ecosystem components. To provide the growing number of participants, the energy throughflow increases as well as exergy degradation, which is necessary for the maintenance of the components. Connectivity is low, and therefore external inputs can modify the system easily; the adaptability is high.
- *Fast development* (middle succession, network growth after Jørgensen et al., conservation function after Holling): After a first structure has been established, the successful actors start funneling energy and matter into their own physiology. Due to the mutual adaptation of the winning community, the connectivity of the system increases by additional structural, energetic, and material interrelations and cycling mechanisms. The single species become more and more dependent on each other, uncertainty decreases, and the role of self-regulating processes grows, reinforcing the prevailing structure. Adaptability is reduced.
- *Maturity* (information growth after Jørgensen et al., conservation function after Holling): In this stage, a qualitative growth in system behavior takes place, changing from exploitative patterns to more conservative patterns with high efficiencies of energy and matter processing. Species that easily adapt to external variability (r-selected species) have been replaced by the variability controlling K-strategists; the niche structure is enhanced widely, and loss is reduced. The information content of the system increases continuously. A majority of the captured exergy is used for the maintenance of the system; thus, there is only a small energetic surplus, which can be used for adaptation processes. Sensitivities versus external perturbations have become high, while the system's buffer capacities are much smaller compared with the former stages of the development. These items result in a rise of the system's vulnerability and a decrease of resilience (see Table 7.4). Adaptability has reached minimum values.
- *Breakdown* (release function after Holling, creative destruction after Schumpeter, 1942): Due to the "brittleness" of the mature stages (Holling, 1986), their structure may break down very rapidly due to minor changes of the exterior conditions. Accumulated resources are released, internal control and organization mechanisms are broken, and positive feedbacks provoke the decay of the mature system. Uncertainty rises enormously, hierarchies are broken, and chaotic behavior can occur (Figure 7.3). There are only extremely weak interactions between the system components, nutrients are lost and cycling webs are disconnected. Adaptability and resilience have been exceeded.
- *Reorganization:* During this short period the structural and functional resources can be arranged to favor in new directions, new species can occur and become successful, and—in spite of the inherited memory (e.g., seed bank of the old system and

neighboring influences)—unpredictable developmental traits are possible. There are weak controls, and innovation, novelty, and change can lead to an optimized adaptation on a higher level.
- *Reset*: A new ecosystem succession starts.

The described sequence has been illustrated in Figure 7.6 as a function of the system's internal connectedness and the stored exergy. Starting with the exploitation function, there is a slow development. The trajectory demonstrates a steady increase in mutual interactions as well as an increase in the stored exergy. As has been described above, this energetic fraction can be distinguished into a material fraction (e.g., biomass, symbolizing the growth conception of Ulanowicz, 1986a,b) and the specific exergy that refers to a complexification of the system's structure (development after Ulanowicz). In spite of multiple variability (e.g., annual cycles), the long-term development shows a steady increase up to the mature state. Here the maximum connectivity can be found, which on the one hand is a product of the system's orientation, but which also is correlated with the risk of missing adaptability, which has been nominated as over-connectedness by some authors. After the fast releasing event, the short-term conditions determine the further trajectory of the system. It might turn into a similar trajectory or find a very different pathway.

This figure looks very similar to the well-known four-box model of the Resilience Alliance, which has been depicted in Figure 7.7. The difference between these approaches lies in the definition of the y-axis. While for interdisciplinary approaches and analyses of human–environmental system the special definition of "potential" in the adaptive cycle metaphor seems to be advantageous; from our thermodynamic viewpoint, the key variable

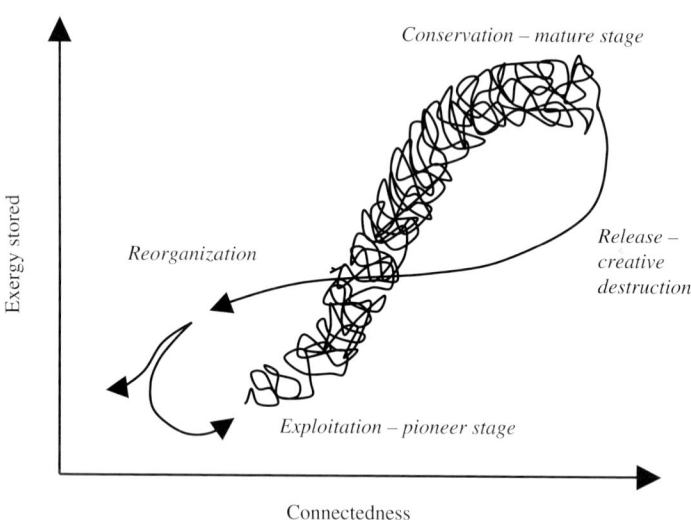

Figure 7.6 Ecosystem succession as a function of structural and functional items.

Chapter 7: Ecosystems have complex dynamics 159

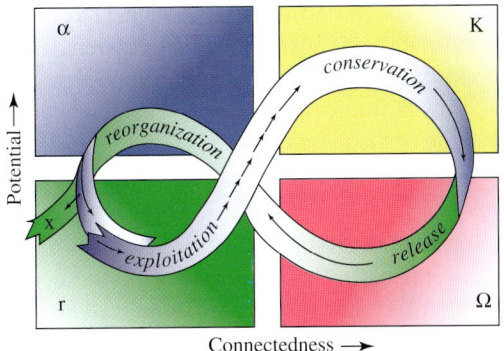

Figure 7.7 Adaptive cycle after Holling, from Gundersson and Holling (2002).

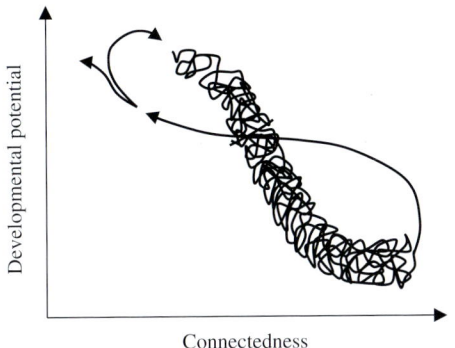

Figure 7.8 Developmental opportunities during the successional cycle from Figure 7.6.

is the total stored exergy, which does not rise again after creative destruction. The nutrients as well as the energetic resources do not grow after the release, but get eroded or leached, and the change of their availability is due to the activities of the organisms, which appear right after the reset of the pioneer stage.

To illustrate the risk discussion from above, Figure 7.8 shows a correlated trajectory of the developmental potential of ecosystems during the adaptive dynamics. This point shows another difference with the concept of the Resilience Alliance, due to another understanding of "potential." Originating from ecosystem theory, we can use the amount of potential trajectories (possibilities, developmental directions) of the system during the whole cycle. As has been described above, there are a high number of developmental possibilities in the beginning during the pioneer phase while thereafter the prevailing interactions are limiting the degrees of freedom and the adaptability of the system continuously. Self-organizing processes have created internal hierarchical constraints, which reduce the flexibility of the entity. Integrating Figures 7.6 and 7.8 demonstrates the dilemma of the orientor philosophy: The more complex and efficient an ecological system's performance is, the better (and more successful) its "old" adaptation to the environmental conditions has been, the lower is its adaptability against unknown environmental changes, and the higher is the system's vulnerability. Thus, a further adaptation to changing conditions is only possible on the base of a destruction of the old structures.

7.5 ADAPTIVE CYCLES ON MULTIPLE SCALES

With the following argumentation we want to link these concepts with another approach to ecosystem theories: Ecosystems are organized *hierarchically* (see Box 2.2 in Chapter 2). Hereafter, we will assume that throughout complexification periods, the focal processes always are influenced by the lower levels' dynamics and the higher levels' development, forming a system of constraints and dynamics of biological potentials. Thus there are four general hierarchical determinants for ecosystem dynamics:

(i) The *constraints* from higher levels are completely effective for the fate of the focal variable. The constraints operate in certain temporal features, with specific regularities and intervals. Some examples for these temporal characteristics are:

- Day–night dynamics (e.g., determining ecosystem temperature, light, or humidity)
- Tides (e.g., determining organism locations in the Wadden Sea)
- Moon phases (e.g., determining sexual behavior)
- Annual dynamics (e.g., determining production phases of plants)
- Longer climatic rhythms (e.g., sun spots influencing production)
- Dynamics of human induced environmental stress factors
    - Typical periodic land use activities (e.g., crop rotation)
    - Land use change (structural and functional)
    - Emission dynamics and environmental policy (e.g., sulfur emission in Germany and their effects on forests)
    - Global change and greenhouse gas emissions (e.g., temperature rise)
    - Continuous climate change
- Biome transitions

These constraints are interacting and constantly changing; therefore, the maximum degree of mutual adaptation is a dynamic variable as well. This is a focal reason why the orientor approach is nominated as a "very theoretical outline" only. As ecosystems "always are recovering from the last disturbance," the orientor dynamics often are practically superseded by the interacting constraints dynamics.

(ii) The *dynamics of the focal variables* themselves exhibit certain natural frequencies. As in the patch dynamics concepts, there can be internal change dynamics on the observed level itself. For example, we can observe the undisturbed succession on the base of biological processes—from a lake to a fen. The system changes enormously due to its internal dynamics. Throughout this process often a limited

Chapter 7: Ecosystems have complex dynamics

number of species become dominant, e.g., stinging nettles in secondary successions on abandoned agricultural systems. This leads to an interruption of orientor dynamics because the dominant organisms do not allow competitors to rise.

(iii) The *biological potential* of the lower levels results from mostly filtered, smoothened, and buffered variables with high frequencies. They can only become effective if the system exceeds certain threshold values. This can happen if disturbances unfold their indirect effects, as has been described above.

(iv) *Disturbances* primarily meet elements that operate on similar spatial and temporal scales. Only after these components have been affected, indirect effects start influencing the interrelated scales and thus can provoke far-reaching changes.

Summarizing these points, we can state that ecosystems under steady state conditions are regulated by a hierarchy of interacting processes on different scales. The slow processes with large extents build up a system of constraints for the processes with high dynamics. Thereby limiting their degrees of freedom, steady states can be characterized by relatively low variability of low-level processes (O'Neill et al., 1986). Furthermore, under steady state conditions, these high dynamic processes cannot influence the system of constraints, resulting in a rather high resilience. Thus, the question arises, what will happen during disturbances?

This can be depicted by the concept of stability landscapes (see Walker et al., 2004) or hypothetical potential functions. In Figure 7.9 the system state is plotted on the x-axis, the z-axis represents the parameter values (may also be taken as a temporal development with changing parameter loadings), and the potential function is plotted on the y-axis. This function can be regarded as the slope of a hill, where the bottom of the valley

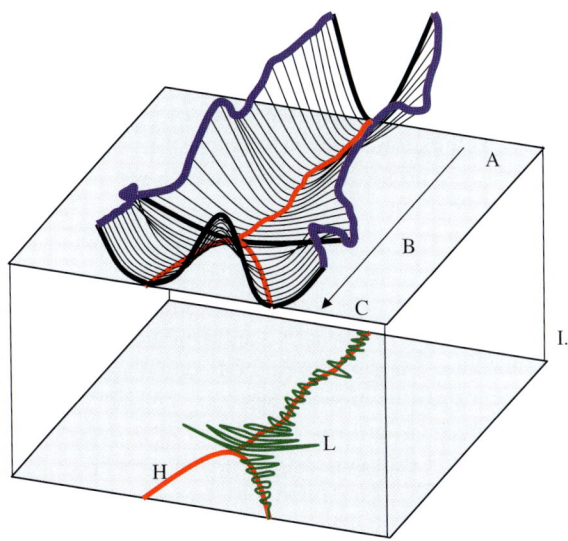

Figure 7.9 Hypothetical potential function of a hierarchical system.

represents steady state conditions. If we throw a marble into this system, then it will find its position of rest after a certain period of time at the deepest point of the curve. If the parameter values change continuously (A → B → C), then a set of local attractors appear, symbolized by the longitudinal profile of the valley, or the broadscale bifurcation line (H) at level I. This manifold sketches a sequence of steady states referring to different parameter values. In Figure 7.9, the straight line below on level I may be interpreted as the sequence of a parameter of a high hierarchical level while the oscillating parameter value line L indicates the states of a lower level holon. The return times of this holon to its different steady states will be different if the states A, B, and C are compared: The steeper the slope the more rapidly a local steady state will be reached, and smaller amplitudes will be measured. When the parameter value is changed continuously within long-term dynamics we will find small variations near state A. As our parameter shifts from A via B toward C, the potential curve's slopes decrease, finding a minimum at B. In this indifferent state the amplitudes of the low-level holon will be very high (see level I). If there is a further change of the parameter value, a first-order phase transition takes place. The state can be changed radically passing the bifurcation point B before a more stable state is achieved again, finally reaching C. Passing B there are two potential states the system can take, and the direction our holon takes is determined by all levels of the broken hierarchy, including the high frequent (small scale) dyna-mics. This process is accompanied by temporal decouplings, by a predominance of positive feedbacks, and by autocatalytic cycles.

This makes it possible for ecosystems to operate at the edge of chaos, but frequently avoid chaos and utilize all the available resources at the same; see also Box 7.4.

Box 7.4   Chaos in ecosystem models

The prevailing conditions including the abundance of other species determine which growth rate is optimal. If the growth rate is too high, then the resources (food) will be depleted and the growth will cease. If the growth rate is too low, then the species does not utilize the resources (food) to the extent that it is possible. The optimal growth rate also yields the highest system exergy. If, in a well-calibrated and validated eutrophication model—state variables include phytoplankton, nitrogen, phosphorus, zooplankton, fish, sediment nitrogen, and sediment phosphorus—the zooplankton growth rate is changed, then exergy will show a maximum at a certain growth rate (which is frequently close to the value found by the calibration and approved by the validation). At both lower and higher growth rates, the *average* exergy is lower because the available phytoplankton is either not utilized completely or is overexploited. When overexploitation occurs the phytoplankton and zooplankton show violent fluctuations. When the resources are available the growth rate is very high but the growth stops and the mortality increases as soon as the resources are depleted, which gives the resources a chance to recover and so on. At a growth rate slightly higher than the value giving maximum exergy, the model starts to show deterministic chaos. Figure 7.10 illustrates the exergy as function of the zooplankton growth rate in the model referred to above, focusing on the time when the model starts to show deterministic chaos. These results are consistent with Kaufmann's (1993) statement: biological systems tend to operate at the edge of chaos to be able to utilize the resources at the optimum. In response to constraints, systems move away as far as possible from thermodynamic equilibrium under the prevailing conditions, but that implies that the system has a high probability to avoid chaos, although the system is operating close to chaos. Considering the enormous complexity of natural ecosystems, and the many interacting processes, it is surprising that chaos is not frequently observed in nature, but it can be explained by an operation at *the edge* of chaos to ensure a high utilization of the resources—to move as far away as possible from thermodynamic equilibrium under the prevailing conditions.

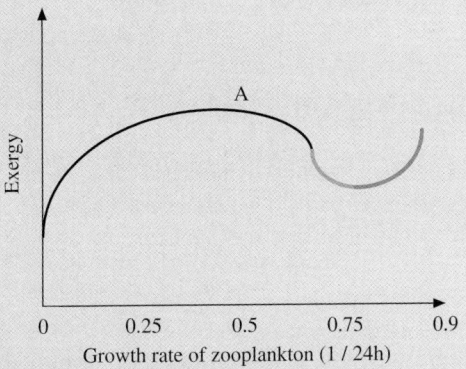

Figure 7.10 Exergy is plotted versus maximum growth rate for zooplankton in a well calibrated and validated eutrophication model. The shaded line corresponds to chaotic behavior of the model, i.e., violent fluctuations of the state variables and the exergy. The shown values of the exergy above a maximum growth rate of about 0.65–0.7 per day are therefore average values. By a minor change of the initial value of phytoplankton or zooplankton in the model, significant changes are obtained after 2 months simulations as an indication of deterministic chaos.

After having elucidated disturbance from the hierarchical viewpoint, one last aspect should be taken into consideration. As we have mentioned above, the adaptive cycle is a metaphor, which can be assigned to a multitude of interacting scales. There is a high normality in disturbance with adaptability as a key function. If this feature cannot reach sufficient quantities by low-scale flexibility, then the breakdown on a higher hierarchical level enables the system to start a reset under the new prevailing conditions. Thus, in the end, disturbance really can be understood as a part of ecosystem growth and development on a higher scale, as indicated in Figure 7.11; disturbance may even be extremely necessary to enable a continuation of the complexifying trajectory of the overall system.

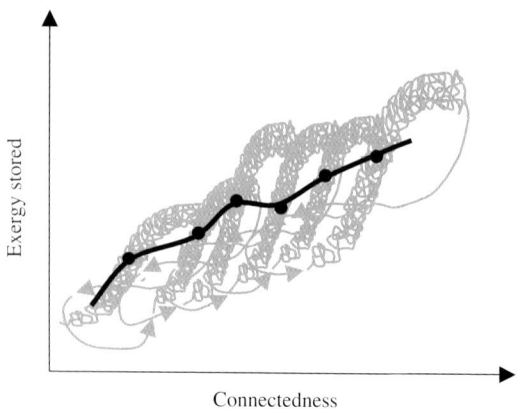

Figure 7.11 Long-term succession of ecosystems, indicated on different scales: small-scale disturbances may support the development of the overall system.

7.6 A CASE STUDY: HUMAN DISTURBANCE AND RETROGRESSIVE DYNAMICS

Up to now, we have focused on "natural dynamics." Thus, in the end of this chapter, we demonstrate human disturbances using a wetland case study. In general, human activities influence disturbance regimes in several mechanisms, such as:

- the rescaling of natural disturbances,
- the introduction of novel disturbances,
- the modification of the reception mechanisms of the disturbed components,
- influences on disturbance rates and intensities,
- the suppression of natural disturbances to ensure the potential of aspired ecosystem services,
- the change of successional pathways due to irreversible changes.

As an example for human pressures and disturbance dynamics, Figure 7.12 describes a case study from ecosystem research in the wetlands of the Bornhöved Lakes District in Northern Germany. Here a holistic indicator system, which has been developed on the basis of the orientor theory (Müller, 2005) has been used to demonstrate the steps of wetland retrogression as provoked by eutrophication and drainage.

Based on field measurement, mappings, and classifications different ecosystem types have been analyzed with the computer-based "digital landscape analysis system" (Reiche, 1996) and the modelling system "Wasmod–Stomod" (Reiche, 1996) which was used to simulate the dynamics of water budgets, nutrient, and carbon fluxes based on a 30-year series of daily data about meteorological and hydrological forcing functions. The model outputs were validated by measured data in some of the systems

Chapter 7: Ecosystems have complex dynamics

(Schrautzer, 2003). The model outputs were extended to include data sets concerning the ecosystem indicators by the following variables:

- Exergy capture: net primary production (NPP)
- Entropy production: microbial soil respiration
- Storage capacity: nitrogen balance, carbon balance
- Ecosystem efficiency: evapotranspiration/transpiration, NPP/soil respiration
- Nutrient loss: N net mineralization, N leaching, denitrification
- Ecosystem structures: Number of plant species (measured values)

The wet grasslands of the Bornhöved Lakes District are managed in a way that includes the following measures: drainage, fertilization, grazing, and mowing in a steep gradient of ecosystem disturbances. The systems have been classified due to these external input regimes, and in Figure 7.12 the consequences can be seen in a synoptic manner. While the farmer's target (improving the production and the yield of the systems), the NPP is growing by a factor of 10, the structural indicator is decreasing enormously throughout the retrogression. Also the efficiency measures (NPP/soil respiration) are going down, and the biotic water flows get smaller. On the other hand, the development

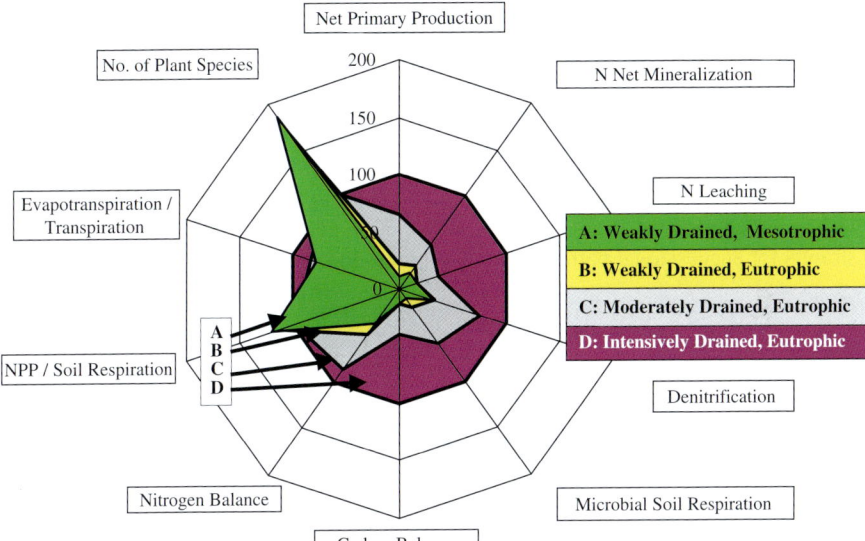

Figure 7.12 Retrogressive ecosystem features at different steps of human intervention, after Müller et al. (i.p.). The figure shows a set of 10 holistic indicators which as a whole represent ecosystem integrity. Starting with the initial state A, drainage and eutrophication of the wet grassland ecosystems affect irreversible changes up to the degraded state D. During that development ecosystem structures (complexity) are reduced, energy and matter efficiencies decrease, and the originally sink ecosystem turns into a source for nitrogen and carbon compounds.

of the carbon and nitrogen balances demonstrates that the system is turning from a sink function into a source, the storage capacity is being reduced, and the loss of carbon and nitrogen compounds (all indicators on the right side of the figure) is rising enormously. With these figures we can state an enormous decrease of ecosystem health, and as many of the processes are irreversible, the capacity for future self-organization is reduced up to a very small degree.

7.7 SUMMARY AND CONCLUSIONS

In this chapter we have discussed the role of destructive processes for ecosystem dynamics. After some examples of destructive events on the organism scale, the population scale and the ecosystem scale, and after a general integration of the disturbance concept into the orientor model, it is shown that especially mature states can suffer from the high risk of reduced adaptability. Therefore, breakdown is the consequent reaction if the living conditions of a community change strongly. Thereafter, new potentials can be realized and the orientor behavior will start again with renewed site conditions. Adopting this argumentation, natural disturbances seem to be crucial for the long-term self-organization, for the ecological creativity, and for the long-term integrity of ecological entities. Destructive processes are focal components of the overall ecosystem adaptability, and they can be found on all relevant scales.

If we follow the ecosystem-based argumentation that integrity and health are relevant variables for ecological evaluation, the potential for self-organizing processes becomes a key variable in environmental management. It is strictly related to the long-term ecosystem adaptability and its buffer capacity. Therefore, human disturbances in fact intervene the natural dynamics: They operate on artificial spatio-temporal scales, they introduce novel qualities and quantities of change, they modify the reception mechanisms of the ecosystems, they often reduce ecosystem adaptability, and—as shown in the case study—they set new constraints for successional pathways, thus suppressing the natural dynamics.

8

Ecosystem principles have broad explanatory power in ecology

THE BEST ANSWER RAISES MOST QUESTIONS

8.1 INTRODUCTION

The criticism that ecology as a whole lacks universal laws and predictive theory is frequent, and there are authors who even argue that theoretical ecology concerned for instance with fitness and natural selection is not scientific (Murray, 2001).

Scientific observations on natural phenomena usually give origin to possible explanations and, furthermore, provide tentative generalizations that may lead to broad-scale comprehension of the available information. Generalizations may be descriptive and inductive, deriving from observations carried out on observable characteristics, or become much more eager, constituting the base of deductive theories. In ecology, we must recognize that there are basically no universal laws (maybe such laws cannot even exist in the same sense as those in physics). In fact, most explanations in ecology are inductive generalizations, without any deductive theory behind them, and as a consequence we may find a large number of non-universal tentative generalizations.

As explained earlier in the book, regarding features such as immense number problem, growth and decay, and network interrelations, ecology is more complex than physics, and it will, therefore, be much more difficult to develop an applicable, predictive ecological theory. Testing explanatory hypotheses by verification instead of by falsification is perhaps the easiest way. But many ecologists probably feel inwards the need for a more general and integrative theory that may help in explaining their observations and experimental results.

In the last 20 or 30 years several new ideas, approaches, and hypotheses appeared in the field of systems ecology, which when analyzed more deeply appear to form a pattern of theories able to explain the dynamics of ecosystems (Jørgensen, 1997, 2002). And in fact, due to the complexity involved, we probably need a number of different complementary approaches to explain ecosystem structure and function (Jørgensen, 1994a; Fath et al., 2001). Such ecosystem theories were only used in a limited way in ecological modeling, namely in the development of non-stationary models, able to take into account the adaptation of biological components (Jørgensen, 1986, 1992b, 1994b, 1997; Jørgensen and de Bernardi, 1997, 1998). It has been argued that to improve substantially the predictive power of ecological models it will probably be necessary to apply theoretical approaches much more widely (Jørgensen and Marques, 2001).

Nevertheless, the question remains: is it possible to develop a theoretical framework able to explain the numerous observations, rules, and correlations dispersed in the ecological literature during the last few decades?

Although we may have no sound answer to this question, it has been argued (Jørgensen and Marques, 2001) that it should at least be possible to propose a promising direction for ecological thinking. The idea in this chapter is to check the compliance of ecosystem principles to a number of ecological rules or laws, and to see if other proposed non-universal explanations provided by different authors about different ecological problems can be further enlightened according to the same ecological principles.

8.2 DO ECOLOGICAL PRINCIPLES ENCOMPASS OTHER PROPOSED ECOLOGICAL THEORIES?: EVOLUTIONARY THEORY

One of the most important, if not the most important, theories in biology is the theory of evolution; so we begin by outlining this theory, with examples and with intent later to show a similarity with it to the ecosystem theories proposed earlier in the book. In biology, *evolution* is the process by which natural populations of organisms acquire and pass on novel characteristics from generation to generation (Darwin and Wallace, 1858; Darwin, 1859), and the theory of evolution by natural selection became decisively established within the scientific community. In the 1930s, work by a number of scientists combined Darwinian natural selection with the re-discovered theory of heredity (proposed by Gregor Mendel) to create the modern evolutionary synthesis. In the modern synthesis, "evolution" means a change in the frequency of an allele within a gene pool from one generation to the next. This change may be caused by a number of different mechanisms: natural selection, genetic drift, or changes in population structure (gene flow).

(a) *Natural selection* is survival and reproduction as a result of the environment. Differential mortality consists of the survival rate of individuals to their reproductive age. Differential fertility is the total genetic contribution to the next generation. The central role of natural selection in evolutionary theory has given rise to a strong connection between that field and the study of ecology.

Natural selection can be subdivided into two categories:

- *Ecological selection* occurs when organisms that survive and reproduce increase the frequency of their genes in the gene pool over those that do not survive.
- *Sexual selection* occurs when organisms that are more attractive to the opposite sex because of their features reproduce more and thus increase the frequency of those features in the gene pool.

Natural selection also operates on mutations in several different ways:

- Purifying or background selection eliminates deleterious mutations from a population.
- Positive selection increases the frequency of a beneficial mutation.

- Balancing selection maintains variation within a population through a number of mechanisms, including:

  ○ Over-dominance or heterozygote advantage, where the heterozygote is more fit than either of the homozygous forms (exemplified by human sickle cell anemia conferring resistance to malaria).
  ○ Frequency-dependent selection, where the rare variants have a higher fitness.

- Stabilizing selection favors average characteristics in a population, thus reducing gene variation but retaining the mean.
- Directional selection favors one extreme of a characteristic; results in a shift in the mean in the direction of the extreme.
- Disruptive selection favors both extremes, and results in a bimodal distribution of gene frequency. The mean may or may not shift.

(b) *Genetic drift* describes changes in allele frequency from one generation to the next due to sampling variance. The frequency of an allele in the offspring generation will vary according to a probability distribution of the frequency of the allele in the parent generation.

Many aspects of genetic drift depend on the size of the population (generally abbreviated as N). This is especially important in small mating populations, where chance fluctuations from generation to generation can be large. Such fluctuations in allele frequency between successive generations may result in some alleles disappearing from the population. Two separate populations that begin with the same allele frequency might, therefore, "drift" by random fluctuation into two divergent populations with different allele sets (e.g. alleles that are present in one have been lost in the other).

The relative importance of natural selection and genetic drift in determining the fate of new mutations also depends on the population size and the strength of selection: when $N \cdot s$ (population size times strength of selection) is small, genetic drift predominates. When $N \cdot s$ is large, selection predominates. Thus, natural selection is 'more efficient' in large populations, or equivalently, genetic drift is stronger in small populations. Finally, the time for an allele to become fixed in the population by genetic drift (i.e. for all individuals in the population to carry that allele) depends on population size, with smaller populations requiring a shorter time to fixation.

The theory underlying the modern synthesis has three major aspects:

(1) The common descent of all organisms from a single ancestor.
(2) The manifestation of novel traits in a lineage.
(3) The mechanisms that cause some traits to persist while others perish.

Essentially, the modern synthesis (or neo-Darwinism) introduced the connection between two important discoveries: the evolutionary units (genes) with its mechanism (selection). It also represents a unification of several branches of biology that previously had little in common, particularly genetics, cytology, systematics, botany, and paleontology.

A critical link between experimental biology and evolution, as well as between Mendelian genetics, natural selection, and the chromosome theory of inheritance, arose from T.H. Morgan's work with the fruit fly *Drosophila melanogaster* (Allen, 1978). In 1910, Morgan discovered a mutant fly with solid white eyes—wild-type *Drosophila* have red eyes—and found that this condition though appearing only in males was inherited precisely as a Mendelian recessive trait. Morgan's student Theodosius Dobzhansky (1937) was the first to apply Morgan's chromosome theory and the mathematics of population genetics to natural populations of organisms, in particular *Drosophila pseudoobscura*. His 1937 work *Genetics and the Origin of Species* is usually considered the first mature work of neo-Darwinism, and works by E. Mayr (1942: systematics), G.G. Simpson (1944: paleontology), G. Ledyard Stebbins (1950: botany), C.D. Darlington (1943, 1953: cytology), and J. Huxley (1949, 1942) soon followed.

According to the modern synthesis as established in the 1930s and 1940s, genetic variation in populations arises by chance through mutation (this is now known to be due to mistakes in DNA replication) and recombination (crossing over of homologous chromosomes during meiosis). Evolution consists primarily of changes in the frequencies of alleles between one generation and another as a result of genetic drift, gene flow, and natural selection. Speciation occurs gradually when geographic barriers isolate reproductive populations. The modern evolutionary synthesis continued to be developed and refined after the initial establishment in the 1930s and 1940s. The most notable paradigm shift was the so-called Williams revolution, after Williams (1966) presented a gene-centric view of evolution. The synthesis as it exists now has extended the scope of the Darwinian idea of natural selection, specifically to include subsequent scientific discoveries and concepts unknown to Darwin such as DNA and genetics that allow rigorous, in many cases mathematical, analyses of phenomena such as kin selection, altruism, and speciation.

Examples

Example 1: Industrial melanism in the peppered moth
Wallace (1858) hypothesized that insects that resemble in color the trunks on which they reside will survive the longest, due to the concealment from predators. The relatively rapid rise and fall in the frequency of mutation-based melanism in populations (Figure 8.1) that occurred in parallel on two continents (Europe, North America), is a compelling example for rapid microevolution in nature caused by mutation and natural selection. The hypothesis that birds were selectively eating conspicuous insects in habitats modified by industrial fallout is consistent with the data (Majerus, 1998; Cook, 2000; Coyne, 2002; Grant, 2002).

Example 2: Warning coloration and mimicry
In his famous book, Wallace (1889) devoted a comprehensive chapter to the topic "warning coloration and mimicry with special reference to the Lepidoptera". One of the most conspicuous day-flying moths in the Eastern tropics was the widely distributed species *Opthalmis lincea* (Agaristidae). These brightly colored moths have developed chemical repellents that make them distasteful, saving them from predation (*Müllerian mimetics*). *O. lincea* (Figure 8.2A) is mimicked by the moth *Artaxa simulans* (Liparidae), which was collected during the voyage of the *Challenger* and later described as a new species (Figure 8.2B). This survival mechanism is called *Batesian mimetics* (Kettlewell, 1965).

A
form: typica

B
form: carbonaria

species: *Biston betularia*

Figure 8.1 Industrial melanism in populations of the peppered moth (*Biston betularia*). Previously to 1850, white moths peppered with black spots (typica) were dominant in England (A). Between 1850 and 1920, as a response to air pollution that accompanied the rise of heavy industry, typica was largely replaced by a black form (carbonaria) (B), produced by a single allele, since dark moths are protected from predation by birds. Between 1950 and 1995, this trend reversed, making form (B) rare and (A) again common. (Adapted from Kettlewell, 1965).

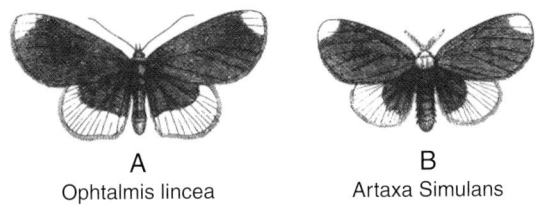

A
Ophtalmis lincea

B
Artaxa Simulans

Figure 8.2 Insects have evolved highly efficient survival mechanisms that were described in detail by A.R. Wallace. One common moth species (*Opthalmis lincea*) (A) contains chemical repellents to make the insects distasteful. This moth is mimicked by a second species (*Artaxa simulans*) (B) From Wallace (1889).

Example 3: Darwin's finches
Darwin's finches exemplify the way one species' gene pools have adapted for long-term survival via their offspring. The Darwin's finches diagram below illustrates the way the finch has adapted to take advantage of feeding in different ecological niches (Figure 8.3).

Their beaks have evolved over time to be best suited to their feeding situation. For example, the finches that eat grubs have a thin extended beak to poke into holes in the ground and extract the grubs. Finches that eat buds and fruit would be less successful at doing this, while their claw like beaks can grind down their food and thus give them a selective advantage in circumstances where buds are the only real food source for finches.

Example 4: The role of size in horses' lineage
Maybe the horses' lineage offers one of the best-known illustrations regarding the role of size, profoundly documented through a very well-known fossil record. In the early Eocene (50–55 million years ago), the smallest species of horses' ancestors had approximately the size of a cat, while other species weighted up to 35 kg. The Oligocene species, approximately

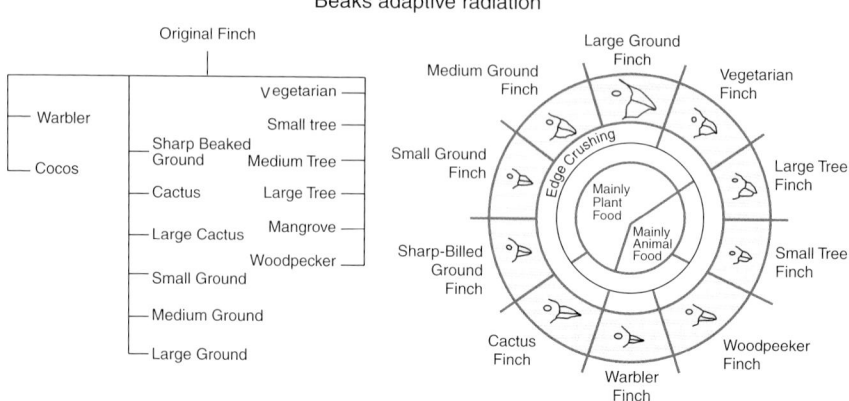

Figure 8.3 Darwin's finches diagram.

30 million years ago, were bigger, probably weighing up to approximately 50 kg. In the middle Miocene, approximately 17–18 million years ago, grazing "horses" of the size up to 100 kg were normal. Numerous fossils have shown that the weight reached approximately 200 kg 5 million years ago and approximately 500 kg 20,000 years ago. Why did this increase in size offer a selective advantage?

Figure 8.4 shows a model in form of a STELLA diagram that has been used to answer this question. The model equations are shown in Table 8.1.

The model has been used to calculate the efficiency for different maximum weights. Heat loss is proportional to weight to the exponent 0.75 (Peters, 1983). The growth rate follows also the surface, but the growth rate is proportional to the weight to the exponent 0.67 (see equations in Table 8.1). The results are shown in Table 8.2 and the conclusion is clear: the bigger the maximum weight, the better the eco-exergy efficiency. This is of course not surprising because a bigger weight means that the specific surface that determines the heat loss by respiration decreases. As the respiration loss is the direct loss of free energy, relatively more heat is lost when the body weight is smaller. Notice that the maximum size is smaller than the supper maximum size that is a parameter to be used in the model equations (see also Table 8.2).

The evolutionary theory at the light of ecosystem principles
Although living systems constitute very complex systems, they obviously comply with physical laws (although they are not entirely determined by them), and therefore in ecological theory it should be checked that each theoretical explanation conforms to basic laws of physics. First, one needs to understand the implications of the three generally accepted laws of thermodynamics in terms of understanding organisms' behavior and ecosystems' function. Nevertheless, although the three laws of thermodynamics are effective in describing system's behavior close to the thermodynamic equilibrium, in far from equilibrium systems, such as ecosystems, it has been recognized that although the

Chapter 8: Ecosystem principles have broad explanatory power in ecology

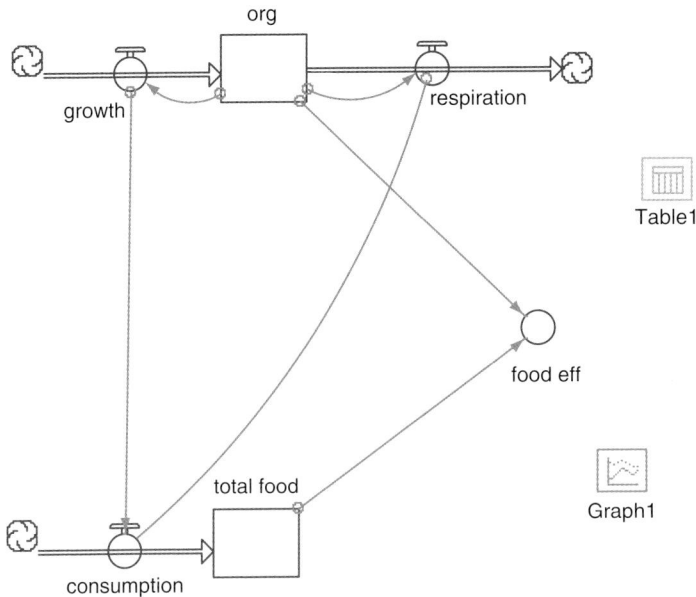

Figure 8.4 The growth and respiration follow allometric principles (Peters, 1983). The growth equation describes logistic growth with a maximum weight. The food efficiency is found as a result of the entire life span, using the β-values for mammals and grass (mostly *Gramineae*). The equations are shown in Table 8.1.

Table 8.1 Model equations

$d(\text{org}(t))/dt = (\text{growth} - \text{respiration})$
INIT org = 1 kg
INFLOWS: growth = $3 \times \text{org}^{(0.67)} \times (1 - \text{org/upper maximum size})$
OUTFLOWS: respiration = $0.5 \times \text{org}^{(3/4)}$
$d(\text{total\_food}(t))/dt = (\text{consumption})$
INIT total_food = 0
INFLOWS: consumption = growth + respiration
food_eff % = $2127 \times 100 \times \text{org}(t)/(200 \times \text{total\_food}(t))$

Note: See the conceptual diagram Figure 8.4.

three basic laws remain valid, they represent an incomplete picture when describing ecosystem functioning. This is the purpose of "irreversible thermodynamics" or "non-equilibrium thermodynamics". A tentative Ecological Law of Thermodynamics was proposed by Jørgensen (1997) as: *If a system has a through-flow of Exergy, it will attempt to utilize the flow to increase its Exergy, moving further away from thermodynamic*

Table 8.2 Eco-exergy efficiency for the life span for different maximum sizes[a]

Maximum size (kg)	Eco-exergy efficiency (percent)	Upper maximum size parameter (kg)
35	1.41	45
50	1.55	65
100	1.84	132
200	2.20	268
500	2.75	690

[a]β-Value for mammals is 2127 and for grass is 200.

equilibrium; If more combinations and processes are offered to utilize the Exergy flow, the organization that is able to give the highest Exergy under the prevailing circumstances will be selected. This hypothesis may be reformulated, as proposed by de Wit (2005) as: *If a system has a throughflow of free energy, in combination with the evolutionary and historically accumulated information, it will attempt to utilize the flow to move further away from the thermodynamic equilibrium; if more combinations and processes are offered to utilize the free energy flow, the organization that is able to give the greatest distance away from thermodynamic equilibrium under the prevailing circumstances will be selected.*

Both formulations mean that to ensure the existence of a given system, a flow of energy, or more precisely Exergy, must pass through it, meaning that the system cannot be isolated. Exergy may be seen as energy free of entropy (Jørgensen, 1997; Jørgensen and Marques, 2001), i.e. energy which can do work. A flow of Exergy through the system is sufficient to form an ordered structure, or dissipative structure (Prigogine, 1980). If we accept this, then a question arises: which ordered structure among the possible ones will be selected or, in other words, which factors influence how an ecosystem will grow and develop? The difference between the formulation by exergy or eco-exergy and free energy has been discussed in Chapter 6.

Jørgensen (1992b, 1997) proposed a hypothesis to interpret this selection, providing an explanation for how growth of ecosystems is determined, the direction it takes, and its implications for ecosystem properties and development. Growth may be defined as the increase of a measurable quantity, which in ecological terms is often assumed to be the biomass. But growth can also be interpreted as an increase in the organization of ordered structure or information. From another perspective, Ulanowicz (1986) makes a distinction between growth and development, considering these as the extensive and intensive aspects, respectively, of the same process. He argues that growth implies increase or expansion, while development involves increase in the amount of organization or information, which does not depend on the size of the system.

According to the tentative Ecological Law of Thermodynamics, when a system grows it moves away from thermodynamic equilibrium, dissipating part of the Exergy in catabolic processes and storing part of it in its dissipative structure. Exergy can be seen as a measure of the maximum amount of work that the ecosystem can perform when it is brought into thermodynamic equilibrium with its environment. In other words, if an ecosystem were in equilibrium with the surrounding environment its exergy would be zero (no free energy), meaning that it would not be able to produce any work, and that all gradients would have been eliminated.

Structures and gradients, resulting from growth and developmental processes, will be found everywhere in the universe. In the particular case of ecosystems, during ecological succession, exergy is presumably used to build biomass, which is exergy storage. In other words, in a trophic network, biomass, and exergy will flow between ecosystem compartments, supporting different processes by which exergy is both degraded and stored in different forms of biomass belonging to different trophic levels.

Biological systems are an excellent example of systems exploring a plethora of possibilities to move away from thermodynamic equilibrium, and thus it is most important in ecology to understand which pathways among the possible ones will be selected for ecosystem development. In thermodynamic terms, at the level of the individual organism, survival and growth imply maintenance and increase of the biomass, respectively.

From the evolutionary point of view, it can be argued that adaptation is a typically self-organizing behavior of complex systems, which may explain why evolution apparently tends to develop more complex organisms. On one hand, more complex organisms have more built-in information and are further away from thermodynamic equilibrium than simpler organisms. In this sense, more complex organisms should also have more stored exergy (thermodynamic information) in their biomass than the simpler ones. On the other hand, ecological succession drives from more simple to more complex ecosystems, which seem at a given point to reach a sort of balance between keeping a given structure, emerging for the optimal use of the available resources, and modifying the structure, adapting it to a permanently changing environment. Therefore, an ecosystem trophic structure as a whole, there will be a continuous evolution of the structure as a function of changes in the prevailing environmental conditions, during which the combination of the species that contribute the most to retain or even increase exergy storage will be selected.

This constitutes actually a translation of Darwin's theory into thermodynamics because survival implies maintenance of the biomass, and growth implies increase in biomass. Exergy is necessary to build biomass, and biomass contains exergy, which may be transferred to support other exergy (energy) processes.

The examples of industrial melanism in the peppered moth and warning coloration and mimicry are compliant with the Ecological Law of Thermodynamics, illustrating at the individual and population levels how the solutions able to improve survival and maintenance or increase in biomass under the prevailing conditions were selected. Also, the adaptations of Darwin's finches to take advantage of feeding in different ecological niches constitute another good illustration at the individual and population levels. Depending on the food resources available at each niche, the beaks evolved throughout time to be best suited to their function in the prevailing conditions, improving survival, and biomass growth capabilities. Finally, the horses' lineage increase in size illustrates very well how a bigger weight determines a decrease in body specific surface and consequently a decrease in the direct loss of free energy (heat loss by respiration). From the thermodynamic point of view, we may say that the solutions able to give the highest exergy under the prevailing circumstances were selected, maintaining or increasing gradients and therefore keeping or increasing the distance to thermodynamic equilibrium.

8.3 DO ECOLOGICAL PRINCIPLES ENCOMPASS OTHER PROPOSED ECOLOGICAL THEORIES?: ISLAND BIOGEOGRAPHY

In the next section, we consider another important ecological theory, namely island biogeography. Why do many more species of birds occur on the island of New Guinea than on the island of Bali? One answer is that New Guinea has more than 50 times the area of Bali, and numbers of species ordinarily increase with available space. This does not, however, explain why the Society Islands (Tahiti, Moorea, Bora Bora, etc.), which collectively have about the same area as the islands of the Louisiade Archipelago off New Guinea, play host to much fewer species, or why the Hawaiian Islands, ten times the area of the Louisiades, also have fewer native birds.

Two eminent ecologists, the late Robert MacArthur of Princeton University and E.O. Wilson of Harvard, developed a theory of "island biogeography" to explain such uneven distributions (MacArthur and Wilson, 1967). They proposed that the number of species on any island reflects a balance between the rate at which new species colonize it and the rate at which populations of established species become extinct (Figure 8.5). If a new volcanic island were to rise out of the ocean off the coast of a mainland inhabited by 100 species of birds, some birds would begin to immigrate across the gap and establish populations on the empty, but habitable, island. The rate at which these immigrant species could become established, however, would inevitably decline, for each species that successfully invaded the island would diminish by one the pool of possible future invaders (the same 100 species continue to live on the mainland, but those which have already become residents of the island can no longer be classed as potential invaders).

Equally, the rate at which species might become extinct on the island would be related to the number that had become residents. When an island is nearly empty, the extinction rate is necessarily low because few species are available to become extinct. And since the resources of an island are limited, as the number of resident species increases, the smaller

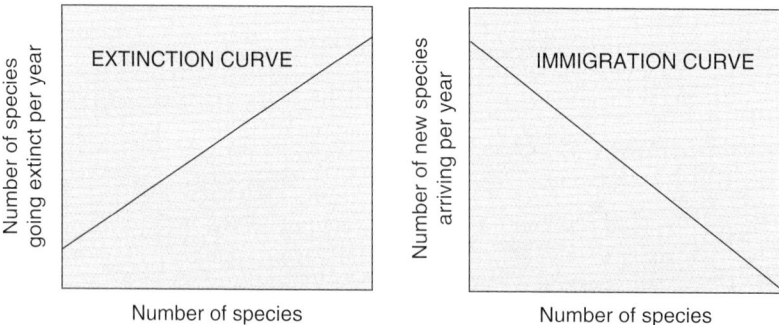

Figure 8.5 Extinction and immigration curves.

Chapter 8: Ecosystem principles have broad explanatory power in ecology 177

and more extinction prone their individual populations are likely to become. The rate at which additional species will establish populations will be high when the island is relatively empty, and the rate at which resident populations go extinct will be high when the island is relatively full. Thus, there must be a point between 0 and 100 species (the number on the mainland) where the two rates are equal, and therefore the input from immigration balances output from extinction. That equilibrium in the number of species (Figure 8.6) would be expected to remain constant as long as the factors determining the two rates did not change. But the exact species present should change continuously as some species go extinct and others invade (including some that have previously gone extinct), so that there is a steady turnover in the composition of the fauna.

Examples
Example 1: Krakatau Island
One famous "test" of the theory was provided in 1883 by a catastrophic volcanic explosion that devastated the island of Krakatau, located between the islands of Sumatra and Java. The flora and fauna of its remnant and of two adjacent islands were completely exterminated, yet within 25 years (1908) 13 species of birds had re-colonized what was left of the island. By 1919–1921 28-bird species were present, and by 1932–1934, 29. Between the explosion and 1934, 34 species actually became established, but five of them went extinct. By 1951–1952 33 species were present, and by 1984–1985, 35 species. During this half century (1934–1985), a further 14 species had become established, and 8 had become extinct. As the theory predicted, the rate of increase declined as more and more species colonized the island. In addition, as equilibrium was approached there was some turnover. The number in the cast remained roughly the same while the actors gradually changed.

The theory predicts other things, too. For instance, everything else being equal, distant islands will have lower immigration rates than those close to a mainland, and equilibrium

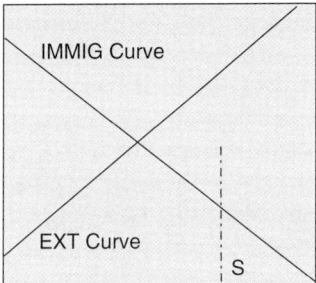

Number of species

Figure 8.6 The equilibrium number of species. Any particular island has a point where the extinction (EXT curve) and immigration curves (IMMIG curve) intersect. At this point the number of new immigrating species to the island is exactly matched by the rate at which species are going extinct.

will occur with fewer species on distant islands (Figure 8.7). Close islands will have high immigration rates and support more species. By similar reasoning, large islands, with their lower extinction rates, will have more species than small ones—again everything else being equal (which frequently is not, for larger islands often have a greater variety of habitats and more species for that reason).

Island biogeography theory has been applied to many problems, including forecasting faunal changes caused by fragmenting previously continuous habitat. For instance, in most of the eastern United States only patches of the once-great deciduous forest remain, and many species of songbirds are disappearing from those patches. One reason for the decline in birds, according to the theory, is that fragmentation leads to both lower immigration rates (gaps between fragments are not crossed easily) and higher extinction rates (less area supports fewer species).

Example 2: Connecticut forest re-establishing
Indications of such changes in species composition during habitat fragmentation were found in studies conducted between 1953 and 1976 in a 16-acre nature preserve in Connecticut in which a forest was re-establishing itself. During that period development was increasing the distance between the preserve and other woodlands. As the forest grew back, species such as American Redstarts that live in young forest colonized the area, and birds such as the Field Sparrow, which prefer open shrub lands, became scarce or disappeared. In spite of the successional trend toward large trees, however, two bird species normally found in mature forest suffered population declines, and five such species went extinct on the reserve. The extinctions are thought to have resulted from lowering immigration rates caused by the preserve's increasing isolation and by competition from six invading species characteristic of suburban habitats.

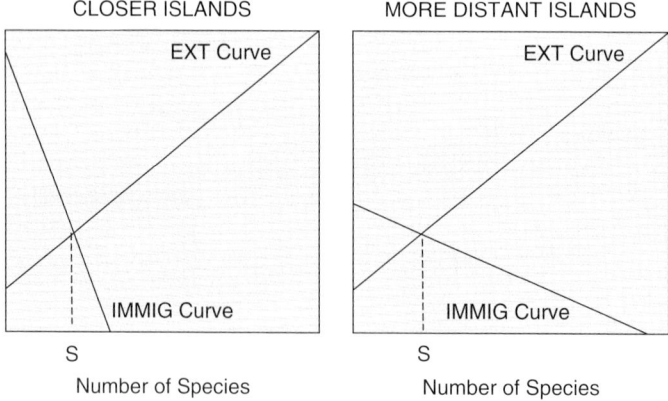

Figure 8.7 The influence of distance of an island from the source on the equilibrium number of species. EXT curve, Extinction curve. IMMIG curve, Immigration curve.

Chapter 8: Ecosystem principles have broad explanatory power in ecology

Example 3: Bird community in an oak wood in Surrey, England
Long-term studies of a bird community in an oak wood in Surrey, England, also support the view that isolation can influence the avifauna of habitat islands. A rough equilibrium number of 32 breeding species were found in that community, with a turnover of three additions and three extinctions annually. It was projected that if the woods were as thoroughly isolated as an oceanic island, it would maintain only five species over an extended period: two species of tits (same genus as titmice), a wren, and two thrushes (the English Robin and Blackbird).

Island biogeography theory can be a great help in understanding the effects of habitat fragmentation. It does not, however, address other factors that can greatly influence which birds reside in a fragment. Some of these include whether nest-robbing species are present in such abundance that they could prevent certain invaders from establishing themselves, whether the fragment is large enough to contain a territory of the size required by some members of the pool of potential residents, or whether other habitat requirements of species in that pool can be satisfied. To take an extreme example of the latter, Acorn, Nuttall's, Downy, or Hairy Woodpeckers would not colonize a grass-covered, treeless habitat in California, even if they were large, and all four woodpeckers are found in adjacent woodlands.

Island biogeography theory at the light of ecosystem principles
In general terms, the Island Biogeography Theory explains therefore why, if everything else is similar, distant islands will have lower immigration rates than those close to a mainland, and ecosystems will contain fewer species on distant islands, while close islands will have high immigration rates and support more species. It also explains why large islands, presenting lower extinction rates, will have more species than small ones. This theory forecasts effect of fragmenting previously continuous habitat, considering that fragmentation leads to both lower immigration rates (gaps between fragments are not crossed easily) and higher extinction rates (less area supports fewer species).

The Ecological Law of Thermodynamics equally provides a sound explanation for the same observations. Let us look in first place to the problem of the immigration curves. In all the three examples, the decline in immigration rates as a function of increasing isolation (distance) is fully covered the concept of openness introduced by Jørgensen (2000a). Once accepted the initial premise that an ecosystem must be open or at least non-isolated to be able to import the energy needed for its maintenance, islands' *openness* will be inversely proportional to its distance to mainland. As a consequence, more distant islands have lower possibility to exchange energy or matter and decreased chance for information inputs, expressed in this case as immigration of organisms. The same applies to fragmented habitats, the smaller the plots of the original ecosystem the bigger the difficulty in recovering (or maintaining) the original characteristics. After a disturbance, the higher the openness the faster information and network (which may express as biodiversity) recovery will be.

The fact that large islands present lower extinction rates and more species than small ones, as well as less fragmented habitats in comparison with more fragmented ones, also complies with the Ecological Law of Thermodynamics. All three examples can be interpreted in this light. Actually, provided that all the other environmental are similar, larger islands offer more available resources. Under the prevailing circumstances, solutions able to give the highest exergy will be selected, increasing the distance to thermodynamic equilibrium not only in terms of biomass but also in terms of information (i.e. network and biodiversity). Moreover, after a disturbance, like in the case of the Krakatau Island, the rate of re-colonization and ecosystem recovery will be a function of system's openness.

8.4 DO ECOLOGICAL PRINCIPLES ENCOMPASS OTHER PROPOSED ECOLOGICAL THEORIES?: LATITUDINAL GRADIENTS IN BIODIVERSITY

On a global scale, species diversity typically declines with increasing latitude toward the poles (Rosenzweig, 1995; Stevens and Willig, 2002). Although the latitude diversity gradient is the most striking biodiversity pattern, the dynamics that generate and maintain this trend remain poorly understood. The latitudinal diversity gradient is commonly viewed as the net product of in situ origination and extinction, with the tropics serving as either a generator of biodiversity (the tropics-as-a-cradle hypothesis), or an accumulator of biodiversity (the tropics-as-museum hypothesis), eventually both.

The causes for latitudinal gradients in biodiversity (http://www.ecology.info/gradients-biodiversity.htm).

The determinant of biological diversity is not latitude *per se*, but the environmental variables correlated with latitude. More than 25 different mechanisms have been suggested for generating latitudinal diversity gradients, but no consensus has been reached yet (Gaston, 2000).

One factor proposed as a cause of latitudinal diversity gradients is the area of the climatic zones. Tropical landmasses have a larger climatically similar total surface area than landmasses at higher latitudes with similarly small temperature fluctuations (Rosenzweig, 1992). This may be related to higher levels of speciation and lower levels of extinction in the tropics (Rosenzweig, 1992; Gaston, 2000; Buzas et al., 2002). Moreover, most of the land surface of the Earth was tropical or subtropical during the Tertiary, which could in part explain the greater diversity in the tropics today as an outcome of historical evolutionary processes (Ricklefs, 2004).

The higher solar radiation in the tropics increases productivity, which in turn is thought to increase biological diversity. However, productivity can only explain why there is more total biomass in the tropics, not why this biomass should be allocated into more individuals, and these individuals into more species (Blackburn and Gaston, 1996). Body sizes and population densities are typically lower in the tropics, implying a higher number of species, but the causes and the interactions among these three variables are complex and still uncertain (Blackburn and Gaston, 1996).

Higher temperatures in the tropics may imply shorter generation times and greater mutation rates, thus accelerating speciation in the tropics (Rohde, 1992). Speciation may also be accelerated by a higher habitat complexity in the tropics, although this does not apply to freshwater ecosystems. The most likely explanation is a combination of various factors, and it is expected that different factors affect differently different groups of organisms, regions (e.g. northern vs. southern hemisphere) and ecosystems, yielding the variety of patterns that we observe.

Chapter 8: Ecosystem principles have broad explanatory power in ecology

Examples

Example 1: Geographic range of marine prosobranch gastropods
Roy et al. (1998) have assembled a database of the geographic ranges of 3916 species of marine prosobranch gastropods living in waters shallower than 200 m of the western Atlantic and eastern Pacific Oceans, from the tropics to the Artic Ocean. They have found that Western Atlantic and eastern Pacific diversities were similar, and that the diversity gradients were strikingly similar despite many important physical and historical differences between the oceans. Figure 8.8 shows the strong latitudinal diversity gradients that are present in both oceans.

The authors have found that one parameter that did correlate significantly with diversity in both oceans was solar energy input, as represented by average sea surface temperature. More, the authors continued saying that if that correlation was causal, sea surface temperature is probably linked to diversity through some aspect of productivity. They defend that if that is the case, diversity is an evolutionary outcome of trophodynamics processes inherent in ecosystems, and not just a by-product of physical geographies.

Example 2: Latitudinal trends in vertebrate diversity (http://www.meer.org/chap3.htm)
Amphibians, absent from arctic regions, are well represented in the mid-latitudes (Figure 8.9A). Forty-seven species of amphibians are found in California (Laudenslayer and Grenfell, 1983). As might be expected given the warmth and humidity of much of the tropics and the inability of amphibians to thermoregulate, this group reaches its greatest diversity here. In fact, one of the three orders (groups of related families) of the class Amphibia, called caecilians (160 species of worm-like creatures), is restricted in its distribution to the tropics.

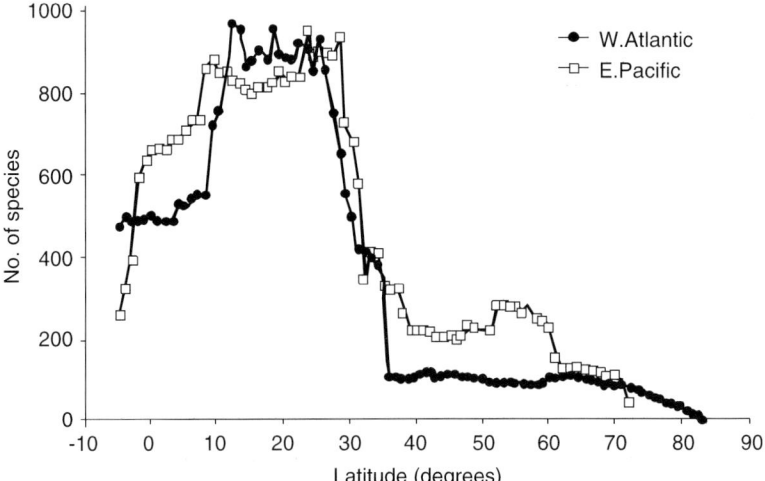

Figure 8.8 Latitudinal diversity gradient of eastern Pacific and western Atlantic marine prosobranch gastropods, binned per degree of latitude. The range of a species is assumed to be continuous between its range endpoints, so diversity for any given latitude is defined as the number of species whose latitudinal ranges cross that latitude.

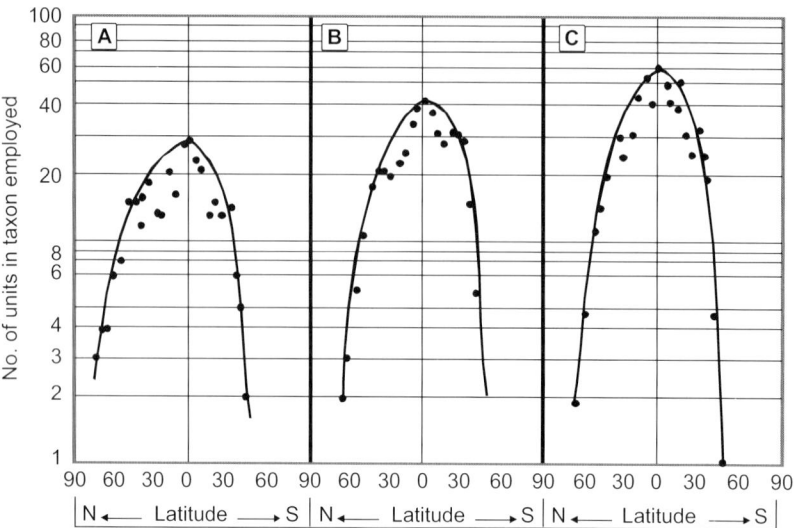

Figure 8.9 Diversity vs. latitude plots for three groups of terrestrial poikilotherms, showing what appear to be latitude-related anomalies in the region 15–30° that are probably a response to the less favorable conditions prevailing in the desert regions often found in those latitudes. Data show the highest number of genera in 5° latitude classes. Solid curve smoothed through points indicating highest diversity, dotted curve following the points suggestive of persistent anomaly. (A) Genera of amphibians. (B) Genera of lizards. (C) Genera of snakes.

Reptiles, too, are represented by more species in the temperate latitudes. The diversity of lizards is shown in Figure 8.9B and for snakes is shown in Figure 8.9C. Both of these figures show slight decreases in diversity for these groups between 15 and 30° latitude. These are the latitudes at which most of the world's deserts are found. There are 77 species of reptiles in California (Laudenslayer and Grenfell, 1983). The two major groups of terrestrial reptiles, lizards and snakes, are represented by more species in the tropics than in higher latitudes. The pattern is even more pronounced for turtles.

Birds really increase in diversity in temperate latitudes. For example, at least 88-bird species breed on the Labrador Peninsula of northern Canada (55° N), 176 species breed in Maine (45° N), and more than 300 species can be found in Texas (31° N; Peterson, 1963). The total number of bird species found in California exceeds 540 (Laudenslayer and Grenfell, 1983); the total for all of North America is roughly 700 (Welty, 1976).

An indication of the latitudinal trend in mammalian diversity was provided by Simpson (1964) for continental North American mammals. Here again, species diversity is apparent with decreasing latitude. This analysis also shows that, superimposed on the latitudinal trend, is an effect due to elevation such that mountainous regions have more species of mammals than lowlands. There are 214 species of mammals in California (Laudenslayer and Grenfell, 1983).

A majority of all *fish* species are found in tropical waters. It is possible to get an indication of the diversity of fish in the tropics by considering two examples, one freshwater and one marine. The first example is provided by the dazzling array of coral reef fish. Something on the order of 30–40% of all marine fish species are in some way associated with tropical reefs and more than 2200 species can be found in a large reef complex (Moyle and Cech, 1996). Second, the Amazon River of South America, huge in comparison to most other river systems—3700 miles long, drains a quarter of the South American continent—has over 2400 species of fish. The Rio Negro, a tributary of the Amazon, contains more fish species than all the rivers of the United States combined.

Example 3: Trends within plant communities and across latitude
Niklas et al. (2003) examined how species richness and species-specific plant density—number of species and number of individuals per species, respectively—vary within community size frequency distributions and across latitude. 226 forested plant communities from Asia, Africa, Europe, and North, Central, and South America were studied (60°4′N–41°4′S) using the Gentry database. An inverse latitudinal relationship was observed between species richness and species-specific plant density. Their analyses showed that the species richness increased toward the tropics and the reverse trend was observed for average species-specific plant density.

Example 4: Trends within marine epifaunal invertebrate communities
Witman et al. (2004) tested the effects of latitude and the richness of the regional pool on the species richness of local epifauna invertebrate communities by sampling the diversity of local sites in 12-independent biogeographic regions from 62°S to 63°N. Both regional and local species richness displayed significant unimodal patterns with latitude, peaking at low latitudes and decreasing toward high latitudes (Figure 8.10).

Latitudinal gradients in biodiversity at the light of ecosystem principles
Latitudinal gradients in biodiversity are easily interpretable at the light of the Ecological Law of Thermodynamics. Obviously, the higher solar radiation in the tropics increases productivity, which in turn is thought to increase biological diversity. In fact, Blackburn and Gaston (1996) found that one parameter that did correlate significantly with diversity in both oceans was solar energy input, as represented by average sea surface temperature. Moreover, these authors claim that if that correlation was causal, sea surface temperature is probably linked to diversity through some aspect of productivity. However, they could not establish the causal nexus, considering that productivity could only explain why there is more total biomass in the tropics, not why this biomass should be allocated into more individuals, and these individuals into more species.

This apparent inconsistency can nevertheless be explained within the frame of ecosystem principles. In fact, Jørgensen et al. (2000), proposed that ecosystems show three growth forms:

I. Growth of physical structure (biomass), which is able to capture more of the incoming energy in the form of solar radiation but also requires more energy for maintenance (respiration and evaporation).
II. Growth of network, which means more cycling of energy or matter.

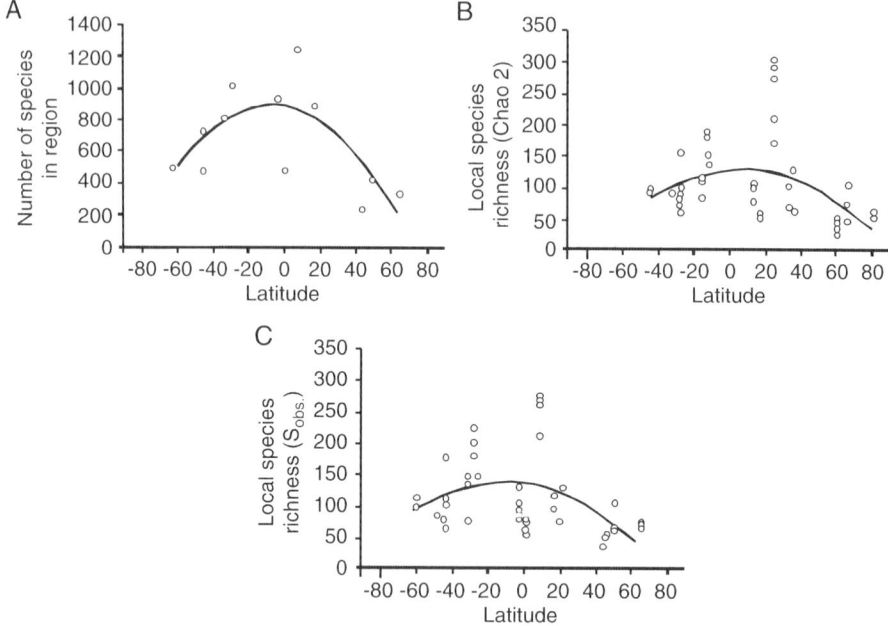

Figure 8.10 Species richness as a function of latitude. (A) Regional species richness (standard regional pool). (B) Local species richness based on the Chao2 estimate. (C) Local species richness as S_{obs}. Lines represent significant, best fits to second-order polynomial equations.

III. Growth of information (more developed plants and animals with more genes), from r-strategists to K-strategists, which waste less energy but also usually carry more information.

This was experimentally confirmed by Debeljak (2002) examining managed and virgin forest in different development stages (e.g. pasture, gap, juvenile, optimum forest). Accordingly, these three growth forms may be considered an integration of Odum's attributes, which describe changes in ecosystems associated with development from the early stage to the mature stage. Clearly, Blackburn and Gaston (1996) were considering only growth form I.

8.5 DO ECOLOGICAL PRINCIPLES ENCOMPASS OTHER PROPOSED ECOLOGICAL THEORIES?: OPTIMAL FORAGING THEORY

Researchers have long pursued theories to explain species' diversity. These theories have focused on quantifying adaptation, fitness, and natural selection through observing an animal's feeding behaviors. The assumption is that feeding behaviors are reflections of these internal processes. Using behavior as a mechanism of adaptation in a feedback loop creates an interactive system between an animal's phenotype and its environment.

MacArthur and Pianka (1966) first proposed an optimal foraging theory, arguing that because of the key importance of successful foraging to an individual's survival, it should be possible to predict foraging behavior by using decision theory to determine the behavior that would be shown by an "optimal forager"—one with perfect knowledge of what to do to maximize usable food intake. In their paper, a graphical model of animal feeding activities based on costs versus profits was developed. A forager's optimal diet was specified and some interesting predictions emerged. Prey abundance influenced the degree to which a consumer could afford to be selective because it affected search time per item eaten. Diets should be broad when prey are scarce (long search time), but narrow if food is abundant (short search time) because a consumer can afford to bypass inferior prey only when there is a reasonably high probability of encountering a superior item in the time it would have taken to capture and handle the previous one. Also, larger patches should be used in a more specialized way than smaller patches because travel time between patches (per item eaten) is lower.

Succinctly, this heavily referenced paper in evolutionary biology presented three concepts:

(1) How long a predator will forage in a specific area?
(2) Influence of prey density on the length of time a predator will forage in an area.
(3) Influence of prey variety on a predator's choice of acquired prey.

These concepts describe a predator's behavior as a function of its relationship with the prey it acquires. Fundamental conditions in these concepts influencing the predator–prey relationship are time foraging and prey availability. Within these concepts, MacArthur and Pianka embodied the study of differential land and resource use in a specific field of study: optimal foraging theory.

Examples
Example 1: Rufous Hummingbirds
Carpenter et al. (1983) found a way to test optimal foraging theory, using Rufous Hummingbirds. These hummers establish feeding territories during stops on their 2000-mile migration between their breeding grounds in the Pacific Northwest and their wintering habitat in southern Mexico. They zealously guard those territories, driving off hawkmoths, butterflies, other hummers, and even bees that might compete for the nectar. In addition, they deplete the nectar resources around the periphery of their territories as early in the day as they can, in order to out-compete other nectar-sippers that might try to sneak a drink at the territory edge.

When half of the flowers in a territory were covered with cloth so the birds could not drain them, Carpenter and her co-workers found that the resident hummer increased its territory size. This showed that territoriality was tied to the availability of nectar, and that the bird could in some way assess the amount of nectar it controlled. Then, by substituting a sensitive scale topped by a perch for the territory-holder's traditional perch, they were able to measure the bird's weight each time it alighted. The researchers found that the hummers optimized their territory size by trial and error, making it larger or smaller until their daily weight gain was at a maximum. In this case of migrant-territorial hummers, theory accurately predicted how a bird behaves in nature.

Example 2: Optimal clam selection by northwestern crows (Alcock, 1997)
Richardson and Verbeek (1986) noticed that crows in the Pacific Northwest often leave littleneck clams uneaten after locating them. The crows dig the clams from their burrows, but they often leave the smaller ones on the beach and only bother with the larger ones, which they drop on the rocks and eat. Their acceptance rate increases with prey size: they open and eat only about half of the 29-mm-long clams they find, while consuming all clams in the 32–33 mm range. The two researchers determined that the most profitable clams were the largest, not because they broke more easily, but because they contained more calories than smaller clams. By considering the caloric benefits from clams of different sizes and the costs of searching for, digging up, opening, and feeding on clams, the authors were able to construct a mathematical model based on the assumption that crows would select an optimal diet, in this case one that maximized their caloric intake. The model predicted that clams approximately 28.5 mm in length would be opened and eaten half the time, given the search costs required to find clams of different sizes; the crows behavior shows that they agree with researchers' match (Figure 8.11). Their work was based on optimal foraging theory.

The optimal foraging theory at the light of ecosystem principles
The optimal foraging theory clearly complies with the Ecological Law of Thermodynamics. The fact that prey abundance influences consumers' selectivity and that diets are broad when prey are scarce and narrow if food is abundant, as a function of search for food time, is clearly translated by "… *If more combinations and processes are*

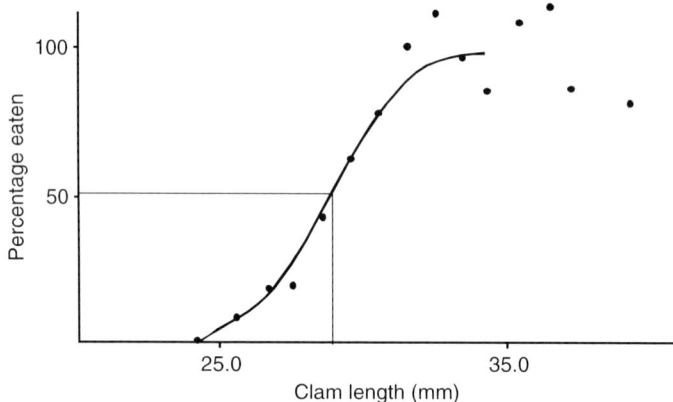

Figure 8.11 Optimality model of prey selection in relation to prey size. The curve represents the predicted percentages of small to large clams that crows should select for consumption after locating, based on the assumption that the birds attempt to maximize the rate of energy gain per unit of time spent foraging for clams. The solid circles represent the actual observations, showing the model's predictions were supported (Richardson and Verbeek, 1986).

offered to utilize the Exergy flow, the organization that is able to give the highest Exergy under the prevailing circumstances will be selected or by if more combinations and processes are offered to utilize the free energy flow, the organization that is able to give the greatest distance away from thermodynamic equilibrium under the prevailing circumstances will be selected". Both examples can, therefore, be easily explained by the same ecosystem principles.

8.6 DO ECOLOGICAL PRINCIPLES ENCOMPASS OTHER PROPOSED ECOLOGICAL THEORIES?: NICHE THEORY

Hutchinson (1957, 1965) suggested that an organism's niche could be visualized as a multidimensional space, or hypervolume, formed by the combination of gradients of each single environmental condition to which the organism was exposed. The N-environmental exposure conditions form a set of N-intersecting axes within which one can define an N-dimensional niche hypervolume unique to each species. The niche hypervolume is comprised of all combinations of the environmental conditions that permit an individual of that species to survive and reproduce indefinitely (*Huthchinson's Fundamental niche*). Hutchinson distinguished the fundamental niche, defined as the maximum inhabitable hypervolume in the absence of competition, predation, and parasitism, from the realized niche, which is a smaller hypervolume occupied when the species is under biotic constraints. Hutchinson also defined the niche breadth for an organism as the habitable range, between the maximum and the minimum, for each particular environmental variable. Thus, the niche breadth is the projection of the niche hypervolume onto each individual environment.

Following Hutchinson's distinction, niche refers to the requirements of the species and habitat refers to a physical place in the environment where those requirements can be met (Figure 8.12).

To interpret Figure 8.12 with regard to the distribution of species 1 and 2, one must understand Hutchinson's emphasis on the fundamental importance of competition as a force influencing the distribution of species in nature. Hutchinson argued that in the face of competition, a species will not utilize its entire fundamental niche, but rather the *realized niche* actually used by the species will be smaller, only consisting of those portions of the fundamental niche where the species is competitively dominant. As a result of competitive exclusion, according to Hutchinson, the realized niche is smaller than the fundamental niche, and a species may frequently be absent from portions of its fundamental niche because of competition with other species. Obviously, the more limited resources two populations have in common (i.e. the more similar their niches are), the greater the impact of competition (all else being equal).

In particular, niche is used to describe and analyze:

(1) Ways in which different species interact (including competition, resource portioning, exclusion, or coexistence).
(2) Why some species are rare and others abundant.
(3) What determines geographical distribution of a given species?
(4) What determines structure and stability of multi-species communities?

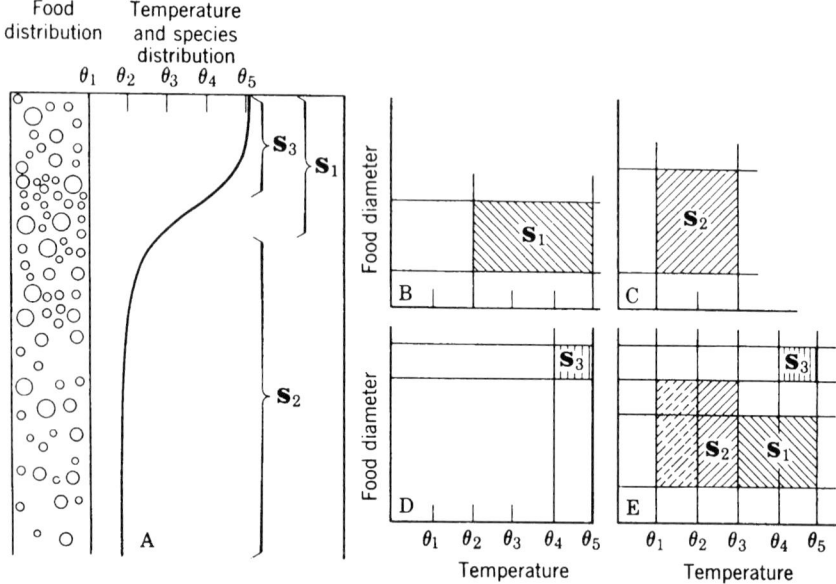

Figure 8.12 Distribution of species 1 and 2.

Consider an extreme case: can two populations occupying the same resource niche coexist in the same environment? (http://courses.washington.edu/anth457/nichelec.htm).

If two populations occupy same resource niche, then by definition, they utilize all the same resources and in the same manner. Common sense tells us there are three possible outcomes to this situation: (1) share resources more or less equally (neither population changes niche); (2) one or both populations alters niche to reduce overlap (*niche partitioning*), and (3) one population loses out completely (*competitive exclusion*). Which outcome will occur? Answer from niche theory = 2 or 3, but not 1. This somewhat counterintuitive finding given the formal name of *competitive exclusion principle* (CEP) (Gause, 1934) states that *no two species can permanently occupy the same niche*: either the niches will differ, or one will be excluded by the other (note: "excluded" here means replaced by differential population growth, not necessarily by fighting or territoriality), and has become a central tenet of modern niche theory. Of course, 100% niche overlap is unlikely if not impossible; but such an extreme case is not necessary for competitive exclusion or other forms of niche change.

Much theory and research in ecology has focused on predicting what actually happens when there is niche overlap and competition: when does exclusion result, when coexistence? How much overlap is possible (a question treated by the "theory of limiting similarity")? How do environmental fluctuations affect this? Why are some species generalists, others specialists? Both possible responses to niche competition (competitive exclusion, and coexistence via reduction in niche overlap) are commonly observed, and their determinants and features have been studied by three means: lab experiments, field observations, and mathematical models or simulations. Competitive exclusion is commonly observed when a species colonizes a habitat and out-competes indigenous species—probably due to absence of parasites and predators adapted to exploit the colonizer (e.g. introduced placentals vs. indigenous marsupials in Australia). Coexistence through niche partitioning is rarely observed directly, but can often be inferred from traces left by "the ghost of competition". The typical means of doing so is to examine two populations that overlap spatially, but only partially, and then to compare the niche of each population in the area of overlap versus the area of non-overlap. In such a case, we often observe that in areas where competitors coexist, one or both have narrower niche range (e.g. diet breadth) than in areas where competitor is absent; this is because competition "forces" each competing population to specialize in those resources—or other niche dimensions—in which it has a competitive advantage, and conversely to "give up" on those in which the other population out competes it. Thus, in absence of competitors a given species will often utilize a broader array of resources, closer to its fundamental niche, than it will in competitor's presence; this phenomenon is termed "competitive release".

When niche shift involves an evolutionary change in attributes ("characteristics") of competing populations, it is termed *character displacement*. This is some of the strongest evidence for the role of competition in shaping niches because it is unlikely to have alternative explanation. A classic example of character displacement is change in length or shape of beaks in ecologically similar bird species that overlap geographically.

Example 1 of the competitive exclusion principle or Gause's principle: two spp. of Paramecium

Gause (1934) placed two species of *Paramecium* into flasks containing a bacterial culture that served as food. Thus, in this artificial laboratory system both species of paramecium were forced to have the same niche. Gause counted the number of *Paramecium* each day and found that after a few days (Figure 8.13) one species always became extinct because it apparently was unable to compete with the other species for the single food resource.

However, extinction is not the only possible result of two species having the same niche. If two competing species can co-exist for a long period of time, then the possibilities exists that they will evolve differences to minimize competition; that is, they can evolve different niches.

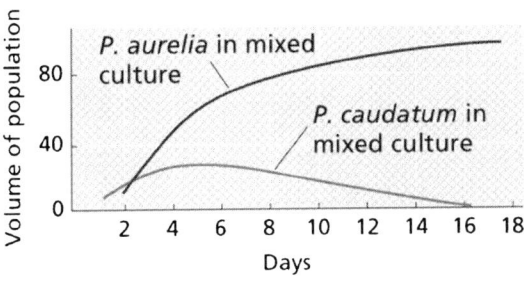

Figure 8.13 Competition between two laboratory populations of *Paramecium* with similar requirements.

Example 2 of the competitive exclusion principle or Gause's principle: Geospiza spp.
We can go to Darwin's finches again to see examples of character displacement on the Galapagos Islands. Members of the genus *Geospiza* are wide spread among the islands. *Geospiza fortis*, for example, is found alone on Daphne Island, while *G. fulginosa* is found alone on Crossman Island. Both ground-feeding birds are about the same size. On Charles and Chatham Islands, on the other hand, the species co-exist. Although *G. fortis* is about the same size as their relatives on Daphne, *G. fulginosa* is considerably smaller than their neighbors on Crossman. The shift in size allows the *G. fortis* to feed on smaller seeds, thus avoiding competition with the larger *G. fortis* on Charles and Chatham. This and other documented cases of character displacement suggest that competition is important in shaping ecosystems. Displacement is interpreted as evidence of historical competition.

Example 3 of the competitive exclusion principle or Gause's principle: squirrels in England (http://www.saburchill.com/IBbiology/chapters02/035.html)
The red squirrel (*Sciurus vulgaris*) is native to Britain but its population has declined due to competitive exclusion, disease, and the disappearance of hazel coppices and mature conifer forests in lowland Britain. The grey squirrel (*Sciurus carolinensis*) was introduced to Britain in approximately 30 sites between 1876 and 1929. It has easily adapted to parks and gardens replacing the red squirrel. Today's distribution is shown below in Figure 8.14.

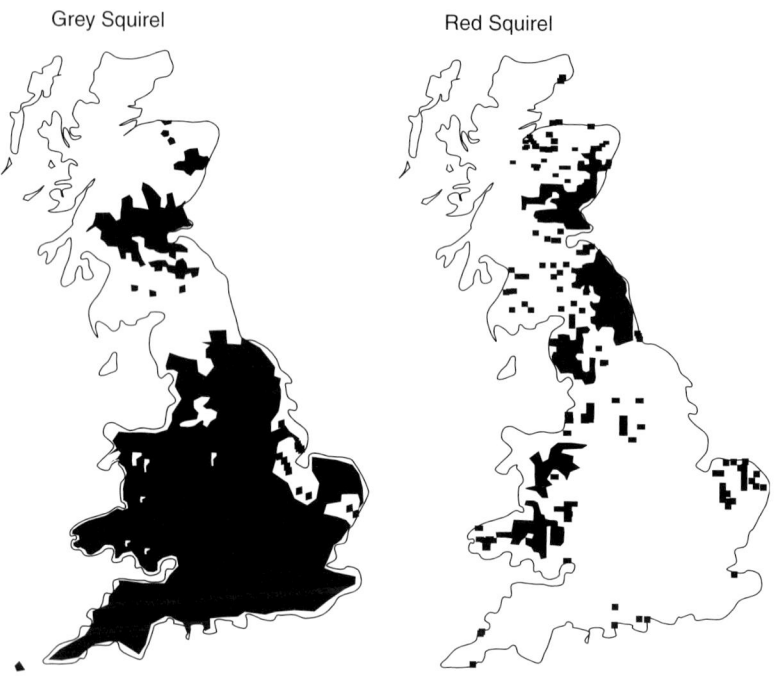

Figure 8.14 Actual distribution of two *Sciurus* species in Britain.

The niche theory at the light of ecosystem principles
In general terms, Hutchinson's niche theory considers that the fundamental niche (theoretical) of a given species comprises all the combinations of environmental conditions that permit an individual of that species to survive and reproduce indefinitely. But from all these possible combinations, only the ones where the species is competitively dominant will in fact be utilized, constituting the realized niche. There will be of course limits of tolerance, maximum and minimum, of the organisms with regard to each environmental variable, which constitute the niche breadth.

This formulation, designed for use at the species and individual levels, is clearly compliant with the Ecological Law of Thermodynamics. In fact, what is said can be translated as: *under the prevailing circumstances the organisms will attempt to utilize the flow to increase its Exergy, moving further away from thermodynamic equilibrium* (Jørgensen, 1997), or alternatively *in combination with the evolutionary and historically accumulated information, it will attempt to utilize the flow to move further away from the thermodynamic equilibrium* (de Wit, 2005).

If the populations of two species occupy the same resources niche one of the two will become out competed, in accordance with the CEP (Gause, 1934), which states that no two species can permanently occupy the same niche. In trophic terms, in the absence of competitors, a given species will most probably specialize less than it will in competitor's presence (competitive release). This also clearly complies with the Ecological Law of Thermodynamics and can be translated as:

If more combinations and processes are offered to utilize the Exergy flow, the organization that is able to give the highest Exergy under the prevailing circumstances will be selected (Jørgensen, 1997), *or alternatively as if more combinations and processes are offered to utilize the free energy flow, the organization that is able to give the greatest distance away from thermodynamic equilibrium under the prevailing circumstances will be selected* (de Wit, 2005).

All the three examples illustrating the CEP can in fact be explained at the light of the Ecological Law of Thermodynamics.

8.7 DO ECOLOGICAL PRINCIPLES ENCOMPASS OTHER PROPOSED ECOLOGICAL THEORIES?: LIEBIG'S LAW OF THE MINIMUM

Many different environmental factors have the potential to control the growth of a population. These factors include the abundance of prey or nutrients that the population consumes and also the activities of predators. A given population will usually interact with a multitude of different prey and predator species, and ecologists have described these many interactions by drawing food webs. Yet, although a given population may interact with many different species in a food web, and also interact with many different abiotic factors outside the food web, not all of these interactions are of equal importance in controlling that population's growth. Experience shows that "only one or two other species dominate the feedback structure of a population at any one time and place (Berryman, 1993)". The identity of these dominating species may change with time and location, but the number of species that limits a given population (i.e. actively controls its dynamics) is usually only one or two.

Liebig's Law (Liebig, 1840), in its modern form, expresses this idea. It says that *of all the biotic or abiotic factors that control a given population, one has to be limiting* (i.e. active, controlling the dynamics) (Berryman, 1993, 2003). Time delays produced by this limiting factor are usually one or two generations long (Berryman, 1999). Moreover, Liebig's Law stresses the importance of limiting factors in ecology. "A factor is defined as limiting if a change in the factor produces a change in average or equilibrium density" (Krebs, 2001).

To summarize, "the functioning of an organism is controlled or limited by that essential environmental factor or combination of factors present in the least favorable amount. The factors may not be continuously effective but only at some critical period during the year or only during some critical year in a climatic cycle".

The Liebig's law of the minimum at the light of ecosystem principles

The Liebig's Law of the minimum may be seen as a deductive consequence of the principle of increasing ascendency (Ulanowicz and Baird, 1999). Let us see why.

Increasing ascendency also implies greater exergy storage, as can be demonstrated by two propositions in sequence:

Proposition 1: Longer biomass retention times contribute to increasing ascendency.

Let B_i represent the amount of biomass stored in the *i*th compartment of the ecosystem. Similarly, let T_{ij} be the amount of biomass that is transferred from compartment *i* to compartment *j* within a unit of time.

Information is now the measure of change in a probability assignment (Tribus and McIrvine, 1971). The two distributions in question are usually the *a priori* and *a posteriori* versions of a given probability, which in the present case is the probability that a quantum of biomass will flow from *i* to *j*. As the *a priori* estimate that a quantum of biomass will leave *i* during a given interval of time, one may use the analogy from the theory of mass–action that the probability can be estimated as $(B_i/B.)$, where $B.$ represents the sum of all the B_i. In strictly similar manner, the probability that a quantum enters some other compartment *j* should be proportional to the quotient $(B_j/B.)$. If these two probabilities were completely independent, then the joint probability that a quantum flows from *i* to *j* would become proportional to the product $(B_i B_j/B^2)$.

Of course, the exit and entrance probabilities are usually coupled and not entirely independent. In such case the *a posteriori* probability might be measured by empirical means in terms of the T_{ij}. That is the quotient $(T_{ij}/T..)$ would be an estimate of the *a posteriori* joint probability that a quantum leaves *i* and enters *j*.

Kullback (1959) provides a measure of information that is revealed in passing from the *a priori* to the *a posteriori*. It is called the Kullback–Leibler information measure, which is given by:

$$I = \sum_{i,j} p(a_i,b_j) \log\left(\frac{p(a_i,b_j)}{p(a_i)p(b_j)}\right)$$

where $p(a_i)$ and $p(b_j)$ are the *a priori* probabilities of event a_i and b_j, respectively, and $p(a_i,b_j)$ is the *a posteriori* probability that a_i and b_j happen jointly. Substituting the probabilities as estimated in the preceding paragraphs, one obtains the form for the Kullback–Leibler information of biomass flow in a network as:

$$I = \sum_{i,j} \frac{T_{ij}}{T_{..}} \log\left(\frac{T_{ij} B_{..}^2}{T_{..} B_i B_j}\right)$$

Following the lead of Tribus and McIrvine, as in Ulanowicz (1980), one may scale this information measure by the total activity ($T_{..}$) to yield the storage-inclusive ascendency (Ulanowicz and Abarca-Arenas, 1997) as:

$$A = \sum_{i,j} T_{ij} \log\left(\frac{T_{ij} B_{..}^2}{T_{..} B_i B_j}\right)$$

Biomass storage: Odum (1969) proposed 24 trends to be expected as ecosystems develop and mature. These could be grouped under increases in species richness, trophic specificity, cycling, and containment. It happens that, other things being equal, an increase in any of these attributes will result in an increase of systems ascendency. As a result, Ulanowicz (1980, 1986) proposed as a phenomenological principle describing ecosystem development, "in the absence of major perturbations, ecosystems exhibit a tendency to increase in ascendency". Those factors that lend to an increasing ascendency, therefore, should be considered as significant contributors to ecosystem development.

From a mathematical point of view, one can elucidate how a system gains in magnitude by calculating what contributes to positive gradients in ascendency. So, for example, if one wishes to know what changes in the B_k foster increases in ascendency, one would want to study the partial derivatives, $(\partial A/\partial B_k)$. After rather tedious algebraic manipulation, the results reduce to:

$$\frac{\partial A}{\partial X_k} = 2\left(\frac{T_{..}}{B_{..}} - \frac{1}{2}\frac{T_{k.} + T_{.k}}{B_k}\right)$$

This formula has a straightforward meaning. The first term in parentheses is the overall throughput rate. The second quotient is the average throughput rate for compartment k. That is, the sensitivity of the biomass-ascendency is proportional to the amount by which the overall throughput rate exceeds that of the compartment in question. If the throughput of compartment is smaller than the overall rate, ascendency is abetted. In other words, increasing ascendency is favored by slower passage (longer storage) of biomass through compartment k., i.e. biomass storage favors increased ascendency.

Proposition 2: When several elements flow through a compartment, that element flowing in the least proportion (as identified by Liebig (1840)) is the one with the longest retention time in the compartment.

We begin by letting T_{ijk} be the amount of element k flowing from component i to component j. We then consider the hypothetical situation of ideally balanced growth (production). In perfectly balanced growth, the elements are presented to the population in exactly the proportions that are assimilated into the biomass. This can be stated in quantitative fashion: for any arbitrary combination of foodstuff elements, p and q, used by compartment j,

$$\frac{T^*_{.jp}}{T^*_{.jq}} = \frac{B_{jp}}{B_{jq}}$$

where an asterisk is used to indicate a flow associated with balanced growth. Now we suppose that one and only one element, say p without loss of generality, enters j in excess of the proportion needed. That is, $T_{.jp} = T^*_{.jp} + e_p$ where e_p represents the excess amount of p presented to j. Under these conditions we have the inequality

$$\frac{T_{.jp} + e_p}{T_{.jp}} > \frac{B_{jp}}{B_{jq}}$$

Multiplying both sides of this inequality by the ratio $T^*_{.jp}/B_{jp}$ yields

$$\frac{T_{.jp} + e_p}{B_{jp}} > \frac{T^*_{.jq}}{B_{jq}}$$

In words, this latter inequality says that the input rate of p into j is greater (faster) than that of any other element by the 'stoichiometric' amount e_p/B_{jp}. Over a long enough interval, inputs and outputs must balance, and so we can speak about the input rate and throughput rate as being one and the same. (This does not weaken our argument, as there is an implied steady-state assumption in the Liebig statement as well.)

Now we suppose that only two of the elements flowing into j are supplied in excess. Again, without loss of generality, we call the second element q. It is immediately apparent that if $e_p/B_{jp} > e_q/B_{jq}$, then the throughput rate of p exceeds that of q, and vice versa. That is, a slower throughput rate indicates that one is closer to stoichiometric proportions. This last result can be generalized by mathematical induction to conclude that the element having the slowest throughput rate is being presented in the least stoichiometric proportion, i.e. it is limiting in the sense of Liebig.

8.8 DO ECOLOGICAL PRINCIPLES ENCOMPASS OTHER PROPOSED ECOLOGICAL THEORIES?: THE RIVER CONTINUUM CONCEPT (RCC)

Vannote et al. (1980) proposed the RCC (Figure 8.15). For the authors, from headwaters to mouth, the physical variables within a river system present a continuous gradient of physical conditions. This gradient should elicit a series of responses within the constituent populations resulting in a continuum of biotic adjustments and consistent patterns of loading, transport, utilization, and storage of organic matter along the length of a river. Based on the

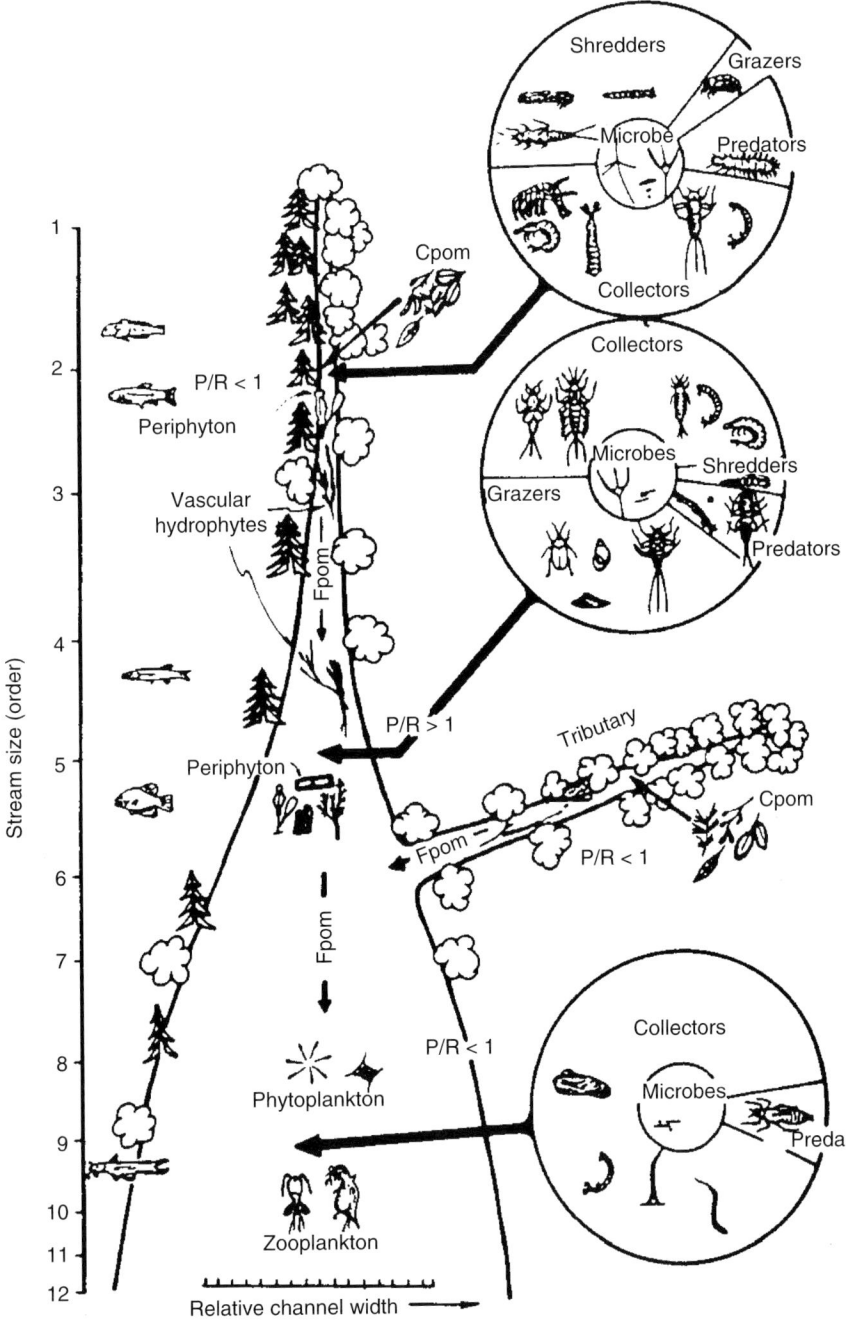

Figure 8.15 A proposed relationship between stream size and the progressive shift in structural and functional attributes of lotic communities (Vannote et al., 1980).

energy equilibrium theory of fluvial geo-morphologists, they hypothesize that the structural and functional characteristics of stream communities are adapted to conform to the most probable position or mean state of the physical system. They reason that producer and consumer communities characteristic of a given river reach become established in harmony with the dynamic physical conditions of the channel. In natural stream systems, biological communities can be characterized as forming a temporal continuum of synchronized species replacements.

This continuous replacement functions to distribute the utilization of energy inputs over time. Thus, the biological system moves toward a balance between a tendency for efficient use of energy inputs through resource partitioning (food, substrate, etc.) and an opposing tendency for a uniform rate of energy processing throughout the year. The authors theorize that biological communities developed in natural streams assume processing strategies involving minimum energy loss.

Downstream communities are fashioned to capitalize on upstream processing inefficiencies. Both the upstream inefficiency (leakage) and the downstream adjustments seem predictable. Finally, they propose that this RCC provides a framework for integrating predictable and observable biological features of lotic systems.

The river continuum theory in the light of ecosystem principles
The river continuum theory can almost be seen as a different wording of the Ecological Law of Thermodynamics applied to rivers, since it is fully compliant with it. Along a continuous gradient of changing environmental conditions, what river communities do under the prevailing conditions is in fact *"attempt to utilize the flow to increase its exergy, moving further away from thermodynamic equilibrium"*. Changing conditions along the gradient determine different constrains and therefore other processing strategies, because *If more combinations and processes are offered to utilize the Exergy flow, the organization that is able to give the highest Exergy under the prevailing circumstances will be selected* (Jørgensen, 1997), or alternatively as *if more combinations and processes are offered to utilize the free energy flow, the organization that is able to give the greatest distance away from thermodynamic equilibrium under the prevailing circumstances will be selected* (de Wit, 2005).

8.9 DO ECOLOGICAL PRINCIPLES ENCOMPASS OTHER PROPOSED ECOLOGICAL THEORIES?: HYSTERESIS IN NATURE

Numerous examples in nature show that that are combinations of environmental factors (external constraints) that may give rise to two equally viable community structures, i.e. they may provide the same degree of support (possibilities to grow) to different sets of internal constraints. In such cases, a hysteresis relationship exists between the dominant external constraining factors and the community structure (internal constraints relieving the external constraints). For instance, in freshwater shallow lakes ecosystems there are references to such type of scenarios:

(1) For concentrations between approximately 50 μg P/l and 120–140 μg P/l a plankton community structure dominated by zooplankton and carnivorous fish has the same

probability to occur than structure dominated by planktivorous fish and phytoplankton (de Bernardi and Giussani, 1995).
(2) For concentrations between approximately 100 and 250 µg P/l shallow lakes can be dominated either by submerged vegetation or by phytoplankton (Scheffer, 1998).

In both cases is the system history that will determine which one of the two possible community structures will actually occur. Once the community installed, within the indicated ranges, a shift in its structure will only take place in case the community (the internal constraints) is changed by external factors (forcing functions). In the first case, this might mean that the planktivorous fish are physically removed and replaced by more carnivorous fish, and in the second case that phytoplankton is removed and submerged vegetation planted. In fact, such interventions are called bio-manipulation and the experience has shown that it only works in the indicated ranges of nutrients concentrations. It can furthermore be shown in the two referred cases that the relief—indicated as the growth measured by eco-exergy, Jørgensen et al., 2000—is the same for the two possible community structures within the indicated ranges (Figure 8.16) (Jørgensen and de Bernardi, 1997). The hysteresis occurrence can thus be explained in the light of the maximum eco-exergy principle.

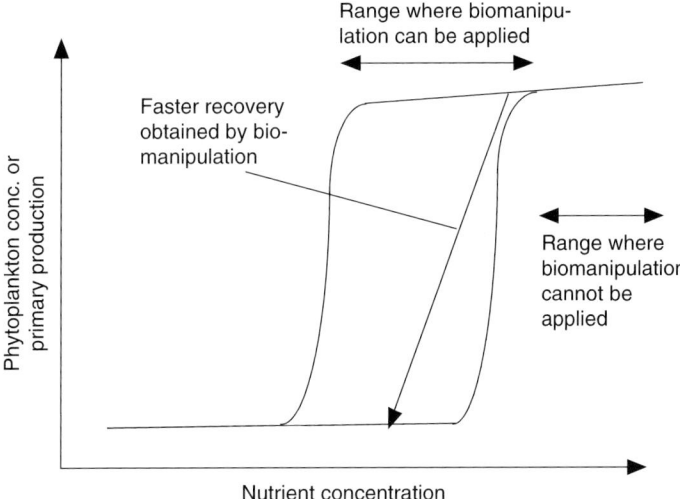

Figure 8.16 The hysteresis relation between nutrient level and eutrophication measured by the phytoplankton concentration is shown. The possible effect of bio-manipulation is shown. An effect of bio-manipulation can hardly be expected above a certain concentration of nutrients, as indicated on the diagram. The bio-manipulation can only give the expected results in the range where two different structures are possible. The range for a change from zooplankton–carnivorous fish control to planktivorous–phytoplankton control is approximately 50–120/140 µg P/l and for a change from dominance by submerged vegetation–phytoplankton from approximately 100 µg P/l to approximately 250 µg P/l.

8.10 CONCLUSIONS

The idea in this chapter was to check the compliance of ecosystem principles to the evolutionary theory, and to see if one could use these same principles to provide an explanation to different ecological problems usually addressed by different theoretical approaches, assuming that to make real progress in the field of Ecology requires a more general and integrative theory. Does this mean that the other theories have no explanatory power? Definitely no. It just means that they are not universal, and therefore they can only be utilized to explain a relatively narrow number of observations, being in most cases specific for a given type of system. It has been demonstrated that ecosystem principles, namely translated in the Ecological Law of Thermodynamics, are fully compliant with the evolutionary theory and, furthermore, can encompass some of the most well-known non-universal ecological theories.

9

Ecosystem principles have applications

Tempus item per se non est, sed rebus ab ipsis consequitur sensus, transactum quid sit in aevo, tumquae res instet, quid porro deinde sequantur.

Time per se does not exist: the sense of what has been done in the past, what is in the present and what will be is embodied in things themselves.
(Lucretius, De Rerum Natura, I, 459–461)

9.1 INTRODUCTION

Orientors, being holistic ecological indicators, can give further information on the state of an ecosystem than can simply reductionistic indicators. Information coming from systematic or analytical approaches should never be neglected but holistic indicators allow us to understand if the system under study is globally following a path that takes the system to a "better" or to a "worse" state. And, we can also compare macroscopic state of different systems, which is impossible to do with isolated reductionistic information. So, advantages of holistic indicators are: additional aggregate information without losing information; ability to compare; ability to compare states of the same system at different times; and possibility of understanding what new data types are needed for this approach.

With indicator concepts like ecosystem health, ecosystem integrity can find operational values, using information coming from approaches like network analysis, eco-exergy, ascendency, emergy evaluation, and the other related indicators. Here, we present several examples in which the systems perspective in ecology has been applied. The types and locations of systems in which they have been applied are very diverse: terrestrial and aquatic ecosystems in Europe, North and South America, and Asia, as are the goals of the research and management questions involved. Regardless of the setting or objective, at its core, holistic indicators always give a broader understanding of the amalgamation of the ecosystem parts into a context of the whole.

9.2 ENTROPY PRODUCTION AS AN INDICATOR OF ECOSYSTEM TROPHIC STATE

References from which these applications of entropy production are extracted:
Aoki I. 1987. Entropy balance in lake Biwa. Ecol. Model. 37, 235–248.
Aoki I. 1995. Entropy production in living systems: from organisms to ecosystems. Thermochim. Acta 250, 359–370.
Aoki I. 2000. Entropy and Exergy principles in living systems. Thermodynamics and Ecological Modelling, Lewis Publishers, New York, NY, pp. 165–190.
Ludovisi A, Poletti A. 2003. Use of thermodynamic indices as ecological indicators of the development state of lake ecosystems. 1. Entropy production indices. Ecol. Model. 159, 203–222.

Entropy flow and entropy production (see Chapter 2) can be quantitatively estimated using physical modelling or calculated from observed energy flow data of biological systems. Here entropy production in lake ecosystems is examined in detail for three ecosystems located in Japan, USA, and Italy.

Case studies

Lake Biwa is located at 34°58′–35°3′ N, 135°52′–136°17′ E (near Kyoto, Japan) and consists of a northern basin (the main part) and a southern basin (the smaller part). The former is oligotrophic and the latter is nearly eutrophic. Only the northern basin is considered. Data for this study were collected in 1970s. The annual adsorbed solar energy was 4153 MJ while the mean depth of the lake is 44 m. It is possible to identify two zones in the column water: a light one (<20 m) and a dark one (between 20 m and 24 m). The average amount of suspended solid (SS) in the light zone was 1.3 [gm^{-3} J] (National Institute for Research Advancement, 1984) while the average amount of dissolved organic carbon (DOC) was 1.6 [gC m^{-3}] (Mitamura and Sijo, 1981). The average amount of total plankton plus zoobenthos in the whole water column was 0.16 [gC m^{-3}] (Sakamoto, 1975).

Lake Mendota is located at 43°04′ N, 89°24′ W (near Madison, Wisconsin, USA) and is a eutrophic lake. Its energy budget was investigated by Dutton and Bryson (1962) and Stewart (1973). The annual adsorbed solar energy was 4494 MJ while the mean depth of the lake is 12.2 m. Two zones of the water column were identified: the euphotic one (until 9 m) and the aphotic one (the last 3.2 m). The average amount of SS in the light zone was 1.9 [gm^{-3}] (National Institute for Research Advancement, 1984) while the average amount of DOC was 3.3 [gC m^{-3} J] (Brock, 1985). The average amount of total plankton plus zoobenthos in the whole water column was 0.62 [gC m^{-3}] (Brock, 1985).

Lake Trasimeno is the largest lake in peninsular Italy (area 124 km^2); it is shallow (mean depth 4.7 m, maximum 6.3 m), and accumulation processes are favored. The water level of the lake showed strong fluctuations with respect to meteorological conditions; hydrological crises occur after several years with annual rainfall <700 mm. Lake Trasimeno can be considered homogeneous for chemical and physical parameters (Maru, 1994) and very sensitive to meteorological variability or human impact. According to the Vollenweider–OECD classification (Giovanardi et al., 1995), Lake Trasimeno is mesotrophic, whereas by using the annual phosphorus loading estimation method (Maru, 1994) and the Hillbrich–Ilkowska method (Hamza et al., 1995), the lake is classified as eutrophic.

Entropy production indices for waterbodies

The quantities necessary to estimate entropy production (see Aoki, 1989, 1990) can be obtained from experimentally observed data. Entropy production plotted against adsorbed solar radiation energy for Lake Biwa and Lake Mendota are shown in Figures 9.1 and 9.2, respectively. The monthly entropy production per unit of volume (S_p) of the Trasimeno Lake was calculated by simple division of entropy production per surface units (S_{prod}) by monthly mean values of water depth; the annual values were calculated as the sum of monthly values and are given in Table 9.1.

Entropy production is expressed in MJ m^{-2} month^{-1} K^{-1}, while solar radiation in MJ m^{-2} month^{-1}. According to Aoki, entropy production in month j (denoted as $(\Delta_i S)_j$) is a linear function of the absorbed solar radiation energy in month j (denoted as Q_j):

$$(\Delta_i S)_j = a + bQ_j \tag{9.1}$$

According to Ludovisi (2003) the definition of the b index as a ratio of S_p (in units MJ m^{-3} year^{-1} K^{-1}) and the solar energy absorbed by the lake surface (Q_s) (MJ m^{-2} per year K^{-1}) in a year is not proper, because entropy and energy flows do not refer to the same

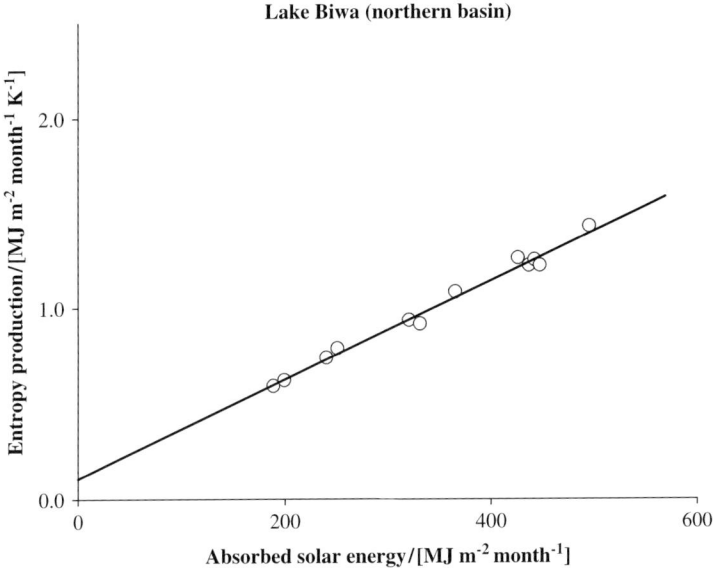

Figure 9.1 Monthly entropy production (S_{prod}) in the northern basin of Lake Biwa per m^2 of the lake surface plotted against monthly solar radiation energy absorbed by 1 m^2 of the lake surface (Qs). The circles represent, from left to right, the months: December, January, November, February, October, September, March, April, June, July, August, May.

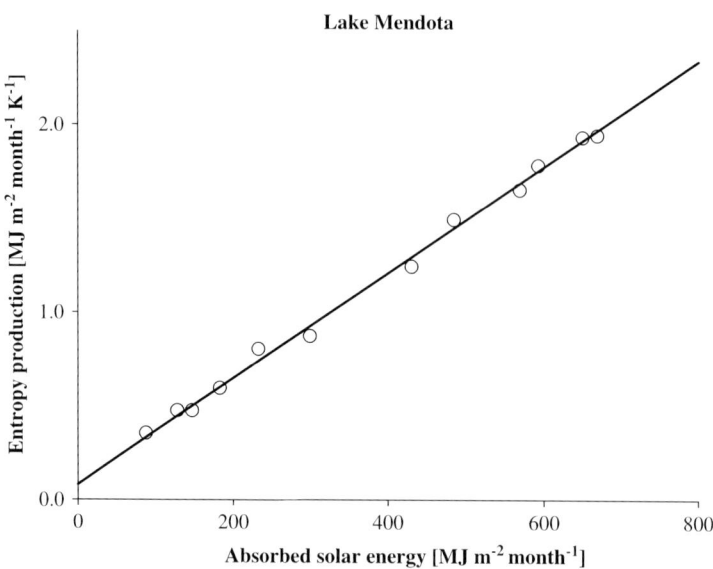

Figure 9.2 Monthly entropy production (S_{prod}) in Lake Mendota per m² of the lake surface plotted against monthly solar radiation energy absorbed by 1 m² of the lake surface (Qs). The circles represent, from left to right, the months: January, February, December, November, March, October, September, April, August, May, June, July.

spatial unit. This fact introduces an artificial dependence on the water depth. Partially following Aoki's indices, a set of new ones (c, d, d') analogous to the a, b, and b' were proposed by Ludovisi (2003) on the basis of the relationship between the S_{prod} and Q_s. The index d' does not demonstrate any significant trend during the years 1988–1996 (Table 9.1).

A good linear correlation between the monthly entropy production (S_{prod}) per surface unit of Lake Trasimeno and the monthly Q_s has been found on a monthly time scale (Figure 9.3) and the regression coefficients of the curve (c, intercept and d, slope) can be compared with the analogous Aoki's indices a, b (Table 9.2).

The comparison of c, d (regression coefficients of the curve Figure 9.3 intercept and slope), d' (the ratio between the annual S_{prod} and Q_s) values (Table 9.2) calculated for Lake Mendota and the northern basin of Lake Biwa significantly distinguishes the eutrophic Lake Mendota from the oligotrophic Lake Biwa, and attributes to Lake Trasimeno higher values of d and d' than both other lakes.

Regarding Equation 9.1, the second term on the right-hand side is the entropy production dependent on solar radiation energy, which is caused by the conversion into heat of the solar energy absorbed by water, by dissolved organic matter, and by SS (negligible are the contributions from photosynthesis and light respiration of phytoplankton). The first term on the right-hand side of Equation 9.1 is the entropy production independent of solar radiation energy and it is caused by respiration of organisms in the lake.

For Lake Biwa and Lake Mendota total and solar energy-dependent entropy productions (per year, per MJ of absorbed solar radiation energy per m³ of the lake water), and

Table 9.1 Annual values of S_{prod} (MJ m^{-2} year^{-1} K^{-1}), S_p (MJ m^{-3} year^{-1} K^{-1}), and the indices b' (10^{-4} m^{-1} K^{-1}), d' (10^{-4} K^{-1}), calculated for Lake Trasimeno in the years 1988–1996

Year	S_{prod}	S_p	b'	d'
1988	16.02	3.20	6.2	31.0
1989	15.60	3.34	6.4	29.9
1990	15.72	3.65	7.3	31.4
1991	15.57	3.74	7.4	30.8
1992	15.42	3.54	7.1	30.8
1993	15.62	3.68	7.1	30.1
1994	16.40	3.91	7.4	30.8
1995	15.60	3.93	7.6	30.2
1996	15.62	4.17	8.0	29.8
Average	15.73	3.69	7.2	30.6

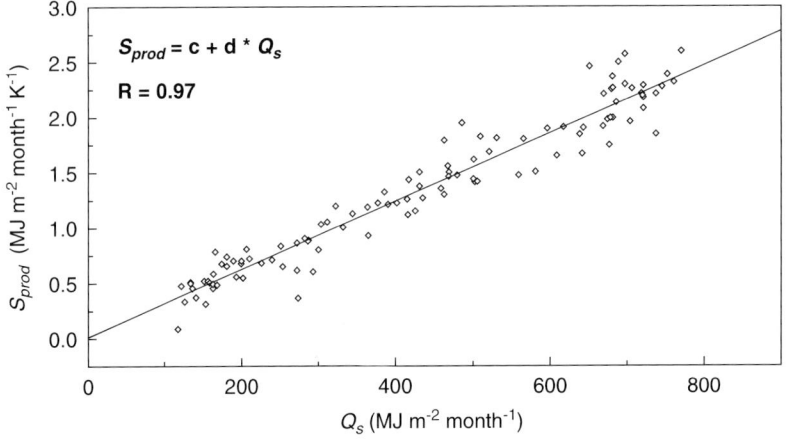

Figure 9.3 Linear regression between the monthly entropy production (S_{prod}) per surface unit of Lake Trasimeno and the monthly solar energy absorbed by the lake (Q_s).

entropy productions independent of solar radiation energy (per year, per m^3 of the lake water) are shown in Table 9.3. The values of entropy production dependent on solar radiation in the light zone (euphotic zone) are related to the amount of dissolved organic matter and SS per m^3 of lake water in the light zone. The ratio of the amount of SS in Lake Mendota to that in Lake Biwa (1:5) and the ratio of DOC in Lake Mendota to that in Lake Biwa (2:1) are consistent with the ratio of entropy production dependent on solar radiation between Lake Mendota and Lake Biwa (Table 9.3). Thus, the greater the amount of SS and DOC, the more the entropy production is dependent on solar radiation. The entropy production dependent on solar radiation gives a kind of physical measure for the

Table 9.2 Environmental parameters, TSI values, and values of trophic indices (a, b, b') proposed by Aoki (1995) and those of the new set of indices c, d, d' for Lake Mendota, Lake Biwa, and Lake Trasimeno

Parameter	Lake Biwa	Lake Mendota	Lake Trasimeno
Mean depth (m)	44	12.2	4.7
Residence time (year)	5.5	3.1–8.8	>20
Transparency (secchi depth in m)	5.2	2.9	1.2
Chlorophyll α ($\mu g\,l^{-1}$)	5	32	8
Total phosphorus ($mg\,l^{-1}$)	0.01	0.07	0.05
TSI (SD)[1]	36	45	58
TSI (Chlα)[1]	46	65	51
TSI (TP)[1]	37	65	59
TSI (average)[1]	39	58	56
Trophic classification[2]	Oligotrophic	Hyper-eutrophic	Eutrophic
a ($MJ\,m^{-3}\,month^{-1}\,K^{-1}$)	0.002	0.006	
b ($10^{-4}\,m^{-1}\,K^{-1}$)	0.6	2.3	
b' ($10^{-4}\,m^{-1}\,K^{-1}$)	0.6	2.4	7.23[3]
c ($MJ\,m^{-2}\,month^{-1}\,K^{-1}$)	0.070	0.070	0.014
d ($10^{-4}\,K^{-1}$)	26.7	27.9	31.0
d' ($10^{-4}\,K^{-1}$)	26.4	29.3	30.7[3]

[1]Trophic state index calculated by using Carlson (1977) equations
[2]Based on the Kratzer and Brezonik (1981) classifcation system
[3]Average value of the years 1988–1996

Table 9.3 Comparison of entropy productions in Lake Biwa and Lake Mendota

Lake	Total (in whole water column)	Solar energy dependent (in light zone)	Solar energy independent (in whole water column)
Lake Biwa	0.07	0.13	19
Lake Mendota	0.24	0.31	69
Lake Mendota/Lake Biwa	3:7	2:3	3:6

Note: Total and solar energy-dependent entropy productions (per year per MJ of absorbed solar radiation energy per m^3 of the lake water) are shown, respectively, in the first and in the second column, and entropy productions independent of solar radiation energy (per year m^3 of the lake water) are in the third column. Units are ($kJ\,K^{-1}\,m^{-3}\,year^{-1}$). Ratios of the values for the two lakes are shown in the last row.

amount of dissolved organic matter and SS in the lake water by means of reactions to incident solar radiation.

The entropy production independent of solar radiation energy (Table 9.3) is the measure of activity of respiration of organisms distributed over the whole water column. The ratio of the amount of plankton plus zoobenthos in Lake Mendota with respect to Lake

Biwa is 3:9 and is consistent with the ratio of entropy production independent of solar radiation (3:6). The larger the amount of organisms, the more the entropy production is independent of solar radiation. The entropy productions in eutrophic Lake Mendota are larger than those in oligotrophic Lake Biwa in any of the categories considered (i.e., due to light absorption, respiration, and total).

Figure 9.4 reports the linear regression curves between d and TSI, TSI (SD) (Carlson, 1977) and the mean depth (because of the little data available, the regression curves cannot

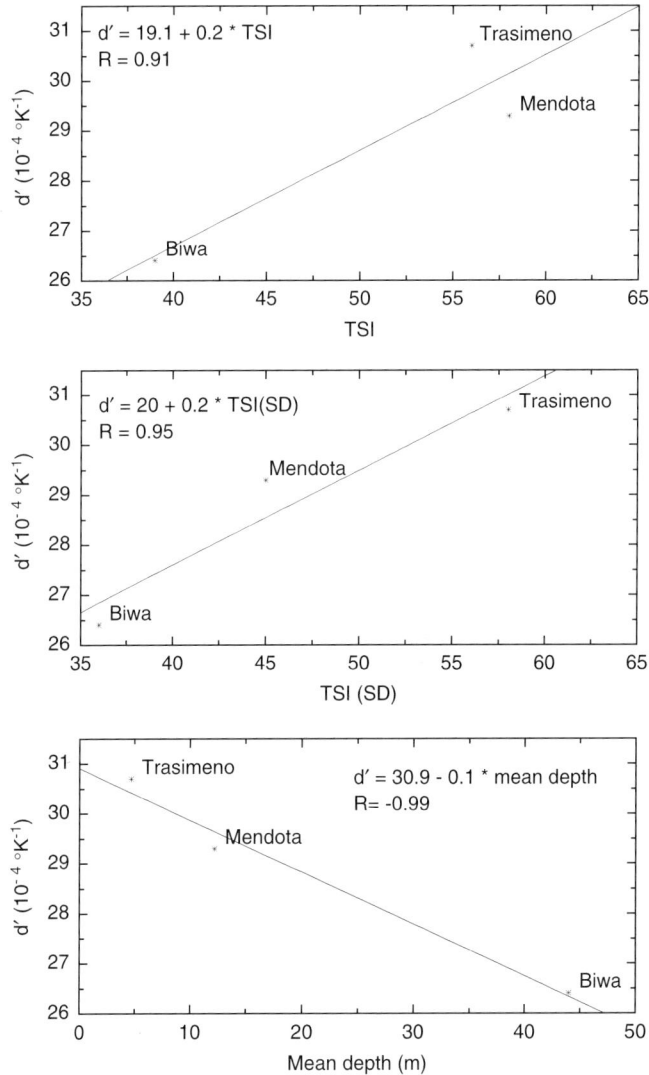

Figure 9.4 Linear regression between the entropy production index d' and TSI, TSI (SD), the mean water depth for Lake Biwa, Lake Mendota, and Lake Trasimeno.

be considered highly significant). As can be seen, d' is positively correlated to TSI, although the relation is not very sharp, because of the similarity of TSI for Lake Trasimeno and Lake Mendota. The index d' shows a good negative linear correlation with the lake's mean depth: the intercept value given by the linear regressions (30.9×10^{-4} K^{-1}) could approach the higher values for d' at the limits of existence of an aquatic ecosystem, which is reached at a rate of 0.1×10^{-4} K^{-1}m^{-1}.

The indices d and d' could be considered measures of the ability of the ecosystems to dissipate the incoming solar energy into the system; the positive correlation between these indices and the trophic state of the lakes indicates that they could account for the influence of the biological productivity on the whole entropy production of the system. As high nutrient concentrations increase the whole biological production as well as the energy flow through an ecosystem, an increase in d and d' values with eutrophication is expected because of the irreversibility of the biological processes.

Furthermore, the efficiency of the energy transfer between the trophic levels in eutrophic systems was found to be lower than in oligotrophic systems (Jonasson and Lindegaard, 1988). In ecological terms, this should mean that a higher nutrient availability in more eutrophic systems induces the achievement of a biological community possessing a better ability to dissipate energy, following a development strategy based on the maximization of the productivity, rather than optimization of the energy exploitation.

Conclusions
The entropy production of the three categories (total entropy production, dependent entropy production, and independent entropy production) can be proposed to be larger in a eutrophic lake than in an oligotrophic lake. Natural processes tends to proceed with time from oligotrophy to eutrophy in most of present lake ecosystems surrounded by the environment full of organic matter; the entropy production of the three categories in a lake will increase with time accompanying the process of eutrophication (Aoki, 1989, 1990).

These entropy production indices can be useful tools for characterizing the trophic status of a water body; however, their ecological interpretation might need more investigation as they depend on the successional stage (Margalef, 1977; Reynolds, 1984) or on the "prevailing condition" the system is following.

9.3 THE USE OF ECOLOGICAL NETWORK ANALYSIS (ENA) FOR THE SIMULATION OF THE INTERACTION OF THE AMERICAN BLACK BEAR AND ITS ENVIRONMENT

Reference from which these applications of ENA are extracted:
  Patten BC. 1997. Synthesis of chaos and sustainability in a nonstationary linear dynamic model of the American black bear (*Ursus americanus* Pallas) in the Adirondack Mountains of New York. Ecol. Model. 100, 11–42.

Here an application of a dynamic model is used to show the importance of indirect effects (see chapter 5) even within a linear approach.

There are many examples of indirect relationships in natural systems, some of them involving the global one—the biosphere. The majority of these relationships remain either overlooked or poorly understood (Krivtsov et al., 2000). To model such systems requires the use of many integrated submodels, due to the complexity of processes involved.

The knowledge that all species in nature are complexly interconnected directly and indirectly to all other biotic and abiotic components of their ecosystems is slow in being translated into models and even more in management practice.

An example for such a synthesis is the simulation model of a wildlife population, the American black bear (*Ursus americanus* Pallas) on the 6000 ha Huntington Wildlife Forest in the central Adirondack Mountain region of upper New York State, USA (Costello, 1992). The model was designed to be conceptually complex but mathematically simple, so its behavior would derive more from biology and ecology than from mathematics. The STELLA II (High Performance Systems, Hanover, NH) model of the Adirondack black bear is linear, donor controlled, nonstationary, and phenomenological (Patten, 1983).

The model's purposes are to express black bear biology as a population system inseparable from its ecosystem and to demonstrate how chaos and sustainability can be realistically incorporated into models, minimizing the use of inappropriate mathematics that, though traditional or classical, may not be well chosen due to an inadequate rationale.

If *envirograms* for all the taxa and significant abiotic categories of the Huntington Wildlife Forest could be formed, then the centrum of each would account for one row and one column of an $n \times n$ interconnection matrix for the whole ecosystem. The centrum of each black bear envirogram for a life history stage would then represent one such row and column within the ecosystem matrix and from these indirect connections between bear and ecosystem compartments could be determined. Of course the forest ecosystem model does not exist, but the rationale for embedding the bear subsystem within it is clear, and the purpose of the envirograms was to implement this in principle by way of organizing relevant information for modeling.

A further criterion was that all the direct interactions between the bear compartments and the environment would be by mass energy transactions, enabling the conservation principle to be used in formulating system equations. The envirograms prepared for this model are depicted in Simek (1995) and were then used to construct a quantitative difference equation model employing STELLA II.

Quantification of the model is still approximate, based on general data and knowledge of the bear's life history, reproductive behavior, environmental relationships, and seasonal dynamics as known for the Huntington Forest and the Adirondack region. The equations are all linear, and donor controlled, with details of temporal dynamics introduced by nonstationary (time-varying) coefficients rather than by nonlinear state variables and constant coefficients.

The model's behavior is here described in detail only for the cub compartment and selected associated parameters (Figure 9.5). The other compartments behave with similar realism.

A baseline simulation was achieved which generated 33–64 individuals 6000 ha during a typical model year; this is consistent with a mean of about 50 animals typically considered to occur on the Huntington property. Yearling M/F sex ratios generated by the

208   A New Ecology: Systems Perspective

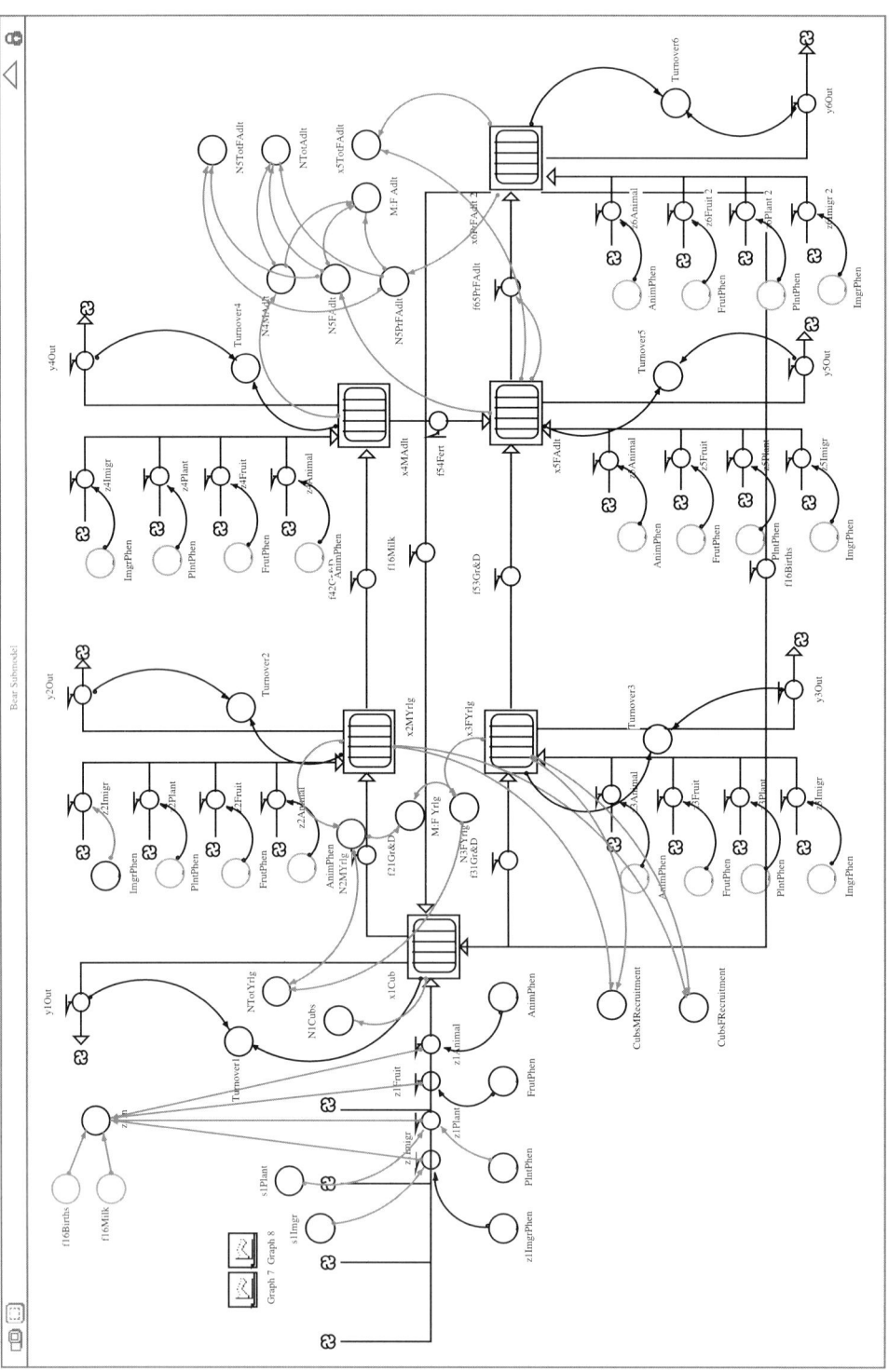

Figure 9.5 Submodel layer depiction of the cub compartment of the black bear model.

Chapter 9: Ecosystem principles have applications

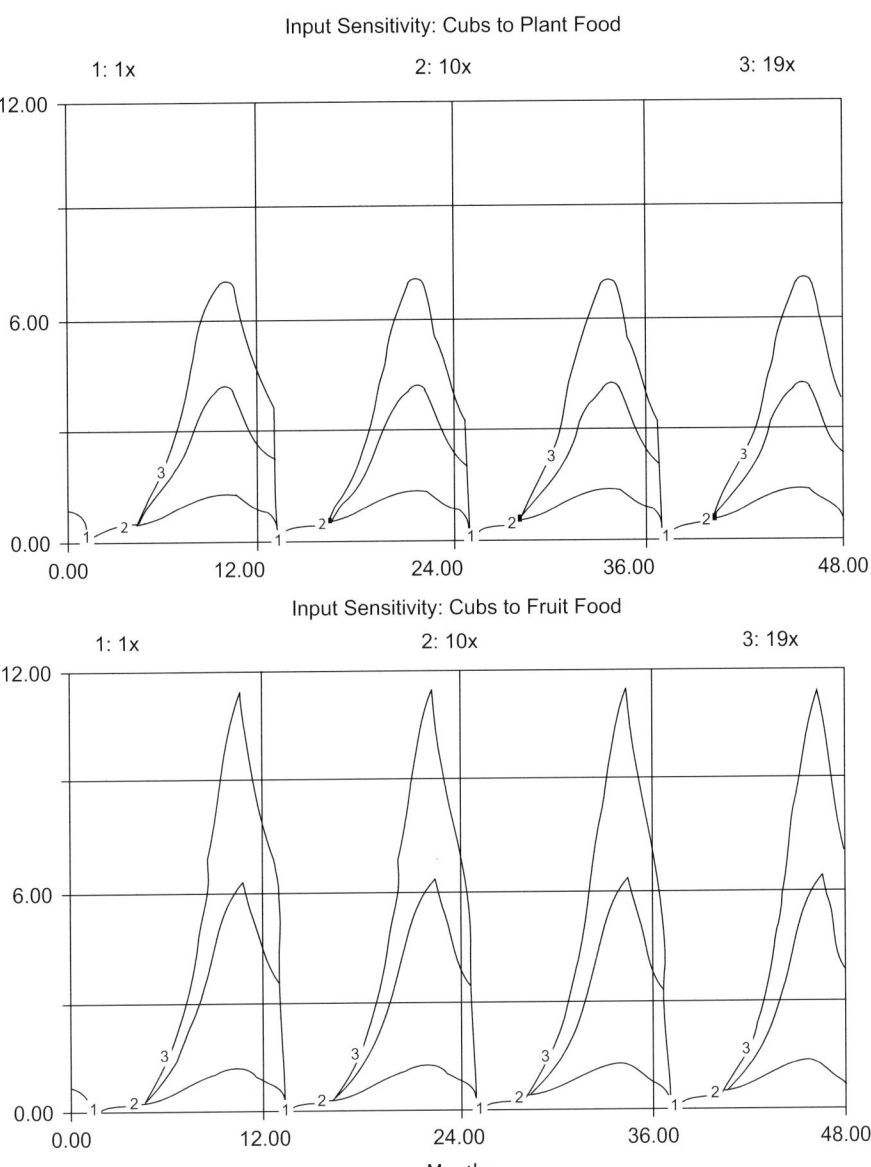

Figure 9.6 Sensitivity of cubs to plant food and fruit. Plant food, principally leaves, fruits, and tubers, comprise 90% of their diets. Fruit is a late-season resource (after July) whereas plant food availability began in May–June. Fruit production occurs when they are approaching going into negative energy balance.

model varied slightly around 0.85, compared to 0.6 observed during 1989–1994. Besides the baseline simulation, model parameters were manipulated to investigate sensitivity relationships. The compartments were indicated to be more sensitive to inputs and less sensitive to outputs. The sensitivity relationships described for cubs generally hold true also for the other age classes in the model.

Conclusions
In descending order, the most sensitive inputs were maternal milk (cubs), fruit production, and plant food availability (Figure 9.6); relatively insensitive inputs were immigration, animal food, and recruitment (to yearlings and adults). Sensitivities to outputs, lower than for inputs, were, in descending order, respiration, egestion, accidental mortality, emigration, parasitic infection, predation (on cubs), harvest, and sickness. Since the model is linear, it can be considered to represent near steady-state dynamics, but its realism suggests that the neighborhood of applicability may actually be very broad around steady state.

9.4 APPLICATIONS OF NETWORK ANALYSIS AND ASCENDENCY TO SOUTH FLORIDA ECOSYSTEMS

Reference from which these applications of ascendency are extracted:
  Heymans JJ, Ulanowicz RE, Bondavalli C. 2002. Network analysis of the South Florida Everglades graminoid marshes and comparison with nearby cypress ecosystems. Ecol. Model. 149, 5–23.

Ascendency (see Chapter 4) is used to compare a cypress system and a graminoids one and to discern the degree of maturity shown by the two systems.

Case studies
The Everglades ecosystem in southern Florida occupies a 9300 km^2 basin that extends from the southern shore of Lake Okeechobee south and southwest to the Gulf of Mexico (Hoffman et al., 1990). Currently, the basin can be divided into three sections: Everglades agricultural area, water conservation areas, and the southern Everglades, the latter of which includes the marshes south of Tamiami Trail and the Shark River Slough. There are two distinct communities in the graminoid system that are differentiated according to short and long hydroperiod areas (Lodge, 1994) and occur in areal ratio of approximately 3:1. Short hydroperiod areas flank both sides of the southern Everglades, and are occupied by a low sawgrass community of plants with a high diversity (100 species) (Lodge, 1994). Typically, vegetation in the short hydroperiod marsh is less than 1 m tall (Herndorn and Taylor, 1986). Long hydroperiod, deeper marsh communities are developed over peat soil (Goodrick, 1984). The long hydroperiod community occurs more commonly in the central Everglades where they typically are straddled between sawgrass marshes and sloughs. These inundated areas are important for fish and aquatic invertebrates, such as prawns. Long hydroperiod areas provide an abundant reserve of prey for wading birds toward the end of the dry season (March–April).

The freshwater marshes of the Everglades are relatively oligotrophic and have been typified as not being very productive—averaging only about $150 \, \text{g m}^{-2}$ per year in wet prairie areas according to DeAngelis et al. (1998). Graminoid ecosystems provide valuable habitat for a wide range of animals, including species listed by the U.S. Fish and Wildlife Service as endangered, threatened, or of concern.

The cypress system is a 295,000 ha wetlands of the Big Cypress Natural Preserve and the adjacent Fakahatchee Strand State Preserve. Both areas cover a flat, gently sloping limestone plain (Bondavalli and Ulanowicz, 1999) with many strands and domes of cypress trees. The cypress swamp does not have a distinct fauna, but shares many species with the adjacent communities (Bondavalli and Ulanowicz, 1999).

The network models of the ecosystems
A model of the freshwater graminoid marshes was constructed by Heymans et al. (2002) and consists of 66 compartments, of which three represent nonliving groups and 63 depict living compartments (see reference for details). The three nonliving compartments include sediment carbon, labile detritus, and refractory detritus, all of which are utilized mainly by bacteria and microorganisms in the sediment (living sediment) and in the water column (living POC—Particulate Organic Carbon). The primary producers include macrophytes, periphyton, *Utricularia*, and other floating vegetation.

Lodge (1994) suggested that: "the Everglades does not have a great diversity of freshwater invertebrates due to its limited type of habitat and its nearly tropical climate, which many temperate species cannot tolerate." The source of most fauna in South Florida is from temperate areas further north. Accordingly, the invertebrate component of the graminoid marshes are broken down into eight compartments, consisting of apple snails (*Pomacea paludosa*), freshwater prawns (*Palaemonetes paludosus*), crayfish (*Procambarus alleni*), mesoinvertebrates, other macroinvertebrates, large aquatic insects, terrestrial invertebrates, and fishing spiders. Loftus and Kushlan (1987) described an assemblage of 30 species of fish in the freshwater marshes, of which 16 species are found in the sawgrass marshes.

The Everglades assemblage of herpetofauna consists of some 56 species of reptiles and amphibians. Nine compartments of mammals were identified for the graminoid marshes. Approximately 350 species of birds have been recorded within the Everglades National Park, and just slightly less than 300 species are considered to occur on a regular basis (Robertson and Kushlan, 1984). Sixty percent of these birds are either winter residents, migrating into South Florida from the north, or else visit briefly in the spring or fall. The remaining 40% breed in South Florida (Lodge, 1994), but of these only eight groups nest or breed in the graminoids. Various species of wading and terrestrial birds roost or breed in the cypress wetlands and feed in the graminoid marshes including anhingas, egrets, herons, wood storks, and ibises. These birds are explicit components of the cypress network. They feed on the aquatic and terrestrial invertebrate members of the graminoid wetland; however, this capture of prey is represented as an export from the graminoid system and an import into the cypress swamp. Waders were not included as explicit components in the graminoid network.

The cypress swamp model consists of 68 compartments and similar to the graminoid system, the cypress model has three nonliving compartments (refractory detritus, labile detritus, and vertebrate detritus) and two microbial compartments (living POC and living sediment). Ulanowicz et al. (1997), Bondavalli and Ulanowicz (1999) give a breakdown of the construction of the model. The primary producers are more diverse than those found in the graminoids and are represented by 12 compartments, seven of which are essentially terrestrial producers: understory, vines, hardwood leaves, cypress leaves, cypress wood, hardwood, and roots (Bondavalli and Ulanowicz, 1999). These seven compartments ramify the spatial dimension of the ecosystem in the vertical extent—an attribute not shared by the graminoid marshes. Other primary producer compartments include phytoplankton, floating vegetation, periphyton, macrophytes, and epiphytes (Bondavalli and Ulanowicz, 1999).

According to Bondavalli and Ulanowicz (1999), cypress swamps do not possess a distinct faunal assemblage, but rather share most species with adjacent plant communities. Most fauna spend only parts of their lives in the swamp. Benthic invertebrates form the heterotrophic base of the food chain. A high diversity of invertebrates has been recorded in cypress domes and strands, but a lack of data at the species level mandated that they resolve the invertebrates into only five compartments (Bondavalli and Ulanowicz, 1999). Similarly, the fish component of this model could not be resolved into more than three compartments, two containing small fish and a third consisting of large fish (Bondavalli and Ulanowicz, 1999).

The herpetofauna compartments of the cypress model were similar to those of the graminoids. The bird community of the cypress swamps was much more diverse than that in the graminoids. The increased diversity can be traced to the inclusion of wading birds in the cypress model. The wading birds do not roost or nest in the graminoids, although they do feed there; therefore, it was assumed that an export of energy and carbon flowed from the graminoids into the cypress. The 17 bird taxa in the cypress include five types of wading birds, two passerines collections, and various predatory birds (Bondavalli and Ulanowicz, 1999). The mammals of the cypress include all the mammalian compartments of the graminoids, as well as some terrestrial mammals unique to the cypress [shrews, bats, feral pigs, squirrel, skunks, bear, armadillos, and foxes (Bondavalli and Ulanowicz, 1999)]. These species are found mostly in the cypress trees and cypress domes, which extend the spatial extent of the ecosystem into the third dimension.

Ascendency, redundancy, and development capacity
Information theory is employed to quantify how well "organized" the trophic web is (expressed in terms of an index called the system's "ascendency"), how much functional redundancy it possesses (what is termed the "overhead"), what its potential for development is, and how much of its autonomy is encumbered by the necessary exchanges with the external world (Ulanowicz and Kay, 1991).

According to the "total system throughput (TST)", the graminoid system is far more active than the cypress system (Table 9.4). Its TST (10,978 g C m^{-2} per year) is fourfold that of the cypress system (2952 g C m^{-2} per year). The development capacity of an

Table 9.4 Information indices for both the graminoid and cypress systems

Index	Cypress Index	Cypress % of C	Graminoids Index	Graminoids % of C
Total system throughput (TST) (gC m^{-2} per year)	2952.3		10,978	
Development capacity = C (gC-bits m^{-2} per year)	14,659		39,799	
Ascendancy (gC-bits m^{-2} per year)	4026.1	34.3	20,896	52.5
Overhead on imports (gC-bits m^{-2} per year)	2881.6	19.7	3637	9.1
Overhead on exports (gC-bits m^{-2} per year)	75.4	0.5	606	1.5
Dissipative overhead (gC-bits m^{-2} per year)	2940	20.1	4932	12.4
Redundancy (gC-bits m^{-2} per year)	3735.8	25.5	9728	24.4
Internal capacity (gC-bits m^{-2} per year)	5443.4		18,122	
Internal ascendancy (gC-bits m^{-2} per year)	1707.5	31.4	8394	46.3
Redundancy (gC-bits m^{-2} per year)	3735.8	68.6	9728	53.7
Connectance indices				
Overall connectance		1.826		1.586
Intercompartmental connectance		3.163		1.807
Foodweb connectance		2.293		

ecosystem is gauged by the product of the diversity of its processes as scaled by the TST. The development capacity of the graminoid system (39,799 gC bits m^{-2} per year) is significantly higher than that of the cypress (14,659 gC bits m^{-2} per year), a difference that one might be inclined to attribute to the disparity in the scalar factor (TST) between the systems. When one regards the normalized ascendancy, however, (ascendancy is a measure of the constraint inherent in the network structure), one notices that the fraction of the development capacity that appears as ordered flow (ascendancy/capacity) is 52.5% in the graminoids. This is markedly higher than the corresponding fraction in the cypress system (34.3%).

The graminoid system has been stressed by a number of modifications to the patterns of its hydrological flow, which have resulted in the loss of transitional glades, reduced hydroperiods, unnatural pooling, and over-drainage (Light and Dineen, 1994). In comparison with the cypress community, however, the system has exhibited fewer changes in its faunal community and is sustained by an abundance of flora and microbiota. The cypress ecosystem, like that of the graminoids, is limited by a dearth of phosphorus, which remains abundant in marine and estuarine waters and sediments. The graminoid system compensates for this scarcity of nutrients with a profusion of periphyton. Periphyton exhibits a high P/B ratio, even under oligotrophic conditions.

The natural stressors that affect the cypress ecosystem appear to have far greater impacts, in that they modulate the rates of material and energy processing to a far greater extent in that system. This analysis is phenomenological and there is no clear reason why the modulation of rates of material and energy occur in the cypress. Thus, even though these systems are (1) adjacent to one another, (2) share many of the same species, and

(3) some of the heterotrophs of the cypress feed off the graminoid system, the characteristic indices of the graminoid system remain distinct from those of the cypress community.

Calculating and ranking "relative sensitivities" proves to be an interesting exercise. For example, when the average trophic levels of the 66 compartments of the graminoid wetland ecosystem were calculated, lizards, alligators, snakes, and mink were revealed to be feeding at trophic levels higher than some of the "charismatic megafauna," such as the snail kite, nighthawk, Florida panther, or bobcat (Table 9.5).

The relative contributions to ascendency by the latter actually outweighed those of the former, however. The relative values of these sensitivities thus seemed to accord with most people's normative judgments concerning the specific "value" of the various taxa to the organization of the system as a whole (Table 9.5).

Similarly, in the cypress system white ibis, large fish, alligators, and snakes feed at high effective trophic levels, but the system performance seemed to be enhanced more by the activities of the vultures, gray fox, bobcat, and panthers (Table 9.5).

In comparing the component sensitivities in the graminoid and cypress systems, one discovers numerous similarities between the taxa of the two systems (Table 9.5). For example, the avian and feline predators ranked high in both systems. The contributions of snail kites and nighthawks to the performance of the graminoid system were highest (at *ca.* 14 bits), while that of the bobcat and panther were highest in the cypress (at *ca.* 13 bits). Both bobcat and panther seem to be more sensitive in the cypress than in the graminoids.

The low sensitivity of crayfish (0.99 bits) in the graminoids was not repeated in the cypress, although aquatic invertebrates generally had a low sensitivity in that system, too (2.01 bits). The sensitivity of labile detritus was similar in both systems (around 1.5 bits), while refractory detritus was more sensitive in the graminoid (1.59 bits), indicating a greater importance in that system. The sensitivities of the primary producers are lower in the cypress (1.51 bits) than in the graminoids (1.66 bits) and are uniform within both systems, except for *Utricularia* in the graminoids. *Utricularia* are carnivorous plants, and, therefore, both its effective trophic level and its sensitivities are higher than those of the other primary producers (Table 9.5). *Utricularia* can exhibit an interesting example of positive feedback in ecosystems; indeed, it harnesses the production of its own periphyton via intermediary zooplankton grazers. This subsidy to the plant apparently allows it to drive in oligotrophic environments that would stress other macrophytes with similar direct uptake rates. As ambient nutrient level rise, however, the advantage gained by positive feedback wanes, until a point is reached where the system collapses (Ulanowicz, 1995).

The cypress system exhibits an additional spatial dimension in comparison with that of the graminoids. The third, vertical (terrestrial) dimension of cypress vegetation provides both additional habitat and food for the higher trophic levels. In the cypress, the appearance of terrestrial vegetation affords increased herbivory by terrestrial fauna such as mammals, birds, and terrestrial invertebrates. Furthermore, much of what is produced by the bacteria is consumed by the higher trophic levels, and less production is recycled back into the detritus. With the addition of the arboreal dimension in the cypress, one would expect that system to be more productive than its graminoid counterpart, and that the total systems throughput (and, consequently, other systems properties) would be higher in the cypress as well. This is not the case, however. In fact, the throughput of the graminoids exceeds that

Table 9.5 Ascendency sensitivity coefficients (Sens. in bits) and effective trophic levels (ETL) for both the graminoid and cypress systems

	Graminoids			Cypress		
	Compartment	ETL	Sens.	Compartment	ETL	Sens.
1	Crayfish	2.14	0.99	Liable detritus	1.00	1.42
2	Mesoinvertebrates	2.15	1.12	Refractory detritus	1.00	1.45
3	Other macroinvertebrates	2.12	1.15	Phytoplankton	1.00	1.51
4	Flagfish	2.00	1.27	Floating vegetation	1.00	1.51
5	Poecilids	2.20	1.47	Periphyton macroalgae	1.00	1.51
6	Labile detritus	1.00	1.55	Macrophytes	1.00	1.51
7	Refractory detritus	1.00	1.59	Epiphytes	1.00	1.51
8	Apple snail	2.12	1.60	Understory	1.00	1.51
9	Tadpoles	2.03	1.63	Vine leaves	1.00	1.51
10	Periphyton	1.00	1.66	Hardwood leaves	1.00	1.51
11	Macrophytes	1.00	1.66	Cypress leaves	1.00	1.51
12	Floating vegetation	1.00	1.66	Cypress wood	1.00	1.51
13	Utricularia	1.03	1.69	Hardwood wood	1.00	1.51
14	Lizards	3.83	1.79	Roots	1.00	1.51
15	Freshwater prawn	2.27	2.12	Aquatic invertebrates	2.20	2.01
16	Ducks	2.20	2.32	Tadpoles	2.16	2.29
17	Bluefin killifish	2.57	2.34	Anseriformes	2.05	2.38
18	Other small fishes	2.48	2.44	Crayfish	2.26	2.46
19	Sediment carbon	1.00	2.44	Terrestrial invertebrates	2.00	2.55
20	Living sediments	2.00	2.58	Living sediment	2.00	2.64
21	Mosquitofishes	2.47	2.64	Squirrels	2.00	2.72
22	Living POC	2.00	2.80	Apple snail	2.26	2.74
23	Chubsuckers	2.50	2.86	Prawn	2.26	2.91
24	Shiners and minnows	2.68	3.60	Rabbits	2.00	2.97
25	Gruiformes	2.01	3.76	White tailed deer	2.00	2.97
26	Muskrats	2.00	3.83	Living POC	2.00	3.08
27	W-T deer	2.00	3.83	Black bear	2.26	3.30
28	Terrestrial inverts	2.00	3.91	Small herb and omni fish	2.60	3.48
29	Rabbits	2.00	5.10	Galliformes	2.33	3.58
30	Killifishes	2.81	5.13	Mice and rats	2.37	3.77
31	Turtles	2.74	2.57	Wood stork	3.24	3.82
32	Large aquatic insects	2.96	5.63	Raccoon	2.74	3.84
33	Salamander larvae	2.57	5.64	Great blue heron	3.24	3.85
34	Grebes	2.63	5.79	Egrets	3.23	3.90
35	Other centrarchids	3.02	6.59	Hogs	2.44	3.96

(*continued*)

Table 9.5 (*Continued*)

	Graminoids				Cypress		
	Compartment	ETL	Sens.		Compartment	ETL	Sens.
36	Rats and mice	2.27	6.66		Other herons	3.21	4.10
37	Raccoons	2.59	6.72		White ibis	3.58	4.19
38	Opossum	2.45	6.77		Turtles	2.82	4.28
39	Pigmy sunfish	3.09	6.79		Wood peckers	2.52	4.43
40	Bluespotted sunfish	3.09	6.83		Omnivorous passerines	2.53	4.45
41	Dollar sunfish	3.09	6.87		Hummingbirds	2.53	4.45
42	Seaside sparrow	2.57	7.10		Small carnivorous fish	3.07	5.56
43	Passerines	2.96	7.16		Opossum	2.35	5.61
44	Topminnows	3.10	7.47		Kites and hawks	3.37	6.10
45	Redear sunfish	3.13	9.09		Owls	3.36	6.10
46	Catfish	3.11	9.21		Mink	3.25	6.21
47	Spotted sunfish	3.16	9.32		Otter	3.25	6.23
48	Warmouth	3.21	9.42		Medium frogs	3.21	6.24
49	Mink	3.41	9.53		Small frogs	3.21	6.24
50	Snakes	3.32	9.66		Salamanders	3.28	6.32
51	Otter	3.34	9.71		Large frogs	3.32	6.38
52	Bitterns	3.25	9.75		Gruiformes	3.35	6.53
53	Alligators	3.39	9.96		Armadillo	2.90	6.54
54	Large frogs	3.29	10.19		Pelecaniformes	3.40	6.61
55	Small frogs	3.17	10.33		Large fish	3.42	6.99
56	Other large fishes	3.27	10.69		Lizards	3.00	7.64
57	Largemouth bass	3.24	10.92		Caprimulgiformes	3.00	7.64
58	Medium frogs	3.16	10.93		Bats	3.00	7.64
59	Gar	3.45	10.96		Predatory passerines	3.00	7.64
60	Cichlids	3.22	10.98		Shrews	3.00	7.65
61	Fishing spider	3.27	11.77		Alligators	3.78	8.30
62	Bobcat	3.02	12.01		Snakes	3.79	8.58
63	Salamanders	3.32	12.29		Salamander larvae	3.20	8.62
64	Panthers	3.17	12.33		Vertebrates detritus	1.00	8.82
65	Snailkites	3.13	14.38		Vultures	2.00	10.03
66	Nighthawks	3.00	14.69		Gray fox	3.41	10.29
67					Bobcat	3.03	12.96
68					Florida panther	3.36	13.48

of the cypress by some fourfold. Although the total biomass in the cypress is three times greater than that in the graminoids, the cypress system's P/B ratio is four times lower there than in the graminoids, thereby yielding the greater throughput in the graminoids.

The increase in throughput in the graminoids increases its development capacity and ascendency. The relative ascendency, which excludes the effects of the throughput, is perhaps a better index with which to compare these two systems. The relative ascendency of the graminoids is exceptionally high. For example, Heymans and Baird (2000) found that upwelling systems have the highest relative ascendency of all the systems they compared (which were mostly estuarine or marine in origin), but the relative ascendency of 52% for the graminoids is higher than any such index they had encountered. The relative ascendency of 34% reported for the cypress is lower than most of the relative ascendencies reported by Heymans and Baird (2000).

Some reasons behind the higher relative ascendency of the graminoids can be explored with reference to the relative contributions of the various components to the community ascendency (Table 9.5). The highest such "sensitivity" in the cypress is more than one bit lower than its counterpart in the graminoids, and, on average, most higher trophic level compartments that are present in both models exhibit higher sensitivity in the graminoids than in the cypress. It is also noteworthy that 41 compartments in the cypress show sensitivities of less than 5 bits, while only 28 compartments lie below the same threshold in the graminoids. The higher sensitivities in the graminoids owe mainly to the greater activity among the lowest trophic compartments, which causes the other compartments to seem rare by comparison. Thus, in the graminoids, community performance seems sensitive to a larger number of taxa, which accords with the analysis of dependency coefficients and stability discussed in Heymans et al. (2002). Pahl-Wostl (1998) suggested that the organization of ecosystems along a continuum of scales derives from a tendency for component populations to fill the envelope of available niche spaces as fully as possible. This expansive behavior is seen in the cypress system, where the arboreal third dimension of the cypress trees fills with various terrestrial invertebrates, mammals, and birds not present in the graminoids. The graminoid system, however, appears to be more tightly organized (higher relative ascendency) than the cypress in that it utilizes primary production with much higher turnover rates. This confirms Kolasa and Waltho (1998) suggestion that niche space is not a rigid structure but rather coevolves and changes in mutual interaction with the network components and the dynamical pattern of the environment. The graminoid system is more responsive, because it utilizes primary producers with higher turnover rates, and has, therefore, been able to track more closely environmental and anthropogenic changes. The cypress system, on the other hand, should have more resilience over the long term due to its higher overhead, especially its redundancy (Table 9.4).

Conclusions
According to Bondavalli et al. (2000), a high value of redundancy signifies that either the system is maintaining a higher number of parallel trophic channels in order to compensate the effects of environmental stress, or it is well along its way to maturity. Even though these authors suggest that the cypress system is not very mature, in comparison to the

graminoids, one would have to conclude that the cypress is a more mature system. A slower turnover rate, as one observes in arboreal systems such as the cypress, is indicative of a more mature ecosystem. Furthermore, the third dimension of terrestrial vegetation affords the system a greater number of parallel trophic channels to the higher trophic levels, compared with the mainly periphyton dominated graminoid system. Although the graminoid system has a large throughput of carbon and a substantial base of fast-producing periphyton, it appears relatively fragile in comparison to the cypress system, which is more resilient over the long run and has more trophic links between the primary trophic level and the heterotrophs. In conclusion, according to ascendency indices, scale—in the guise of the vertical dimension, of the cypress makes that system more resilient as a whole, and less sensitive with respect to changes in material processing by many of its composite species.

9.5 THE APPLICATION OF ECO-EXERGY AS ECOLOGICAL INDICATOR FOR ASSESSMENT OF ECOSYSTEM HEALTH

Reference from which these applications of eco-exergy used as ecosystem health indicator are extracted:

Zaldívar JM, Austoni M, Plus M, De Leo GA, Giordani G, Viaroli P. 2005. Ecosystem Health Assessment and Bioeconomic Analysis in Coastal Lagoon. Handbook of Ecological Indicator for Assessment of Ecosystem Health. CRC Press, pp. 163–184.

In this paragraph an application of Eco-Exergy is reported (see Chapters 2 and 7) to assess the ecosystem health of a coastal lagoon.

Coastal lagoons are subjected to strong anthropogenic pressure. This is partly due to freshwater input rich in organic and mineral nutrients derived from urban, agricultural, or industrial effluent and domestic sewage, but also due to the intensive shellfish farming.

The Sacca di Goro is a shallow water embayment of the Po Delta. The surface area is 26 km^2 and the total water volume is approximately 40×10^6 m^3. The catchment basin is heavily exploited for agriculture, while the lagoon is one of the most important clam (*Tapes philippinarum*) aquaculture systems in Italy. The combination of all these anthropogenic pressures call for an integrated management that considers all different aspects, from lagoon fluid dynamics, ecology, nutrient cycles, river runoff influence, shellfish farming, macro-algal blooms, and sediments, as well as the socio-economical implication of different possible management strategies. All these factors are responsible for important disruptions in ecosystem functioning characterized by eutrophic and dystrophic conditions in summer (Viaroli et al., 2001), algal blooms, oxygen depletion, and sulfide production (Chapelle et al., 2000). Water quality is the major problem. In fact from 1987 to 1992 the Sacca di Goro experienced an abnormal proliferation of macroalga *Ulva* sp. This species has become an important component of the ecosystem in Sacca di Goro. The massive presence of this macroalga has heavily affected the lagoon ecosystem and has prompted several interventions aimed at removing its biomass in order to avoid anoxic crises, especially during the summer, when the *Ulva* biomass starts decomposing. Such crises are responsible for considerable damage to the aquaculture industry and to the ecosystem functioning.

To carry out such an integrated approach a biogeochemical model, partially validated with field data from 1989 to 1998, has been developed (Zaldívar et al., 2003). To analyze its results it is necessary to utilize ecological indicators, using not only indicators based on particular species or component (macrophytes or zooplankton) but also indicators able to include structural, functional, and system-level aspects. Eco-exergy and specific eco-exergy are used to assess the ecosystem health of this coastal lagoon. Effects of *Ulva*'s mechanical removal on the lagoon's eutrophication level are also studied with specific exergy (Jørgensen, 1997) and cost–benefit analysis (De Leo et al., 2002). Three scenarios are analyzed (for a system with clam production and eutrophication by *Ulva*) using a lagoon model: (a) present situation, (b) optimal strategy based on cost–benefit for removal of *Ulva*, and (c) a significant nutrient loading reduction from watershed. The cost–benefit model evaluates the direct cost of *Ulva* harvesting including vessel cost for day and damage to shellfish production and the subsequent mortality increase in the clam population. To take into account this factor, the total benefit obtained from simulating the biomass increase was evaluated using the averaged prices for clam in northern Adriatic; therefore, an increase in clam biomass harvested from the lagoon will result in an increase of benefit.

The Sacca di Goro model has several state variables for which the exergy was computed: organic matter (detritus), phytoplankton (diatoms and flagellates), zooplankton (micro- and meso-), bacteria, macroalgae (*Ulva* sp.), and shellfish (*Tapes philippinarum*). The exergy and the specific eco-exergy are calculated using the data from Table 9.6 on genetic information content and all biomasses were reduced to $gdw\,l^{-1}$ (grams of dry weight per liter).

Figures 9.7 and 9.8 present the evolution of exergy and specific exergy under the two proposed scenarios: *Ulva* removal and nutrient load reduction, in comparison with the "do nothing" alternative. As it can be seen the eco-exergy and specific eco-exergy of both increase, due to the fact that in our model both functions are dominated by clam biomass.

However, the optimal result from the cost/benefit analysis will considerably improve the ecological status of the lagoon in term of specific exergy.

Table 9.6 Parameters used to evaluate the genetic information content

Ecosystem component	Number of information genes	Conversion factor
Detritus	0	1
Bacteria	600	2.7
Flagellates	850	3.4
Diatoms	850	3.4
Micro-zooplankton	10,000	29.0
Meso-zooplankton	15,000	43.0
Ulva sp.	2000[1]	6.6
Shellfish (bivalves)	–	287[2]

Source: Jørgensen (2000b).
[1] Coffaro et al. (1997).
[2] Marques et al. (1997); Fonseca et al. (2000).

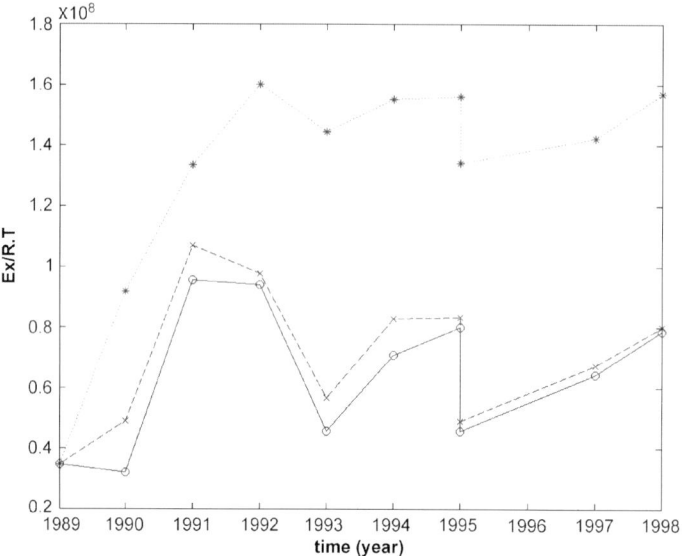

Figure 9.7 Eco-exergy mean annual values: present scenario (continuous line), removal of *Ulva*, optimal strategy from cost–benefit point of view (dotted line), and nutrients load reduction from watershed (dashed line). Reprinted with permission.*

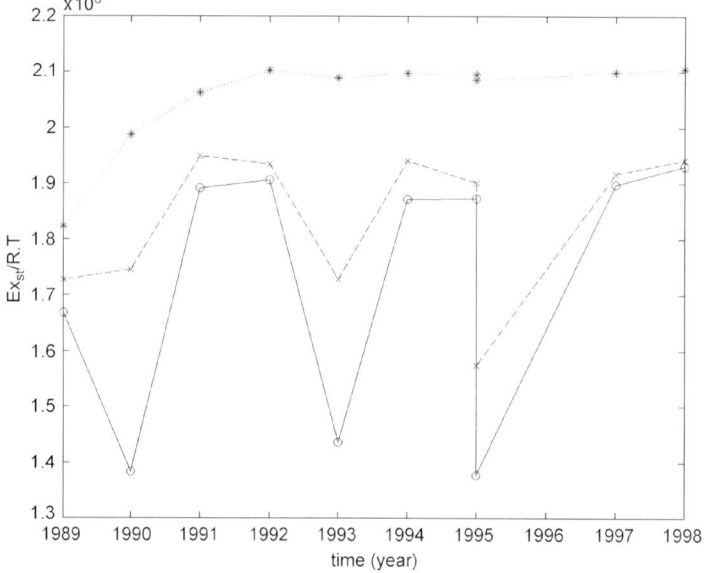

Figure 9.8 Specific eco-exergy mean annual values: present scenario (continuous line), removal of *Ulva*, optimal strategy from cost–benefit point of view (dotted line), and nutrients load reduction from watershed (dashed line). Reprinted with permission.*

---

*Copyright © 2005 Handbook of Ecological Indicators for Assessment of Ecosystem Health, edited by S.E. Jørgensen, F-L Xu, R. Costanza, from chapter by J.M. Zaldívar et al. Two figures reproduced by permission of Taylor & Francis, a division of Informa plc.

Conclusions

The results show that cost–benefit optimal solution for removal of *Ulva* has the highest eco-exergy and specific eco-exergy, followed by a significant removal of nutrients from the watershed. In the case of removal of *Ulva*, specific exergy continues to increase as the number of vessels operating in the lagoon increase. The present situation had the lowest eco-exergy and specific eco-exergy. The result shows that it is a good sustainability policy to take care of natural resources, in this case the clams.

Eco-exergy expresses the system biomass and genetic information embedded in that biomass, while specific eco-exergy tells us how rich in information the system is. These indicators broadly encompass ecosystem characteristics and it has been shown that they are correlated with several important parameters such as respiration, biomass, etc. However it has been pointed out (Jørgensen, 2000b) that eco-exergy is not related to biodiversity, and for example, a very eutrophic system often has a low biodiversity but high eco-exergy.

When a manager has to select between different alternatives, it is difficult to evaluate the optimal solution from an ecological point of view. As eco-exergy and specific exergy are global parameters of the ecosystem, they give an idea of benefits that a measure will produce.

9.6 EMERGY AS ECOLOGICAL INDICATOR TO ASSESS ECOSYSTEM HEALTH

Reference from which these applications of emergy as ecological indicator are extracted:
Howington TM, Brown MI, Wiggington M. 1997. Effect of hydrologic subsidy on self-organization of a constructed wetland in Central Florida. Ecol. Eng. 9, 137–156.

Emergy (see Chapter 6) is used to study and explain theories concerning the effect of an external subsidy on a complex system (constructed wetland) seen by a holistic point of view.

Lake Apopka is a shallow (mean depth = 1.7 m) hypereutrophic lake in Central Florida, with an area of 124 km^2 (Lowe et al., 1989, 1992). In the early 1940s a hurricane removed most of the rooted macrophytes in the lake which led to the early stages of increased nutrient availability and subsequently increased algal productivity (Schelske and Brezonik, 1992). Addressing the nutrient status of this lake, the St. Johns River Water Management District (SJRWMD) constructed a 200 ha freshwater marsh on former agricultural lands with the goal of reducing the nutrient levels in the lake. It was suggested that by pumping enriched lake water through a constructed marsh, filtration of phosphorus and suspended sediments could be maximized. The pump system was turned on in early 1991. The subsidized and unsubsidized marsh maintained similar average water levels (0.76 m) throughout the study period varying yearly by no more than 0.2 m. Theory suggests that an external subsidy should increase the carrying capacity for wildlife of an ecosystem, all other things being equal. The increased capacity for wildlife may be an indirect result of certain self-organizational processes such as changes in vegetative cover. Other factors influencing the relationship between wetland productivity and hydro-period include nutrient inputs, export, nutrient cycling, and decomposition (Carpenter et al., 1985).

This study tested theories concerning the effect of an external subsidy on ecosystem structure and organization. Two newly established marshes (one receiving nutrient-enriched lake water and the other not receiving the subsidy) were the areas under study. The 63 ha subsidized marsh is the first of two cells that constitute the treatment wetland receiving lake water. The unsubsidized marsh, 46 ha, was created as a result of being a borrow pit for building berms around the treatment wetland. Vegetative cover richness and percent cover were determined using aerial photos and GIS, and was calculated using Margalef's (1977) index for species richness. Percent cover provided a further description of the changes in structural complexity of each marsh over time. Also avifauna surveys were conducted. Shannon diversity indexes were used to compare the avian communities found in the surveyed marshes. A synoptic study on the fish population of the subsidized and unsubsidized marshes was also conducted. A model of the marsh system (see Figure 9.9 for energy symbols) was created to

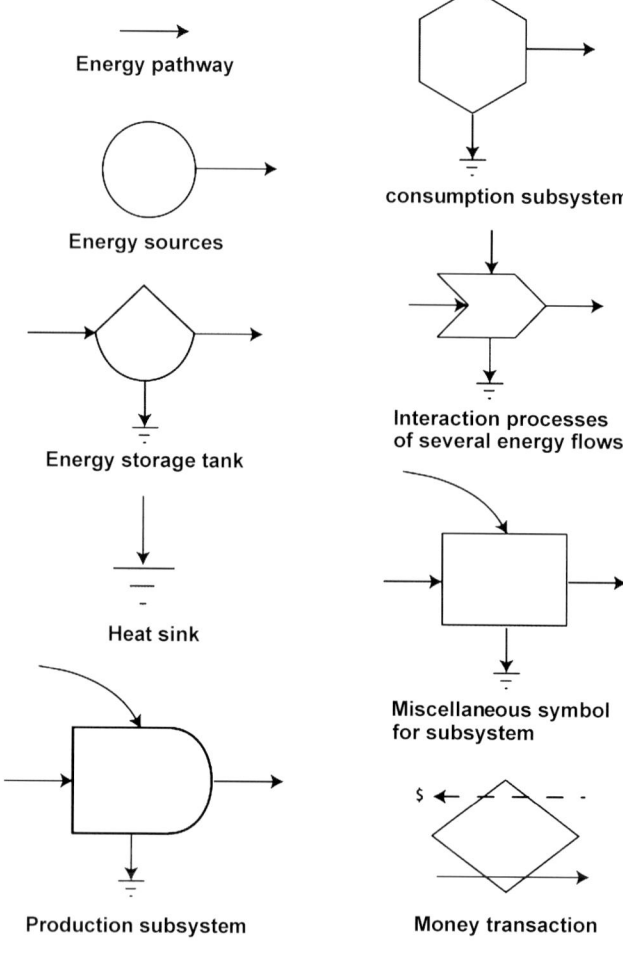

Figure 9.9 Energy symbols used to make an energy diagram.

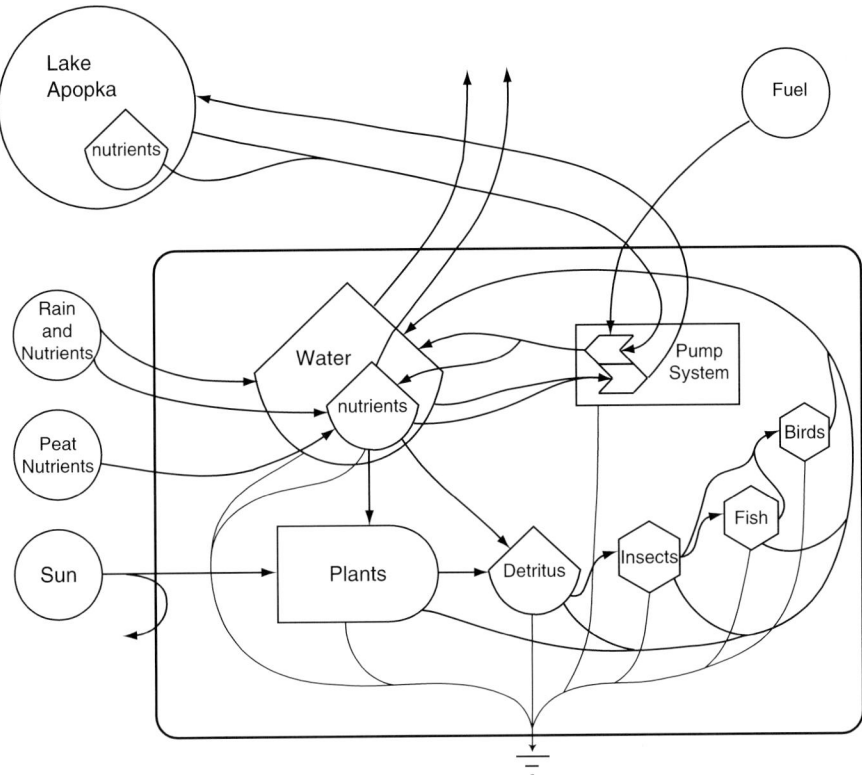

Figure 9.10 Diagram of constructed marsh. Removal of pump system simulated unsubsidized marsh.

describe the role of the most important components and relationships (Figure 9.10). An emergy analysis was performed to evaluate on a common basis (solar energy) the contributions of the various inputs (pumps, water, nutrients, human services, and renewable energies) driving the marshes ecosystems.

Emergy evaluation separates inputs on the basis of the origin (local or purchased) and of their renewability (see also Chapter 6). An environmental loading ratio (ratio of local and exogenous nonrenewable emergy to renewable emergy) and an investment ratio (ratio of exogenous to local emergies) were calculated to compare the quantities and qualities of the energies entering each system. Emergy analysis tables were developed separately in Tables 9.7 and 9.8 for the subsidized and unsubsidized marshes.

The environmental loading ratio showed a large contrast between the two marshes. Investment ratios for the two marshes showed a large difference in the amount of purchased energy necessary to maintain the flows of environmental inputs.

Table 9.9 contains the ratios of free to purchased energy (environmental loading) and nonrenewable energy to renewable energy (investment ratio). Renewable energy sources

Table 9.7 Annual energy, material and dollar flows, and resulting emergy flows supporting 1 ha of the subsidized marsh

Notes		Quantity and unit	Emergy per unit (sej unit^{-1})	Emergy (E + 14 sej)
	Renewables			
1	Sun	5.41E + 09 J	1.00E + 00	0.00
2	Rain-chemical potential	6.27E + 10 J	1.54E + 04	9.66
	Nonrenewables-free			
3	Total nitrogen	7.85E + 05 g	4.21E + 09	33.05
4	Total phosphorus	4.31E + 04 g	6.88E + 09	2.96
5	Phytoplankton	2.12E + 09 J	1.00E + 04	0.00
6	Pumped water-chemical potential	2.39E + 09 J	2.35E + 04	0.56
	Nonrenewables-purchased			
7	Liquid fuel	1.21E + 11 J	6.60E + 04	79.59
8	Construction structure	5.11E + 03 g	6.70E + 09	0.34
9	Construction services	9.34E + 01 $	1.60E + 12	1.49
10	Operation and maintenance	9.55E + 02 $	1.60E + 12	15.28

Table 9.8 Annual energy, material and dollar flows, and resulting emergy flows supporting 1 ha of the unsubsidized marsh

Notes		Quantity and unit (J, g, $)	Emergy per unit (sej unit^{-1})	Emergy (E + 14 sej)
	Renewables			
1	Sun	5.41E + 09 J	1.00E + 00	0.00
2	Rain-chemical potential	6.27E + 10 J	1.54E + 04	9.66
3	Total nitrogen	1.54E + 04 g	4.21E + 09	0.65
4	Total phosphorus	6.35E + 02 g	6.88E + 09	0.04
	Nonrenewables-purchased			
5	Construction-services	3.76E + 01 $	1.60E + 12	0.60

for the subsidized marsh and the unsubsidized marsh were the same in both marshes. Lake water pumped into the subsidized marsh largely increased the emergy of total nitrogen and total phosphorus compared to that entering the unsubsidized marsh. This implies that the emergy flux of free nonrenewable energy sources influencing self-organization contributed 26% of the total emergy flow to the subsidized marsh. Liquid fuel used to operate the hydraulic pumps and the physical structure of the pump system itself are the two nonrenewable purchased energy sources that were included in the subsidized marsh system and that contributed 68% of its total emergy flow.

Table 9.9 Summary of emergy evaluation

	Subsidized (E + 15 sej)	Unsubsidized marsh (E + 15 sej)
Emergy flows		
Renewable emergy	9.7	9.7
Nonrenewable emergy		
Free	36.6	9.7
Purchased	96.7	0.6
Total emergy flux	142.9	19.9
Emergy index		
Environmental Loading Ratio	13.8	0.1
Investment Ratio	2.1	0.1

Table 9.10 Summary of fish community structure

Parameter	Subsidized	Unsubsidized marsh	Significance difference
Fish density	$230\,m^{-2}$	$165\,m^{-2}$	$n = 30$, $df = 5$, $p = 0.01$
Fish biomass	$6.44\,kg\,m^{-2}$	$4.29\,kg\,m^{-2}$	$n = 30$, $df = 5$, $p = 0.03$

Vegetative community richness in the subsidized marsh was lower than that of the unsubsidized marsh. Fish biomass was also significantly different between marshes (Table 9.10).

A dynamic model was used to simulate situations in which the fuel use was increased (0% = unsubsidized marsh; 100% = subsidized marsh). Figure 9.11 shows the changes in biomass carrying capacity with different levels of fuel used. Material and energy balances, as shown in the emergy analysis, were significantly different between the subsidized and unsubsidized marshes.

Due to the nonrenewable energy sources from the lake (e.g. nutrients, phytoplankton) and the pump system itself, the subsidized system had much higher flows of available resources. This is also clearly evident in the higher densities and biomass of the avian and fish communities. On the other hand, the complexity of the subsidized marsh as measured by diversity and community structure was lower. High emergy subsidies may compromise the complexity of the system in favor of high productivity. Community structure and dynamics are likely a result of many processes including demographics, energy cycling, habitat disturbance, and the influence of other populations (Brown and Maurer, 1987; Maurer and Brown, 1988; Weins, 1989). In the case of the subsidized marsh, organization and community dynamics are also controlled by the availability of energy sources with high transformities. The importance of certain high emergy sources is their ability to facilitate the input of additional nonrenewable energies. The nutrient enrichment seemed to speed up self-organizational processes in the subsidized marsh

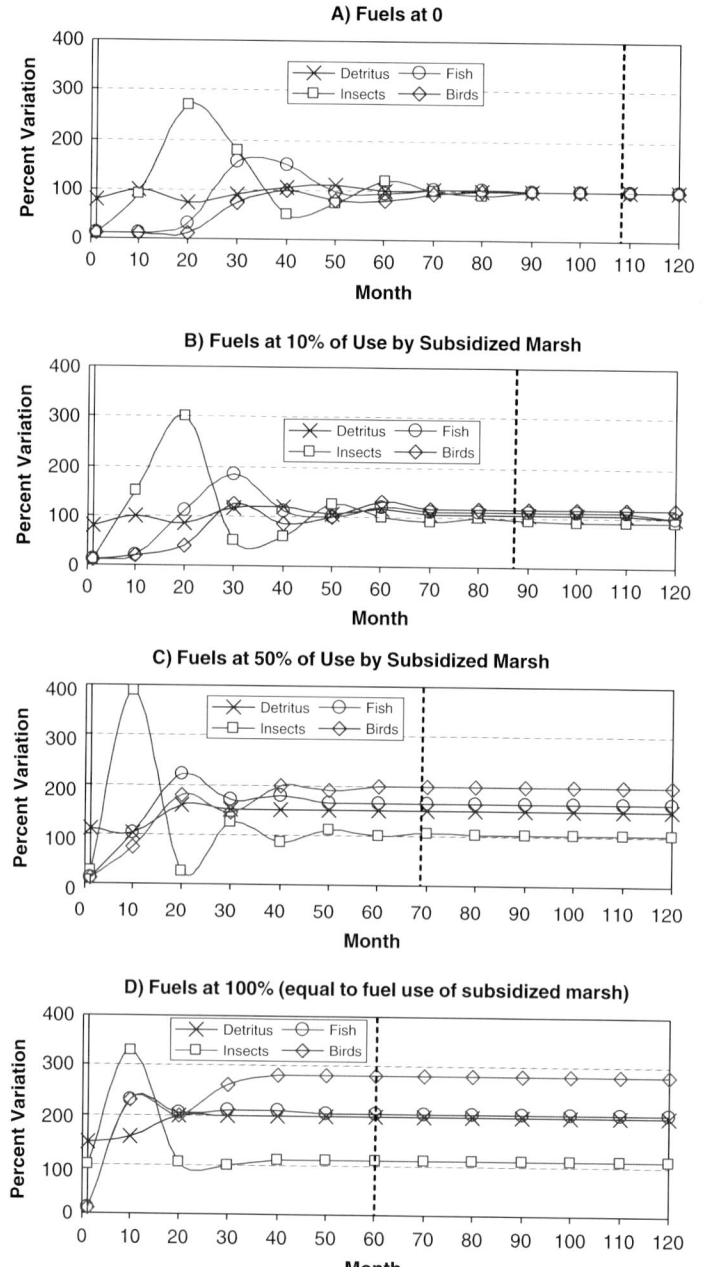

Figure 9.11 Percent variation of biomass over time for different rates of fuel use showing: (A) 0 fuel usage represented by unsubsidized marsh; (B) 10%; (C) 50%; (D) 100% fuel usage relative to actual usage by subsidized marsh. (100% = biomass carrying capacity of unsubsidized marsh). Vertical dashed line marks appropriate time when marsh biomass reaches a steady state.

Chapter 9: Ecosystem principles have applications

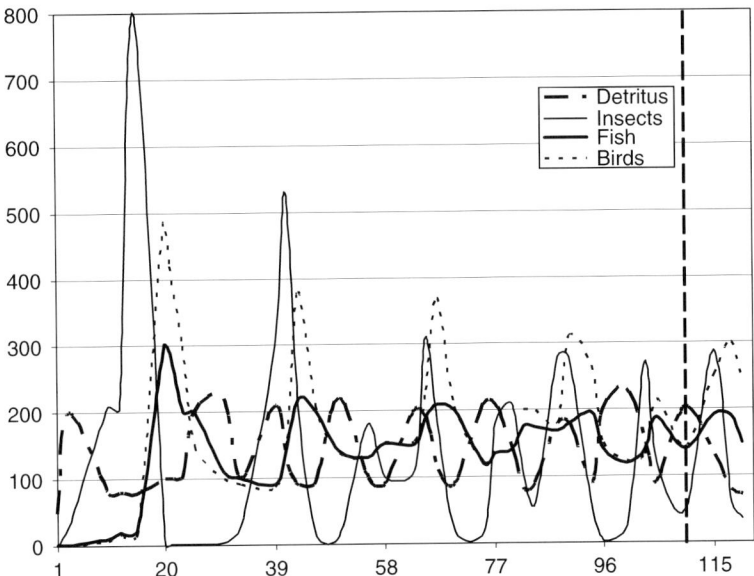

Figure 9.12 Percentage variation of biomass over time as fuel usage changes monthly as sine wave between 0 and 100% fuel usage relative to actual fuel usage of subsidize marsh.

increasing the rate of vegetative coverage of the marsh. Given the higher animal densities and biomass, the external subsidy may have also increased the rate that these components reached their respective carrying capacities.

This theory seemed to be validated by the computer model simulation; it suggests that carrying capacity varies with different levels of external subsidy.

The simulation in Figure 9.12 of changing subsidy reveals that if the subsidy is pulsed, biomass will also pulse; the simulation model is sustained by studies in literature about the relationship between nutrients increase and biomass increase (Kerekes, 1990; Price, 1992). Overall, the external subsidy increased the emergy flux in the subsidized marsh by increasing the input of nonrenewable energy sources. As a result, community parameters such as density and overall biomass also increased in the subsidized marsh, but at a cost of lowered richness, diversity, and evenness.

Conclusions

The emergy analysis reveals that the higher emergy flow in the subsidized marsh is caused by a higher input of nonrenewable energy sources. Community parameters (density and overall biomass) are also higher in the subsidized marsh, but at a cost of lowered richness, diversity, and evenness. Complexity of community structures is not influenced by the external subsidy. External subsidies increase the total emergy flow in an ecosystem and may increase the rate of successional processes in both the vegetative and wildlife communities.

9.7 THE ECO-EXERGY TO EMPOWER RATIO AND THE EFFICIENCY OF ECOSYSTEMS

Reference from which these applications of eco-exergy to empower as ecological indicator are extracted:

Bastianoni S, Marchettini N. 1997. Emergy/exergy ratio as a measure of the level of organization of systems. Ecol. Model. 99, 33–40.

Bastianoni S. 2002. Use of thermodynamic orientors to assess the efficiency of ecosystems: a case study in the lagoon of Venice. Sci. World J. 2, 255–260.

Bastianoni S. 2006. Emergy, empower and the eco-exergy to empower ratio: a reconciliation of H.T. Odum with Prigogine? Int. J. Ecodyn., in press.

The ratio of exergy to emergy flow (empower) has been used in order to assess the efficiency of an ecosystem in transforming available inputs in actual information and organization in eight aquatic ecosystems located in Argentina, Italy, and USA.

The case studies

Eight aquatic ecosystems are used to understand the importance as an indicator of the eco-exergy to empower (emergy flow) ratio. Two of these ecosystems (called "control pond" and "waste pond") are in North Carolina (USA) and are part of a group of similar systems built to purify sewage. Near the town of Morehead City, six artificial lakes were created: three control ponds fed with estuary water and "clean" water from the local sewage treatment plant, and three "waste" ponds fed with estuary water mixed with more "polluted" (i.e., richer in nutrients) effluent (Odum, 1989). Plant and animal species were introduced in and around the lakes to colonize the new areas and eventually produce new ecosystems by natural selection.

The third ecosystem is the Lagoon of Caprolace in Latium, Italy, at the edge of the Circeo National Park. The Lagoon of Caprolace is an ancient natural system fed by rainwater and farmland runoff rich in nitrogen, phosphorus, and potassium.

The fourth ecosystem is Lake Trasimeno. This is the largest lake in peninsular Italy (area 124 km^2), it is shallow (mean depth 4.7 m, maximum 6.3 m) and accumulation processes are favored. The water level of the lake shows strong fluctuations with respect to meteorological conditions; hydrological crises occur after several years with annual rainfall < 700 mm.

The fifth system is the Lagoon of Venice. With a surface area of about 550 km^2, it is the largest Italian lagoon. The sea and the lagoon are connected through three inlets. The average daily volume of water that enters the lagoon from the sea is about 400 million m^3, while 900 million m^3 of fresh water flow into the lagoon every year from the drainage basin.

The sixth system is an artificial one, located in the central part of the Lagoon of Venice, i.e., the Figheri basin. Fish farming basins consist of peripheral areas of lagoon surrounded by banks in which local species of fish and crustaceans are raised. Salt water from the sea and freshwater from canals and rivers are regulated by locks and drains. The fishes of highest demand raised in basins are *Dicentrarchus labrax* (bass) and *Sparus auratus*. Various types of mullet are also raised, as well as eels and mollusks.

Two ecosystems are located within the Esteros del Iberá (northeastern Argentina), one of the most pristine and largest wetlands of South America (13,000 km^2). This subtropical wetland is located between 27°36′–28°57′S and 58°00′–57°30′W. The macrosystem consists of a mosaic of marshes, swamps, and open water bodies. It is located between three large rivers, the Rio Paraná alto, the Rio Paraná medio, and the Rio Uruguay, with a single outlet to the Rio Corrientes that feeds into the Parana Medio (Loiselle et al., 2001; Bracchini et al., 2005).

The Galarza Lagoon is a mesotrophic, round-shaped lake with an area of 14 km^2 and averages 2 m in depth. The lagoon is fed by a small stream that originates in the large marsh area (200 km^2) directly above the lagoon and feeds into another small stream that leads to another large shallow lagoon. The water then flows out of this second lagoon into another large marsh area.

Laguna Iberá (area 58 km^2, mean depth 3.2 m), has a more irregular morphology and an eutrophic status. This lake is divided into two basins by a narrow passage that acts as a barrier reducing the interchange of wave energy and water masses. A small river (Río Miriñay) feeds the southern basin.

Emergy, exergy, and their joint use

Why use emergy flow (empower) and eco-exergy together on the same systems? Emergy and eco-exergy are complementary concepts, the former based on the history of the system (Odum, 1988, 1996) and the latter examining the actual state (Jørgensen, 1992a, 2006). When systems follows a process of selection and organization, we can use the ratio of eco-exergy to emergy flow in order to assess the efficiency of an ecosystem in transforming available inputs in actual information and organization. The higher the ratio, the greater is the efficiency of the ecosystem in transforming available inputs (as emergy flow) into structure and ecosystem organization (as eco-exergy). Its units are $J\,yr\,sej^{-1}$. Since dimensions are those of time, it cannot be regarded as a real efficiency (which is dimensionless), but more as an index of efficiency.

According to Svirezhev (2000), this fact is normal, since this concept resembles that of a relaxation time, i.e., the time necessary to recover from disturbances, so that the exergy to empower ratio should be related with concepts like resilience and resistance of an ecosystem.

The *eco-exergy/empower* ratio indicates the quantity of external input necessary to maintain a structure far from equilibrium: if the eco-exergy/empower ratio tends to increase (apart from oscillations due to normal biological cycles), it means that natural selection is making the system follow a thermodynamic path that will bring the system to a higher organizational level.

As an efficiency indicator the eco-exergy to empower ratio enlarges the viewpoint of a pure exergetic approach as described in Fath et al. (2004), where the exergy degraded and the eco-exergy stored for various ecosystems are compared: using emergy there is a recognition of the fact that solar radiation is the driving force of all the energy (and exergy) flows on the biosphere, important when important "indirect" inputs (of solar energy) are also present in a process.

To compare ecosystems different in size empower and eco-exergy densities were used. Table 9.11 shows empower and eco-exergy density values and the ratio of eco-exergy to empower. The emergy flow to Iberá Lagoon has been underestimated due to lack of data about the release of nutrients from the surrounding rice farms. In a sense this explains the highest value for eco-exergy to empower ratio, while the ecosystem does not seems to be in ideal conditions (Bastianoni et al., 2006). Nonetheless, the important fact is that all the natural systems that are better protected from human influence show very close figures. It seems that there is a tendency common to different ecosystems in different areas and of different characteristics to evolve toward similar thermodynamic efficiencies.

Figheri basin is an artificial ecosystem, but has many characteristics typical of natural systems. This depends partly on the long tradition of fish farming basins in the Lagoon of Venice, which has "selected" the best management strategies (Bastianoni, 2002). The human contribution at Figheri basin manifests as a higher emergy density (of the same order of magnitude as that of artificial systems) than in natural systems. However, there is a striking difference in eco-exergy density, with values of a higher order of magnitude than in any of the other systems used for comparison: Man and Nature are acting in synergy to enhance the performance of the ecosystem. The fact that Figheri can be regarded as a rather stable ecosystem (i.e., quite regular in its behavior) makes this result even more interesting and significant.

It was observed that the "natural" lagoon of Caprolace had a higher eco-exergy/emergy ratio than the control and waste ponds, due to a higher eco-exergy density and a lower emergy density (Bastianoni and Marchettini, 1997). These observations were confirmed by the study of Lake Trasimeno (Ludovisi and Poletti, 2003).

Also, the results on the entire lagoon of Venice confirm the general trend, showing figures in the range of Trasimeno and Caprolace, albeit the differences in the structure of the ecosystems and the huge inputs from the watershed.

Table 9.11 Empower density, eco-exergy density, and eco-exergy to empower ratio for eight different ecosystems

	Control pond	Waste pond	Caprolace Lagoon	Trasimeno Lake	Venice Lagoon	Figheri Basin	Iberá Lagoon	Galarza Lagoon
Empower density (sej year^{-1}l^{-1})	$20.1 \cdot 10^8$	$31.6 \cdot 10^8$	$0.9 \cdot 10^8$	$0.3 \cdot 10^8$	$1.4 \cdot 10^9$	$12.2 \cdot 10^8$	$1.0 \cdot 10^8$	$1.1 \cdot 10^8$
Eco-exergy density (Jl^{-1})	$1.6 \cdot 10^4$	$0.6 \cdot 10^4$	$4.1 \cdot 10^4$	$1.0 \cdot 10^4$	$5.5 \cdot 10^4$	$71.2 \cdot 10^4$	$7.3 \cdot 10^4$	$5.5 \cdot 10^4$
Eco-exergy/empower (10^{-5} J year sej^{-1})	0.8	0.2	44.3	30.6	39.1	58.5	73	50.0

Conclusions
In general, in the more "natural" systems, where selection has acted relatively undisturbed for a long time, the ratio of eco-exergy to empower is higher, and decreases with the introduction of artificial inputs or stress factors that make the emergy flow higher and/or the eco-exergy content of the ecosystem lower.

9.8 APPLICATION OF ECO-EXERGY AND ASCENDENCY AS ECOLOGICAL INDICATOR TO THE MONDEGO ESTUARY (PORTUGAL)

References from which these applications of eco-exergy and ascendency as ecological indicators are extracted:

Jørgensen SE, Marques J, Nielsen SN. 2002. Structural changes in an estuary, described by models and using exergy as orientor. Ecol. Model. 158, 233–240.

Marques JC, Pardal MA, Nielsen SN, Jørgensen SE. 1997. Analysis of the properties of exergy and biodiversity along an estuarine gradient of eutrophication. Ecol. Model. 102, 155–167.

Patricio J, Ulanowicz R, Pardal MA, Marques JC. 2004. Ascendency as an ecological indicator: a case study of estuarine pulse eutrophication. Est. Coast. Shelf Sci. 60, 23–35.

The Mondego Estuary has been used to benefit of the integration of the information derived from different ecological indicators: eco-exergy, specific eco-exergy (Chapter 2 and 7), and ascendency (Chapter 4).

Mondego Estuary: Description
The Mondego River drains a hydrological basin of approximately $6670\,km^2$ at the western coast of Portugal. Urban wastewater is still discharged into the Mondego without treatment, and the estuary supports industrial activities, desalination ponds, and aquaculture. Additionally, the lower Mondego River valley has about 15,000 ha of farming fields (mainly rice paddies), with a significant loss of nutrients to the estuary (Marques, 1989).

The Mondego Estuary is located in a warm/temperate region with a basic mediterranean temperate climate. It consists of two arms, north and south (Figure 9.13) separated by an island. The two arms splits in the estuarine upstream area about 7 km from the sea, and join again near the mouth. These two arms of the estuary present very different hydrographic characteristics. The north arm is deeper (5–10 m during high tide, tidal range about 2–3 m), while the south arm (2–4 m deep, during high tide) is almost filled with silt in the upstream areas, directing most of the freshwater through the north arm. The water circulation in the south arm is controlled by tidal circulation and the relatively small fresh water input from the tributary, the Pranto River, which is controlled by a sluice located 3 km from the confluence with the south arm of the estuary. Due to differences in depth, the tidal excursion is longer in the north arm, causing daily changes in salinity to be much stronger, whereas daily temperature changes are highest in the south arm (Marques, 1989).

Seasonal intertidal macroalgae blooms (mainly of *Enteromorpha* spp.) have been reported in the south arm of the estuary for several years (Marques et al., 1993a, b) and *Zostera noltii* beds, which represent the richest habitat with regard to productivity and biodiversity, are

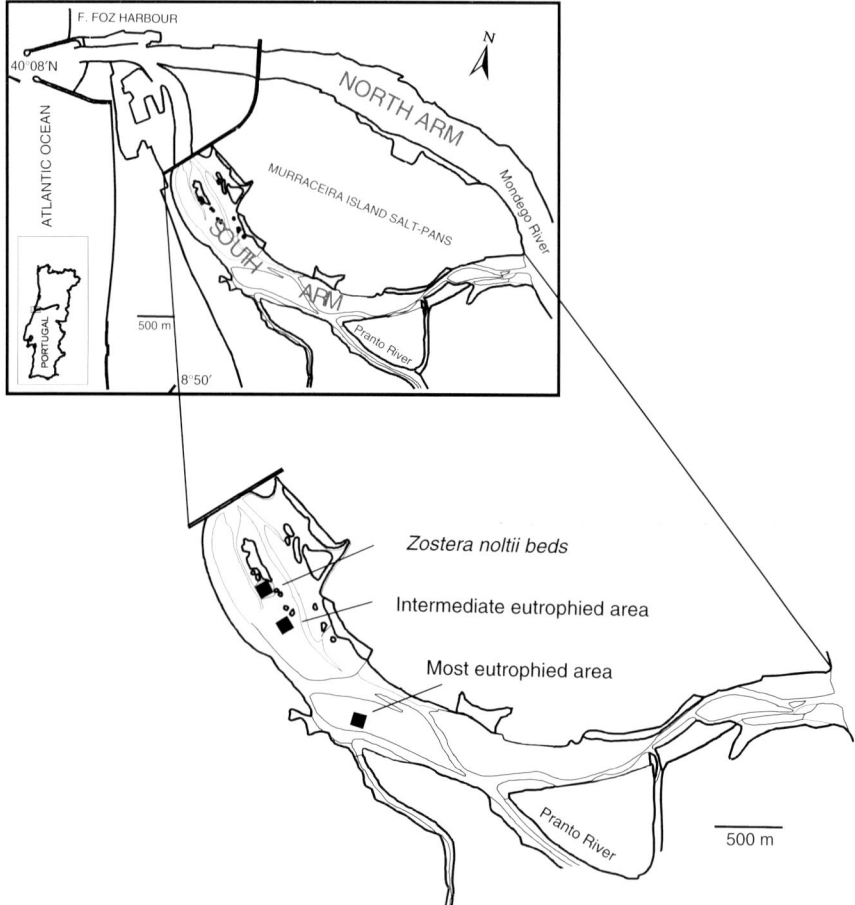

Figure 9.13 Map of Mondego River showing the field areas and the division of the river into the north and south arms.

being drastically reduced in the south arm of the Mondego estuary, presumably out competed by *Enteromorpha* (Rafaelli et al., 1991). The physical data are listed in Table 9.12.

Nutrient loading into the south arm of the estuary was estimated, assuming that the major discharge is through the Pranto River, from the Armazens channel (there is no freshwater discharge but, in each cycle, the tidal wave washes out the channel, where several industries discharge waste waters), and from the downstream communication of the south arm. The only way out of the system is the downstream communication (Figure 9.13). The nutrient inputs from the Pranto River and Armazens channel, and the exchanges (input vs. output) in the downstream communication of the south arm have been monitored from May 1993 to June 1994.

Table 9.12 Physical parameters of Mondego River

Physical parameters	Mondego River
Length and width of estuary (km^2)	10, 0.3
Area (km^2)	3.4
Volume (km^3)	0.0075
Mean depth (m)	1 and 2
Tidal range	0.35–3.3
Drainage area (km^2) (excluding the estuary)	6670
Discharge (m^3 year^{-1})	8.5×10^9
Mean residence time (days) (based on fresh water discharge)	North arm 2
	South arm 9
Temperature range (min., max.)	7–32
Mean Secchi depth in m (May–Sept.)	0.5–1.0
Annual insolation of PAR (400–700 nm) (mol fot m^{-2} year^{-1})	3200–32,000

The annual nitrogen loading to the south arm of the Mondego Estuary was roughly estimated to 134 t (126 t of nitrate and 8 t of nitrite), of which 14 t is still in the system (18 t of nitrate were imported and 4 t of nitrite were exported) and 120 t were transported to the sea. For phosphorus the loading was estimated to be 14 t (1 t was imported to the system, while 15 t were exported to the sea, which means that 14 t were net released from the south arm of the estuary). In Table 9.13 are listed the chemical aspects of the area.

Maximization of eco-exergy to predict the behavior of the system
It is often of interest to determine among several possibilities which structure of an ecosystem will prevail under given environmental circumstances. Here the thermodynamic variable, eco-exergy, was used as an orientor to describe adaptation and changes in the species composition. In the Mondego Estuary two very different types of communities have been observed:

(1) An *Enteromorpha* dominant community with the presence of *Cyathura carinata*, mollusca, and crustacea. The algae community shows often a crash at early summer due to oxygen depletion. This community is found where the salinity is not too low and the nutrient concentration is high.
(2) A *Z. noltii* dominant community with the presence of oligochaeta, polychaeta, mollusca, and crustacea. This ecosystem is found where the nutrient concentration is lower. Mollusca is more abundant in ecosystem (2) than in ecosystem (1), while for crustacea the reverse is valid.

Table 9.13 Chemical parameters of Mondego River

Chemical parameters	Mondego River	Inner part	Outer part
External N loading (tN year^{-1})	126		
External P loading (tP year^{-1})	1		
Salinity	Winter	10	10
	Summer	25	30
DIN (μg l^{-1})	Winter	300	200
	Summer	50	40
DIP (μg l^{-1})	Winter	30	25
	Summer	25	30
Tot-N (mg l^{-1})	Winter	0	0
	Summer	0.25	0.2
Tot-P (mg l^{-1})	Winter	0.30	0.25
	Summer	0.25	0.30
Sediment C (mg l^{-1})		350 g C m^{-2}	500 g C m^{-2}

From an ecological management point of view the *Zostera* dominated community is preferred because the oxygen concentration is higher, the water is clearer, and no crash due to anaerobic conditions takes place.

Starting from the hypothesis that the ecosystem structure having the highest eco-exergy among the possible ones would prevail, two models (one for an *Enteromorpha* dominant community and another one for a *Z. noltii* dominant community) were developed to compare eco-exergy for several conditions, using STELLA. The growth was described as a function of internal concentrations of nutrients, temperature, light, and salinity (Duarte, 1995).

If the hypothesis is correct, comparing the models for the two types of ecosystems, the highest exergy under eutrophied and medium to high-salinity conditions should be found for ecosystem (1), while the highest exergy should be found for ecosystem (2) under low-nutrient and low-salinity conditions.

The models show that if the freshwater with high concentration of nutrients (particularly nitrogen) is discharged during the last part of the year *Enteromorpha* will be dominant. The eco-exergy calculations show that the exergy is approximately the same for the two models, which may be interpreted as the initial value, may be crucial for the final results.

The results of the five simulations suggest that the ecological management of the freshwater discharge is a key factor for the prevailing of the two communities (*Enteromorpha, Z. noltii*). From a management viewpoint at least two possibilities can be considered: artificial control of the freshwater discharges through the use of sluices, increasing the discharge during the first part of the year; or reduction of the nutrient input from freshwater (and if possible also from tide water). The joint use of these two alternatives should give the *Zostera* dominated community better conditions.

Eco-exergy, specific eco-exergy, and diversity

The spatial and temporal variation of eco-exergy, specific eco-exergy, species richness, and heterogeneity were analyzed to examine in what extent these ecological indicators would capture changes in benthic communities along the gradient of eutrophication.

The benthic communities in the Mondego Estuary (Portuguese western coast) were monitored during a yearly cycle. Samples of macrophytes, macroalgae, and associated macrofauna were taken fortnightly at three different sites, during low water, along an estuarine gradient of eutrophication in the south arm of the estuary, from the non-eutrophicated zone, where the *Z. noltii* community is present, up to the heavily eutrophicated zone, in the inner areas of the estuary, where *Enteromorpha* spp. blooms have been observed. An overview of the major taxonomic groups contributing to the exergy in this system is provided in Table 9.14.

With regard to eco-exergy, values were consistently higher in the *Z. noltii* community than in the eutrophicated areas. Also, eco-exergy values were higher in the most heavily eutrophicated area when compared with the intermediate eutrophicated area, especially during spring and early summer. This was related to the intensity of the *Enteromorpha* bloom, which gave rise to much higher values for total biomass in the most eutrophicated area.

Table 9.14 Major contributors to the exergy in the Mondego Estuary benthic communities along the gradient of eutrophication

Contributors	Non-eutrophicated area	Intermediate eutrophicated area	Eutrophicated area before the algae crash	Eutrophicated area after the algae crash
Enteromorpha + *Ulva*	2.099	28.211	264.642	1.273
Other macroalgae	16.141	2.138	6.152	0.165
Z. noltii leafs	128.368	0.000	0.000	0.000
Z. noltii roots	87.975	0.000	0.000	0.000
Z. noltii-total	216.343	0.000	0.000	0.000
Anthozoa	0.003	0.000	0.000	0.000
Sipunculia	0.001	0.001	0.001	0.002
Nemertinea	0.005	0.003	0.005	0.001
Oligochaeta	0.128	0.031	0.010	0.002
Polychaeta	1.254	0.709	0.569	0.846
Mollusca	63.950	14.192	31.195	13.240
Crustacea	1.372	1.088	14.945	3.419
Insecta	0.007	0.006	0.009	0.001
Echinodermata	0.000	0.000	0.000	0.000
Pisces	0.000	0.006	0.034	0.000

Note: For the non-eutrophicated and intermediate eutrophicated areas, the average annual biomass ($g\,m^{-2}$) of each is given. For the eutrophicated area, the average biomass ($g\,m^{-2}$) of each group before and after the algae crash is given.

Specific eco-exergy was found to be consistently higher in the *Z. noltii* community than in the eutrophicated areas until late spring when the picture changed completely and values became higher in the eutrophicated areas. This was due to a macroalgae crash in the eutrophicated areas, which determined not only a drastic reduction of the total biomass but also a change from a primary production-based situation toward a detritus-based food web. Therefore, since total biomass values after late spring consisted basically of animals (consumers), primarily deposit feeders, and detritus feeders (e.g., annelid worms and crustaceans), it is clear that the abrupt increase of specific eco-exergy in the eutrophicated areas after the algae crash do not reflect any augmentation of the structural complexity of the community, but simply the different quality of the biomass involved in the calculations.

Regarding the *Zostera* community (data after 6 July), accounting for the primary producers and the consumers, specific exergy is lower than in the eutrophicated areas. But if we account only for the consumers, it is higher, following the same pattern from before the algae crash. Hence, specific eco-exergy may shift very drastically as a function of yearly dynamics (like in communities dominated by r-strategists), providing a spatial and temporal picture that may not be related with the long-term evolution and integrity of the system. With regard to biodiversity, the variation of species richness and of heterogeneity (species richness + evenness) along the gradient of eutrophication provided quite different pictures. Through time species richness was consistently higher in *Zostera* community, decreasing along the gradient of eutrophication. On the contrary, heterogeneity was always higher in the eutrophicated areas, except for the decrease observed in the most heavily eutrophicated area after an algae crash.

The observed spatial variation of heterogeneity is due to the fact that the Shannon–Wiener's index integrates two components: the number of species (species richness) and their relative abundance (evenness). Therefore, although species richness decreased as a function of increasing eutrophication, as we expected, the dominance of a few species (e.g., *Hydrobia ulvae*, a detritus feeder and epiphytic grazer gastropod, and *Cerastoderma edule*, a filter feeder bivalve) in the *Zostera* community, probably due to the abundance of nutritional resources, decreased species evenness, and consequently heterogeneity values. In this case, lower values of heterogeneity must be interpreted as expressing higher biological activity, and not as a result of environmental stress (Legendre and Legendre, 1984).

Taking into account the yearly data series for each site along the eutrophication gradient (non-eutrophicated, intermediate eutrophicated, eutrophicated), eco-exergy, and specific eco-exergy were significantly correlated ($p \leq 0.05$) providing a similar picture from the system. Values were consistently higher and more stable in the non-eutrophicated area. The comparison of yearly data series (*t*-test, $p \leq 0.05$) showed that using eco-exergy values it was possible to distinguish between the three situations considered, even though differences between the intermediate and eutrophicated areas were not significant, which suggests that eco-exergy, an extensive function, might be more sensitive to detect subtle differences.

Species richness and eco-exergy were significantly correlated ($p \leq 0.05$), following a similar pattern, both decreasing from non-eutrophicated to eutrophicated areas

(Figure 9.14B). On the contrary, heterogeneity and eco-exergy appeared negatively correlated (although not significantly), providing a totally distinct picture of the benthic communities along the eutrophication gradient (Figure 9.14A). This obviously resulted from the properties of the heterogeneity measure, as explained above.

Similar results were obtained comparing the patterns of variation of species richness, heterogeneity, and specific eco-exergy. Species richness and specific eco-exergy appeared clearly positively correlated ($p \leq 0.05$) (Figure 9.14B), while the patterns of variation of heterogeneity and specific eco-exergy showed to be distinct (Figure 9.14A). Moreover, from the comparison of yearly data series (t-test, $p \leq 0.05$), heterogeneity values were not significantly different in the intermediately eutrophicated and eutrophicated areas, and therefore did not permit to discriminate relatively subtle differences.

The hypothesis that eco-exergy and biodiversity would follow the same trends in space and time was validated with regard to species richness, but not for heterogeneity. Actually, eco-exergy, specific eco-exergy, and species richness responded as hypothesized, decreasing from non-eutrophicated to eutrophicated areas, but heterogeneity responded in the opposite way, showing the lowest values in the non-eutrophicated area.

Their range of variation (eco-exergy and specific eco-exergy) through time was smaller in the non-eutrophic area, expressing a more stable situation, while the magnitude of the variations was stronger in the other two areas, but especially in the intermediate eutrophic area (Marques et al., 2003). On the other hand, both eco-exergy and species richness were able to grade situations presenting relatively subtle differences, but specific eco-exergy and heterogeneity appeared to be less sensitive. Although biodiversity may be considered as an important property of ecosystem structure, the relative subjectivity of its measurements and their interpretation constitutes an obvious problem.

The spatial variation of species richness was significant; biodiversity may be seen as the full range of biological diversity from intraspecific genetic variation to the species richness, connectivity, and spatial arrangement of entire ecosystems at a landscape-level scale (Solbrig, 1991). If we accept this biodiversity concept, then eco-exergy, as system-oriented characteristic and as ecological indicator of ecosystem integrity, may encompass biodiversity.

Moreover, eco-exergy implies the existence of the transport information through scales, from the genetic to the ecosystem level, accounting not only for the biological diversity, but also for the evolutionary complexity of organisms, and ecosystem-emergent properties arising from self-organization.

Ascendency calculations

Eutrophication can be described in terms of network attributes as any increase in system ascendency (due to a nutrient enrichment) that causes a rise in TST that more than compensates for a concomitant fall in the mutual information (Ulanowicz, 1986). This particular combination of changes in variables allows one to distinguish between instances of simple enrichment and cases of undesirable eutrophication. Three sampling stations representative of the non-eutrophic area, of the intermediate eutrophic area, and of the

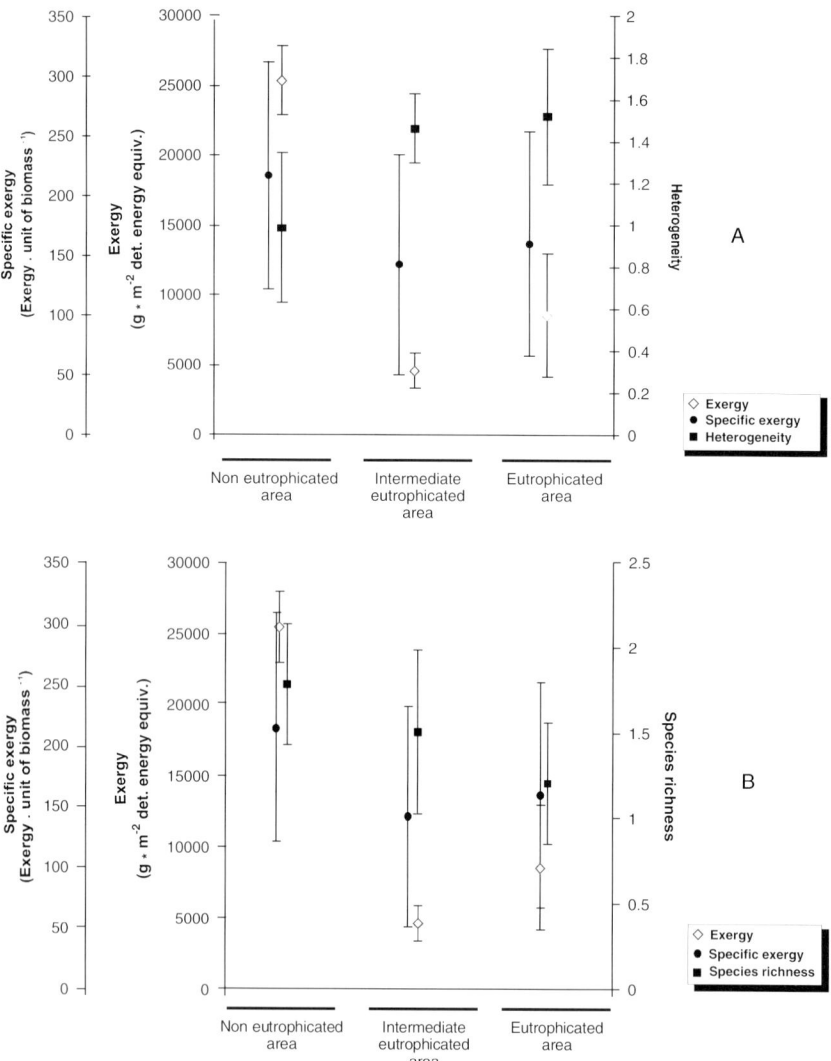

Figure 9.14 Variation of exergy and specific exergy in comparison with heterogeneity (A) and species richness (B) along the gradient of eutrophication gradient. For each situation, respectively, non-eutrophicated (ZC), intermediate eutrophicated (INT), and eutrophicated (EUT), we indicate the average values and the standard deviation, taking into account the entire yearly data set. The spatial variation of exergy and specific exergy was significantly correlated ($r = 0.59$; $p \leq 0.05$). The spatial variation of heterogeneity was not significantly correlated with exergy or specific exergy ($r = -0.48$ and $r = 0.38$, respectively; $p \leq 0.05$).

strongly eutrophic area were chosen. Estuarine food webs were constructed at the three sites and these quantified food webs were examined using network analysis. Taken together with Table 9.15 these provide the measures that were used to characterize the trophic status of the three estuarine ecosystems. Although the three habitats are clearly distinct in physical appearance, network analyses revealed both differences and similarities among their trophic structures that had not been apparent at first glance.

It was possible to observe (Table 9.15) that the *Zostera* dominated community had the highest TST, followed (unexpectedly) by the strongly eutrophic system and finally by the intermediate eutrophic area. The development capacity was highest in the *Zostera* beds and lowest in the intermediately eutrophic area.

The index differed significantly among the three areas. Due to the logarithmic nature of this index, small differences can represent appreciable disparities in structure. The average mutual information (AMI) was slightly higher in the non-eutrophic area, followed closely by the eutrophic area and was lowest in the intermediate eutrophic area. Concerning ascendency, it increased in order from the intermediate eutrophic to the heavily eutrophic zone to the *Zostera* meadows, while redundancy increases in the opposite direction.

The long-term study in the Mondego Estuary indicated that years of low precipitation tended to be associated with reductions in turnover rates and increases in water column stability, temperature, salinity, and light penetration (Martins et al., 2001). These changes in habitat conditions encouraged blooms of macroalgae that gradually replaced the resident macrophytes (Marques et al., 1997; Martins et al., 2001). In the

Table 9.15 Network analysis ecosystems indices for the three areas

Information indices	Non-eutrophic area	Intermediate eutrophic area	Strongly eutrophic area
Total system throughput (TST) (g AFDW m^{-2} y^{-1})	10,852	1155	2612
Development capacity (g AFDW m^{-2} y^{-1}; bits)	39,126	5695	10,831
Ascendency (%)	42.3	30.4	36.7
Overhead on imports (%)	12.3	8.2	6.2
Overhead on exports (%)	1.3	1.5	2.5
Dissipative overhead (%)	17.7	22.1	19.9
Redundancy (%)	26.4	37.8	34.6
Average mutual information (AMI) (bits)	1.52	1.50	1.52
Φ/TST	2.08	3.43	2.62
Connectance indices			
Overall connectance	1.67	2.43	2.1
Inter-compartmental connectance	2.41	3.57	2.63
Finn cycling index (FCI)	0.0575	0.2045	0.1946
Total number of cycles	74,517	15,009	9164

intermediate and strongly eutrophic areas, primary production is largely the result of these macroalgal blooms. Production appears as a strong pulse during this specific time, but remains at very low levels during the rest of the year. This limited temporal interval of primary production results in a significantly lower figure for the cumulative annual primary production and TST in these areas as compared with the corresponding measures in the *Zostera* beds. Comparing the AMI values of the flow structure for the three areas, it is possible to discern a very small decrease in the measure among the three zones, suggesting that, as regards trophic structure, these areas are indeed different.

The three zones appear nevertheless much more distinct by eye than is illustrated by the AMI values. In the light of these results, the network definition of eutrophication appears to be inappropriate for the Mondego estuarine ecosystem. It would be more accurate to describe the enrichment processes occurring in this ecosystem as "pulse eutrophication." This process could be characterized as a disturbance to system ascendency in the form of an intermittent supply of excess nutrients that, when coupled with a combination of physical factors (e.g., salinity, precipitation, temperature, etc.), causes both a decrease in system activity and a drop in the mutual information of the flow structure. Even though a significant rise in the TST occurs during the period of the algal bloom and at that time there is a strong increase of the system ascendency, the annual picture nevertheless suggests that the other components of the intermediate and strongly eutrophic communities were unable to accommodate the pulse in production. The overall result was a decrease in the annual value of the TST and, as a consequence, of the annual ascendency as well.

Regarding the results of the trophic analysis, the *Zostera* community has one more trophic level than those counted in the strongly eutrophic chain, implying that this community possesses a more complex web with additional top consumers. At the same time, the *Zostera* community exhibits lower transfer efficiency at the first trophic level, probably because the production of *Z. noltii* meadows usually cannot be eaten directly, but needs to be decomposed first (Lillebø et al., 1999). Concerning the analysis of cycled materials, the overall percentage of cycled matter increases as the degree of eutrophication rises. This is indicated by the Finn Cycling Index (FCI) that reveals the proportion of TST that is devoted to the recycling of carbon (Finn, 1976).

Eugene Odum (1969) suggested that mature ecosystems recycle a greater percentage of their constituent material and energy than do pioneer or disturbed communities. Hence, according to Odum, the progressive increase in the FCI would suggest the maturation of the ecosystem.

It has been observed, however, that disturbed systems also often exhibit greater degrees of recycling. The speculation is that such an increase in cycling in disturbed systems is the homeostatic response that maintains in circulation resources, which before the perturbation had been stored as biomass in the higher organisms (Ulanowicz and Wulff, 1991). This latter scenario seems consistent with the present results.

When the whole-system properties of the three areas were compared, the measures associated with the system considered to lie between the two extremes in nutrient loading, did not plot intermediate to the other two.

From this viewpoint, the intermediate eutrophic area appears to be the most disturbed of the three areas, since it has the lowest ascendency, AMI, TST, and development capacity values and the highest figures for redundancy and FCI.

9.9 CONCLUSIONS

Chapters 2–8 have shown several ecosystem properties that need indicators to be measured. This chapter has presented the up-to-date level of knowledge about these indicators. All these indicators are used to compare characteristics and performance of different ecosystems, or of an ecosystem in time, more than to give absolute measures. These indicators cover a wide range of important properties of ecosystems, more than those shown in this chapter, for the evaluation of ecosystem health. Here just few examples have been picked among the many to let the reader have at least an idea of what can be done with these indicators, but nowadays a whole literature of papers and books can provide further examples and types of applications. The use of these indicators spans from agricultural to industrial systems, and from ecosystem management to ecological economics.

Especially for management purposes, it is necessary to use these indicators with others, more focused on particular aspects than on the global ecosystem. Nonetheless, the approaches used in this chapter are fundamental to describe ecosystems as "systems" and not just as the sum of singular components and therefore should always be used. They provide information that is sometimes complementary and sometimes overlaps: in this direction more research is needed to clarify the level of overlapping and to fully explore the essence of the *indicandum* provided by every indicator, but their relevance is undoubted.

10

Conclusions and final remarks

10.1 ARE BASIC ECOLOGICAL PROPERTIES NEEDED TO EXPLAIN OUR OBSERVATIONS?

Take a walk on a pleasant May day in a temperate deciduous forest, visit the Serengeti National Park in Tanzania when the wildebeests are emigrating North, paddle a canoe through a North-American wetland, or hike the alpine tundra of Austria, whatever your preference, you will be impressed by the diversity and beauty that nature offers to you. We know that the diversity of nature is enormous. We can find on the order of 10^7 different species on earth and they can be combined in ecological networks in an almost infinite number of ways. We also have a fairly good image of the evolution from 3.8 billion years ago when the first primitive cells emerged to *Homo sapiens* with advanced technology of today: airplanes, computers, Internet, and so on. We could, therefore, turn the question around and ask: which properties do ecosystems have that explain the diversity, adaptability, and beauty of nature and evolution. How can we explain that the interactions between matter, energy, and information lead to the abiotic and biotic web of life on earth, as we can observe? We definitely do not need an intelligent designer to come up with a clear and fully acceptable explanation. This book presents an overview of which systems-based, thermodynamic properties are known to underpin this natural growth and development.

10.2 PREVIOUS ATTEMPTS TO PRESENT AN ECOSYSTEM THEORY

Previously, various attempts have been made to present an ecosystem theory that could be applied to quantitatively explain ecosystem processes and their responses to disturbance and changing impacts. While we cannot cover all the attempts here, we focus on those based on systems perspectives and thermodynamics.

One of the early pioneers in Systems Ecology, Ken Watt, proposed his theory in the important work *Ecology and resource management* in 1968, which opened the way for greater systems thinking in ecology. In the 1970s B.C. Patten edited four volumes with the title *Systems Analysis and Simulation in Ecology*. These four volumes gave the state-of-the-art of systems ecology at that time and was a useful reference in systems ecology. All the volumes formed in a sense an early attempt to develop an ecosystem theory. During the 1980s a number of scientists contributed to an ecosystem theory: H.T. Odum, R. Ulanowicz, B.C. Patten, R. Margalef, and S.E. Jørgensen to mention a few. For example, H.T. Odum's book from this period, *Systems Ecology: An Introduction*, is probably one of the best attempts to make a comprehensive ecosystem theory. The discussion of hierarchy theory, allometry, and scaling problems in the 1980s should also be mentioned. T.F.H. Allen and T.B. Starr in their book *Hierarchy, Perspectives for Ecological Complexity* (1982) and R. O'Neill, D.L. De Angelis, J.B. Waide, and T.F.H. Allen in the book *A Hierarchical Concept of Ecosystems* (1986) presented hierarchy theory and made it an almost fully accepted part of ecosystem theory already 20 years ago. Peters (1983) publication of many allometric principles should also be mentioned in this context. Polunin (1986) edited a book titled *Ecosystem Theory and Application*, which was an early attempt to apply ecological theory to address some of the global environmental issues of the day. The 1980s and early 1990s saw a lot of interest in the Stream Ecosystem Theory (e.g. Cummins et al., 1984; Minshall et al., 1985; Minshall, 1988; Wiley et al., 1990) which focused on streams as open systems, controlled primarily by their allochthonous riparian input.

In 1992, S.E. Jørgensen gave an overview of these contributions in his book *Integration of Ecosystem Theories: A Pattern*. The various contributions to an ecosystem theory were very different, but a closer study of the proposed theories revealed that they actually were different angles and covering different aspects, but largely were consistent, complementary, and formed as the title of the book indicates a pattern. H.T. Odum's theoretical contributions to systems ecology were summarized by C.A.S. Hall (1995) in the book *Maximum Power—the Ideas and Applications of H.T. Odum*. R. Margalef's (1997) book *Our Biosphere* summarized his contributions to systems ecology. It was based on a well-balanced cocktail of thermodynamics and ecology. *Macroecology* by J.H. Brown (1995) presents from this period a quantitative ecological attempt to explain particularly biogeophaphical observations and Reynolds (1997) expands his ideas on theory describing aquatic habitats.

Patten and Jørgensen (1995) edited the book *Complex Ecology—The Part-Whole Relation in Ecosystems* in which 31 systems ecologists contributed, presenting a wide overview of many different approaches and viewpoints, from quantum mechanic considerations (see herein, Chapter 3), to modeling theory, to network theory (Chapter 5), to feedback mechanisms (Chapters 4 and 7), to cybernetics (Chapters 4 and 7), and thermo-dynamics (Chapters 2 and 6). Furthermore, Jørgensen, Patten, and Straškraba have published a series of papers in the journal *Ecological Modelling* under the title "Ecosystems Emerging". The paper subtitles to date are: (1) Introduction, (2) Conservation, (3) Dissipation, (4) Openness, and (5) Growth. The remaining papers include: (6) Constraints, (7) Differentiation, (8) Adaptation, (9) Coherence, and (10) Applications. Similar to this book, these papers are rooted in thermodynamic laws and basic properties of ecosystems.

Coming from a more biogeochemical perspective, Ågren and Bosatta (1996) published *Theoretical Ecosystem Ecology—Understanding Element Cycles*, which put emphasis on the importance of carbon and nitrogen cycling in ecosystems and is a commonly used textbook in this field.

In 2001, Jørgensen and Marques published "Thermodynamics and systems theory, case studies from hydrobiology" (Hydrobiologia, 445: 1–10). The paper claimed that we could develop ecosystem laws and apply them similarly to the application of physical laws in physics. Similarly, the December 2002 issue of the journal *Ecological Modelling* (158: 3) was based on nine papers by different authors invited to show that we could explain theoretically many papers published in ecology, which themselves were presented as observations or rules without any theoretical basis. The nine papers showed successfully that it is possible to explain theoretically much more in ecology than is generally presumed.

In 2004, Jørgensen and Svirezhev published *Towards a Thermodynamic Theory for Ecosystems*. The book covers a major part of the ecosystem theory because thermodynamics is the foundation for understanding many ecosystem processes. Thermodynamics is, however, a difficult scientific discipline to understand, which unfortunately prevents wider application. The presented theory is, however, coherent and is able to explain many ecological observations.

To claim that we have laws providing precise predictive capacity is a simplification in the sense that ecosystem laws inevitably will be different from physical laws due to the complexity of ecosystems compared with physical systems. Expressed differently, it will be much harder to formulate causality in ecology than in physics, but there seems no doubt that ecosystems have some general *properties* that can be applied to make predictions and understand ecosystems' response to perturbations. The focus has, therefore, in this book been on general processes, properties, and patterns.

10.3 RECAPITULATION OF THE ECOSYSTEM THEORY

This theory integrates and extends the above-mentioned initiatives, building on those contributions. As stated in the first chapter and carried throughout the book, the Ecosystem Theory presented here rests on seven basic principles we observe in ecosystems:

(1) Ecosystems have thermodynamic openness,
(2) Ecosystem have ontic openness,
(3) Ecosystem have directed development,
(4) Ecosystems have connectivity,
(5) Ecosystems have hierarchic organization,
(6) Ecosystems have complex dynamics: growth and development, and
(7) Ecosystems have complex dynamics: disturbance and decay.

Physical–chemical systems can usually be described by matter and energy relations, while biological systems in addition need to include information relations. Biological systems can be characterized using three growth forms: structural (biomass) growth, network growth, and information growth. The last two growth forms give biological systems, including ecosystems, possibilities to move further and further from thermodynamic equilibrium and explain also the arrow of evolution. The synergistic effect of networks gives ecosystems the possibility to utilize available resources better (see Chapter 5), and thereby move further away from equilibrium. Information can be copied at almost no energy cost (see Chapter 6) and the increased information yields better utilization of the available resources to move the system still further from equilibrium. The first growth form is conservation limited but the second two are not and are far from their possible limits, as discussed in Chapters 5 and 6.

It was demonstrated in Chapter 9 that the seven ecosystem properties presented in Chapters 2–8 can be applied to explain a number of ecosystem rules and observations.

It can be concluded that we do not have a complete theory (no scientific discipline has a complete theory), but one that is adequate to explain many of our observations. Chapter 10 shows that the theory can be applied to assess ecosystem indicators useful in environmental management.

10.4 ARE THERE BASIC ECOSYSTEM PRINCIPLES?

Jørgensen and Fath (2004) have discussed eight basic principles or propositions of ecosystems, their properties and processes, and Jørgensen (2006) added two more. These propositions include the thermodynamic laws that are underlying all ecosystem functions, in addition to what is implicitly covered by the general properties of ecosystems in Chapters 2–7. To the extent possible, it will be mentioned below for each of the propositions how they are rooted in the seven ecosystem properties presented in Section 10.2. Interpretation of the propositions has, however, to be subject to the recognition that ecosystems are ontically open—too complex to allow accurate and complete predictions in all details. Nevertheless, let us try to set up the propositions because they can together with the properties presented in Chapters 2–7 and applied in Chapters 8 and 9 suggest new avenues to understand ecosystems:

1. *Mass and energy are conserved.* This principle is used again and again in ecology since it allows one to write balance equations at the core of ecosystem modeling, such as with a basic box-and-arrow diagram in which: accumulation = input–output.
2. *All ecosystem processes are irreversible* (this is probably the most useful way to express the second law of thermodynamics in ecology). Evolution and directionality, implicit in autocatalysis, can only be understood in light of the irreversibility principle rooted in the second law of thermodynamics. Evolution is a step-wise development that is based on previously achieved good solutions to survival in a changeable and very dynamic world. Evolution has been proceeded in the direction of ever more complex solutions.
3. *All ecosystems are open systems embedded in an environment from which they receive energy–matter input and discharge energy–matter output.* From a thermodynamic point of view, this principle is a prerequisite for ecosystem processes. If ecosystems were isolated in the physics' sense, then they would inevitably go to thermodynamic equilibrium without gradients and without life. This proposition is of course completely consistent with Chapter 2. It is noticeable that quantification of openness leads to an understanding of many ecological rates rooted in scaling theory and allometric principles.
4. *Thermodynamically, carbon-based life has a viability domain between approximately 250 and 350 K.* It is within this temperature range that there is a good balance between the opposing ordering and disordering processes: decomposition of organic matter and building of biochemically important compounds. At lower temperatures process rates are too slow and at higher temperatures the enzymes catalyzing the biochemical reactions will decompose too rapidly.
5. *Ecosystems have many levels of organization and operate hierarchically.* This principle is used again and again when ecosystems are described: atoms, molecules,

cells, organs, organisms, populations, communities, ecosystems, and the ecosphere. Hierarchy theory has been presented in Chapters 2, 3, and 7, and has been widely used to explain ecological observations.

6. *Carbon-based life on earth has a characteristic basic biochemistry which all organi-sms share.* It implies that many biochemical compounds can be found in all living organisms. They have, therefore, almost the same elemental composition derived from approximately 25 elements (Morowitz, 1968). This principle is widely used when stoichiometric calculations are made in ecology, i.e. an approximate average composition of living matter is applied. The proposition is able to give a biochemical explanation of feedback.

7. *Biological processes use captured energy (input) to move further from thermodynamic equilibrium and maintain non-equilibrium states of low entropy and high exergy relative to surroundings.* This is just another way of expressing that ecosystems can grow. Svirezhev (1992) has shown that eco-exergy of an ecosystem corresponds to the amount of energy that is needed to degrade the system. This proposition is consistent with the properties presented in Chapters 4 and 6.

8. *No ecosystem organism exists in isolation but is connected to other organisms and its abiotic environment.* Simply put, this states that connectivity is a basic property that, through transactions and relations, binds ecosystem parts together as an interacting and often integrated system. It can be shown by observations and ecosystem network calculations that the network has a synergistic effect on the components: the ecosystem is more than the sum of the components (see e.g. Patten, 1991; Fath and Patten, 1998). The proposition is completely consistent with the content of Chapter 5.

9. *After the initial capture of energy across a boundary, ecosystem growth and development is possible by (1) an increase of the physical structure (biomass), (2) an increase of the network (more cycling), and then (3) an increase of information embodied in the system.* All three growth forms imply that the system is moving away from thermodynamic equilibrium (Jørgensen et al., 2000), and all three growth forms are associated with an increase of (1) stored eco-exergy and (2) the energy throughflow in the system (power). When cycling flows increase, the eco-exergy storage capacity, the energy use efficiency and space–time differentiation all increase (Ho and Ulanowicz, 2005). When the information increases, the feedback controls and autocatalysis become more effective, specific respiration decreases, and there is a tendency to replace r-strategist species with K-strategists, which means less energy is wasted on reproduction. When earth systems (physical and biological) capture approximately 75% of available solar energy, it is not possible to increase this capture further. The same is true for limiting elements. Under these conditions ecosystems cannot benefit further from growth Form I and must graduate to growth Forms II and III. Thereby, the efficiency of exergy utilization is increased. This description is in accordance with Margalef (1991, 1995): the first stages proceed rapidly with an apparently wasteful use of available energy; later a higher efficiency along a defined direction occurs, because of competition, in the frame of natural selection. Growth Form I is constrained by the conservation of energy and matter, while the two other growth forms are not following the conservation laws. In ecosystem succession the information is transferred from the present to the future

and the shift is manifested in a historical way that has many aspects. One of these, the production and accumulation of biomass, prevails at the beginning, and this is often described as "bottom up" control. Later, the high trophic levels take more control, and "top-down-control" becomes more apparent. This proposition has been presented in Chapter 6 and partly in Chapters 4 and 7. It has furthermore been applied several times in Chapters 8 and 9.

10. *An ecosystem receiving solar radiation will attempt to maximize eco-exergy storage, ascendency, or maximize power such that if more than one possibility is offered, then in the long-run the one which moves the system furthest from thermodynamic equilibrium will be selected.* Eco-exergy storage increases with all three growth forms—see above. When an ecosystem develops it can, therefore, apply all three growth forms in a continuous Darwinian selection process. It is intuitively obvious why the nested space–time differentiation in organisms optimizes thermodynamic efficiency as expressed in the tenth proposition because it allows the organism to simultaneously exploit equilibrium and non-equilibrium energy transfer with minimum dissipation (Ho and Ulanowicz, 2005). This proposition has been touched on in Chapters 4, 6, and 7 and been applied several times in Chapters 8 and 9.

As seen from this short overview of the ten propositions they may be considered a useful organization of the basic system ecology needed to understand the ecosystem reactions and processes. The organization in propositions are different from the basic properties that were applied as the fundament for the presentation of our "New Ecology—A System Approach" shown in Chapters 1–9. It is, of course, not surprising that we need different descriptions of ecosystem processes and responses. Ecosystems are complex systems as touched on several times throughout the book. So, how many different descriptions do we need to describe ecosystems when we consider that a simple physical phenomenon as light requires two descriptions, as waves and as particles?

We, the nine authors, conclude, however, that we do have an ecosystem theory, that can be presented in different ways but under all circumstances can be used to explain and understand ecological observations, properties, and processes (Chapter 8) and even be applied in environmental management (Chapter 9). The theory should be considered one of the first attempt to present an (almost) complete ecosystem theory. It is most likely that it will changed in the coming years as we gain experience by using it, but it can anyhow today as hopefully demonstrated for our readers be applied to explain ecology and to develop a good environmental strategy for ecosystem management. Further development of an ecosystem theory in system ecology will only be possible by trying to use theoretical approaches in ecology and environmental management. We, therefore, encourage all ecologists and environmental managers to apply the presented theory.

10.5 CONCLUSION

The earth is a non-isolated system. There is almost no exchange of matter with the outer space (the earth loses a little hydrogen and receives meteorites). To be able to utilize the matter many times during the evolution or from one year and decade to the next, cycling is necessary. Cycling implies that the ecosystem components are linked in an interacting network (Chapter 5).

Ecosystems must be, as the earth, non-isolated because otherwise they could not receive the energy needed to maintain the ecosystems far from thermodynamic equilibrium and even move further away from thermodynamic equilibrium. Ecosystems are actually open systems (Chapter 2) because they need to exchange at least water (precipitation and evaporation) with their environment. In addition, it is practical that suitable solutions (for instance species with new emergent properties that facilitate survival under a combination of new and emergent conditions) in one ecosystem can be exported to other ecosystems. Moreover, it is easy to observe that ecosystems *are* open systems.

The flow of energy from the sun to the ecosystems is also limited. It is important that an ecosystem captures as much sunlight as possible to cover its energy needs. Therefore, ecosystems, with increased biomass, can increase net primary productivity. But even the best photosystems can only capture a certain part of the solar radiation, which anyhow is limited to about 10^{17} W on average. Therefore moving further away from thermodynamic equilibrium requires that an ecosystem develop better utilization of the exergy that it is able to capture. Network development, where the components have been fitted together, provides improved exergy efficiencies (Chapters 4 and 5). Another possibility is to increase information in the form of better process efficiencies (Chapter 6). Increased sizes of the organisms imply also that the exergy lost for respiration decreases relatively to the biomass (Chapter 2).

While matter and energy flow limit evolution, the amount of information is far from its limit. It is, therefore, understandable that information embodied in genes and in ecological networks has increased throughout evolution (Chapter 6).

On the one side, we do know the characteristic ecosystem response to changed conditions and can make predictions. On the other side, ecosystems are so complex that very accurate detailed predictions always will be impossible and "surprises" should always be expected; see Chapter 3.

The development of the life forms that we know from the earth has been possible because the earth has the elements that are needed to build the biochemical compounds that explain the life processes. It includes water that is an ideal solvent for biochemical reactions. In addition, the earth has the right temperature range that means that the biochemical reactions proceed with a certain rate and that the decomposition of particularly proteins is moderate. The right balance between formation and destruction of high molecular proteins that are the enzymatic compounds controlling the life processes is thereby obtained.

The life processes take place in cells, because they have a sufficiently high specific surface to allow an exchange rate with the environment that is suitable. Cells are, therefore, the biological units that make up organs and organisms. Nature must, therefore, use a hierarchical construction (Chapters 2, 3, and 7): atoms, molecules, cells, organs, organisms, populations, communities, ecosystems, and the ecosphere. The addition of units in one hierarchical level to form the next level gives the next level new and emergent properties (Chapters 3 and 7).

The variability of the life conditions in time and space is very high (Chapter 6). When an ecosystem has adapted to certain conditions, it can still be disturbed by catastrophic events. Ecosystems have due to their properties (see all the chapters but particularly Chapter 7) the adaptability and flexibility to meet these changed conditions and still maintain the system far from thermodynamic equilibrium. The disturbances call for new and creative solutions for life to survive. Disturbances may therefore also be beneficial in the long term for ecosystems (Chapter 7).

We can explain the ecosystems, their processes and responses, and evolution by the properties presented in this book. The discussion throughout the book has clearly shown that the properties are sufficient and the discussion in this last section has demonstrated that the properties are also necessary. It may, however, not be the only possible explanation to life in general. We cannot exclude that we will find other life forms somewhere else in the universe, for instance based on silica or carbon but with another biochemistry, better suited maybe for a different situation. The properties presented above are, however, very consistent with both direct and indirect observations, which render them a good basis for an ecosystem theory applicable on Earth. Chapters 8 and 9 have, not surprisingly, shown that the ecosystem theory presented in this book can be used to explain other ecological rules and hypotheses and have potential for application in environmental management.

References

Abramson N. 1963. Information Theory and Coding. McGraw-Hill, New York, NY, 201 pp.

Ågren GI, Bosatta E. 1996. Theoretical Ecosystem Ecology—Understanding Element Cycles. Cambridge University Press, Cambridge, MA, 234 pp.

Allen G. 1978. Thomas Hunt Morgan: The Man and His Science. Princeton University Press, Princeton, NJ, 430 pp.

Allen TFH, Starr TB. 1982. Hierarchy, Perspectives for Ecological Complexity. The University of Chicago Press, Chicago, IL, 310 pp.

Allesina S, Bondavalli C. 2004. WAND: an ecological network analysis user-friendly tool. Environ. Model. Software 19, 337–340.

Alcock J. 1997. Animal Behavior: An Evolutionary Approach. Sinauer Associates, 640 p.

Andrewartha HG, Birch LC. 1984. The Ecological Web. University of Chicago Press, Chicago, IL.

Aoki I. 1987. Entropy balance in Lake Biva. Ecol. Model. 37, 235–248.

Aoki I. 1988. Entropy laws in ecological networks at steady state. Ecol. Model. 42, 289–303.

Aoki I. 1989. Ecological study of lakes from an entropy viewpoint—Lake Mendota. Ecol. Model. 49, 81–87.

Aoki I. 1990. Monthly variations of entropy production in lake Biwa. Ecol. Model. 51, 227–232.

Aoki I. 1995. Entropy production in living systems: from organisms to ecosystems. Thermochim. Acta 250, 359–370.

Aoki I. 2000. Entropy and Exergy Principles in Living Systems. Thermodynamics and Ecological Modelling. Lewis Publishers, New York, NY, pp. 165–190.

Barber MC. 1978a. A Markovian model for ecosystem flow analysis. Ecol. Model. 5, 193–206.

Barber MC. 1978b. A retrospective markovian model for ecosystem resource flow. Ecol. Model. 5, 125–135.

Barbieri M. 2001. The Organic Codes. The Birth of Semantic Biology. PeQuod, Ancona, Italy.

Barkmann J, Baumann R, Meyer U, Müller F, Windhorst W. 2001. Ökologische Integrität: risikovorsorge im Nachhaltigen Landschaftsmanagement. Gaia 10(2), 97–108.

Bastianoni S. 2002. Use of thermodynamic orientors to assess the efficiency of ecosystems: a case study in the lagoon of Venice. The Scientific World Journal, 2, pp. 255–260.

Bastianoni S. 2006. Emergy, empower and the eco-exergy to empower ratio: a reconciliation of H.T. Odum with Prigogine? International Journal of Ecodynamics, 1 (3), 226–235.

Bastianoni S, Marchettini N. 1997. Emergy/exergy ratio as a measure of the level of organization of systems. Ecol. Model. 99, 33–40.

Bastianoni S, Pulselli FM, Rustici M. 2006. Exergy versus emergy flow in ecosystems: is there an order in maximizations? Ecol. Indicators. 6(1), 58–62.

Bateson G. 1972. Steps to an Ecology of Mind. Ballantine Books, New York, NY, 517 p.

Baumann R. 2001. Konzept zur Indikation der Selbstorganisationsfahigkeit terrestrischer Ökosysteme anhand von Daten des Ökosystemforschungsprojekts Bornhöveder Seenkette. Dissertation, University of Kiel.

de Bernardi R, Giussani G. (eds.). 1995. Biomanipulation in lakes and reservoirs management. Guidelines of lake management, Vol. 7. ILEC (International Lake Environmental Committee)/UNEP (United Nations Environmental Program), Otsu, Japan.

Berryman AA. 1993. Food web connectance and feedback dominance, or does everything really depend on everything else? Oikos 68, 183–185.

Berryman AA. 1999. Principles of Population Dynamics and their Application. Cheltenham, UK.

Berryman AA. 2003. On principles, laws and theory in population ecology. Oikos 103, 695–701.

Blackburn TM, Gaston KJ. 1996. A sideways look at patterns in species richness, or why there are so few species outside the tropics. Biodiversity Lett. 3, 44–53.

Blum HF. 1951. Time's Arrow and Evolution. Princeton University Press, Princeton, NJ, 222 pp.

Boltzmann L. 1872. Weitere Studien ueber das Waermegleichgewicht unter Gasmolekuelen. Wien. Ber. 66, 275–370.

Boltzmann L. 1905. The Second Law of Thermodynamics (Populare Schriften. Essay No. 3 (Address to Imperial Academy of Science in 1886)). Reprinted in English in: Theoretical Physics and Philosophical Problems, Selected Writings of L. Boltzmann. D. Riedel, Dordrecht.

Bondavalli C, Ulanowicz RE. 1999. Unexpected effects of predators upon their prey: the case of the American alligator. Ecosystems 2, 49–63.

Bondavalli C, Ulanowicz RE, Bodini A. 2000. Insights into the processing of carbon in the South Florida Cypress wetlands: a whole-ecosystem approach using network analysis. J. Biogeography 27, 697–720.

Bonner JT. 1965. Size and Cycle. An Essay on the Structure of Biology. Princeton University Press, Princeton, NJ.

Borrett SR, Patten BC, Fath BD. Pathway proliferation and modularity in ecological networks. J. Theor. Biol. (submitted).

Bracchini L, Cózar A, Dattilo AM, Picchi MP, Arena C, Mazzuoli S, Loiselle SA. 2005. Modelling the components of the vertical attenuation of ultraviolet radiation in a wetland lake ecosystem. Ecol. Model. 186, 43–54.

Brock TD. 1985. A Eutrophic Lake, Lake Mendota. Springer, Wisconsin.

Brown JH. 1995. Macroecology. University of Chicago Press, Chicago, IL.

Brown JH, Marquet PA, Taper ML. 1993. Evolution of body size: consequences of an energetic definition of fitness. Am. Nat. 142, 573–584.

Brown JH, Maurer BA. 1987. Evolution of species assemblages: effects of energetic constraints and species dynamics on the diversification of North American avifauna. Am. Nat. 130, 1–17.

Buzas MA, Collins LS, Culver SJ. 2002. Latitudinal difference in biodiversity caused by higher tropical rate of increase. Proc. Natl. Acad. Sci. 99, 7841–7843.

Capek V, Sheehan DP. 2005. Challenges to the Second Law of Thermodynamics. Theory and Experiment. Fundamental Theories of Physics, vol. 146. Springer, New York, NY, 355 pp.

Carlson RE. 1977. A trophic state index for lakes. Limnol. Oceanog. 22 (2), 361–369.

Carnot S. 1824. Reflections on the Motive Power of Heat (translated 1943). ASME. New York, NY, 107 pp.

Carpenter FL, Paton DC, Hixon MA. 1983. Weight gain and adjustment of feeding territory size in migrant hummingbirds. Proc. Natl. Acad. Sci. 80, 7259–7263.

Carpenter SR, Kitchell JF, Hodgson JR. 1985. Cascading trophic interactions and lake productivity. BioScience 35, 634–638.

Carrington A, McLachlan AD. 1979. Introduction to Magnetic Resonance. Chapman and Hall, London.

Chapelle A, Ménesguen A, Deslous-Paoli JM, Souchu P, Mazouni N, Vaquer A, Millet B. 2000. Modelling nitrogen, primary production and oxygen in a Mediterranean lagoon. Impact of oysters farming and inputs from the watershed. Ecol. Model. 127, 161–181.

Christensen V, Pauly D. 1992. Ecopath II: a software for balancing steady-state ecosystem models and calculating network characteristics. Ecol. Model. 61, 169–185.

Clements FE. 1916. Plant Succession: An Analysis of the Development of Vegetation. Carnegie Institution of Washington, Washington, DC.

Coffaro G., Bocci M, Bendoricchio G. 1997. Structural dynamic application to space variability of primary producers in shallow marine water. Ecol. Model. 102, 97–114.

Cohen JE, Briand F, Newman CM. 1990. Community Food Webs: Data and Theory. Springer-Verlag, Berlin, Germany.

Commoner B. 1971. The Closing Circle. Alfred A. Knopf, New York, NY, 328 pp.

Connell JH, Slayter RO. 1977. Mechanisms of succession in natural communities and their roles in stabilisation and organisation. Am. Nat. 111, 199–1144.

Cook LM. 2000. Changing views on melanic moths. Biol. J. Linn. Soc. 69, 431–444.

Costello CM. 1992. Black bear habitat ecology in the Central Adirondack as related to food abundance and forest management. M.S. Thesis. State University of New York, College of Environmental Science and Forestry, Syracuse, NY.

Cousins SH. 1990. Countable ecosystems deriving from a new food web entity. Oikos, 57, 270–275.

Cowles HC. 1899. The Ecological Relations of the Vegetation on the Sand Dunes of Lake Michigan. Bot. Gaz. 27, 95–117; 167–202; 281–308; 361–391.

Coyne JA. 2002. Evolution under pressure. Nature 418, 19–20.

Cummins KW, Minshaal GW, Sedell JR, Cushing CE, Petersen RC. 1984. Stream ecosystem theory. Verhandlung Internationale Vereinigung Limnologie 22 (3), 1818–1827.

Dame RF, Patten BC. 1981. Analysis of energy flows in an intertidal oyster reef. Mar. Ecol. Progr. Ser. 5, 115–124.

Darlington CD. 1943. Race, class and mating in the evolution of man. Nature 3855, 315–319.

Darlington CD. 1953. Genetics and Man. Penguin Books, New York, NY, 256 pp.

Darwin C. 1859. On the Origin of Species by means of Natural Selection or the Preservation of Favoured Races in the Struggle for Life. John Murray, London, UK.

Darwin C, Wallace A. 1858. On the tendency of species to form varieties; and on the perpetuation of varieties and species by natural means of selection. J. Proc. Linn. Soc. Lond. Zool. 3, 45–62.

DeAngelis DL, Gross LJ, Huston MA, Wolff WF, Fleming DM, Comiskey EJ, Sylvester SM. 1998. Landscape modeling for everglades ecosystem restoration. Ecosystems 1, 64–75.

DeAngelis DL, Post WM, Travis CC. 1986. Positive Feedback in Natural Systems. Springer-Verlag, New York, NY.

Debeljak M. 2001. Application of exergy degradation and exergy storage as indicator for the development of managed and virgin forest. Ph.D. Thesis. Ljubliana University.

Debeljak M. 2002. Characteristics of energetics of managed and virgin forest. Ph.D. Thesis. Ljubliana.

De Leo G, Bartoli M, Naldi M, Viaroli P. 2002. A first generation stochastic bioeconomic analysis of algal bloom control in a coastal lagoon (Sacca di Goro, Pò river Delta). Mar. Ecol. 410, 92–100.

Depew DJ, Weber BH. 1995. Darwinism Evolving: Systems Dynamics and the Genealogy of Natural Selection. MIT Press, Cambridge, MA, 588 pp.

De Wit R. 2005. Do all ecosystem maximize their distance to thermodynamic equilibrium. A Comment to the Ecological Law of Thermodynamics (ELT) proposed by S.E. Jørgensen. Sci. Mar. 69, 427–434.

Di Castri F, Hadley M. 1988. Enhancing the credibility in ecology: interacting along and across hierarchical scales. Geo. J. 17, 5–35.

Dierssen K. 2000. Ecosystems as states of ecological successions. In: Jørgensen SE, Müller F (eds.), Handbook of Ecosystem Theories and Management. Boca Raton, FL, pp. 427–446.

Dobzhansky T. 1937. Genetics and the Origin of Species. Columbia University Press, Boston, MA.

Drury WH, Nisbet ICT. 1973. Succession. J. Arnold Arboretum 54 (3), 331–368.

Duarte P. 1995. A mechanistic model of the effects of light and temperature on algal primary productivity. Ecol. Model. 82, 151–160.

Dunne JA, Williams RJ, Martinez ND. 2002. Food-web structure and network theory: The role of connectance and size. Proc. Natl. Acad. Sci. USA 99, 12917–12922.

Dutton JA, Bryson RA. 1962. Heat flux in Lake Mendota. Limnol. Oceanogr. 7, 80–97.

Ebeling W, Engel A, Feistel R. 1990. Physic der Evoltionsprozesse. Akademie Verlag. Berlin, 374 pp.

Ecosystems and Human Well-being, a Report of the Conceptual Framework Working Group of the Millennium Ecosystem Assessment. 2003.

Eggenberger F, Polya G. 1923. Ueber die Statistik verketteter Vorgaenge. Zeitschrift fuer Angewandte Mathematik Mechanik 3, 279–289.

Egler FE. 1954. Vegetation science concepts. 1. Initial floristic composition, a factor in the old-field vegetation development. Vegetation 4, 412–417.

Eigen M. 1971. Self organization of matter and the evolution of biological macromolecules. Naturwiss 58, 465–523.

Eldredge N. 1998. Life in the Balance: Humanity and the Biodiversity Crisis. Princeton University Press, Princeton, NJ, 224 pp.

Elmquist T, Folke C, Nyström M, Peterson G, Bentsson J, Walker B, Norberg J. 2003. Response diversity, ecosystem change, and resilience. Front. Ecol. Environ. 1(9), 488–494.

Elsasser WM. 1969. A causal phenomena in physics and biology: a case for reconstruction. Am. Sci. 57(4), 502–516.

Elsasser WM. 1978. Memoirs of a Physicist in the Atomic Age. Neale Watson Academic Publications, Inc., New York, NY.

Elsasser WM. 1981. A form of logic suited for biology? In: Rosen R (ed.), Progress in Theoretical Biology, vol. 6. Academic Press, New York, NY, pp. 23–62.

Elsasser WM. 1987. Reflections on a Theory of Organisms. Holism in Biology. Johns Hopkins University Press, Baltimore, MD, 160 pp.

Fath BD. 1998. Network analysis: foundations, extensions, and applications of a systems theory of the environment. Ph.D. Thesis. University of Georgia, Athens, GA, 176 pp.

Fath BD. 2004. Editorial: control of distributed systems and environmental applications. Ecol. Model. 179, 151–152.

Fath BD. 2007. Community-level Relations and Network Mutualism. Accepted in Ecological Modelling.

Fath BD, Borrett SR. 2006. A matlab® function for network environ analysis. Environ. Model. Software 21, 375–405.

Fath BD, Halnes G. Cyclic Energy Pathways in Ecological Food Webs. Submitted to Ecological Modelling.

Fath BD, Jørgensen SE, Patten BC, Straškaba M. 2004. Ecosystem growth and development. BioSystem 77, 213–228.

Fath BD, Patten BC. 1998. Network synergism: emergence of positive relations in ecological systems. Ecol. Model. 107, 127–143.

Fath BD, Patten BC. 1999. Review of the foundations of network environ analysis. Ecosystems 2, 167–179.

Fath BD, Patten BC. 2000. Goal Functions and Network Theory. Presented in Porto Venere May 2000 at the Second Conference on Energy.

Fath BD, Patten BC. 2001. A progressive definition of network aggradation. In: Ulgiati S, Brown MT, Giampietro M, Herendeen RA, Mayumi K (eds.), 2nd International Workshop on Advances in Energy Studies: Exploring Supplies, Constraints and Strategies. Porto Venere, Italy, 23–27 May 2000, pp. 551–562.

Fath BD, Patten BC, Choi JS. 2001. Complementarity of ecological goal functions. J. Theor. Biol. 208, 493–506.

Fenchel T. 1974. Intrinsic rate of natural increase: the relationship with body size. Oecologia. 14, 317–326.

Finn JT. 1976. Measures of ecosystem structure and function derived from analysis of flows. J. Theor. Biol. 56, 363–380.

Finn JT. 1980. Flow analysis of models of the Hubbard Brook ecosystem. Ecology 61(3), 562–271.
Fong P. 1962. Elementary Quantum Mechanics. Addison-Wesley Publishing Company, Reading, MA.
Fonseca JC, Marques JC, Paiva AA, Freitas AM, Madeira VMC, Jørgensen SE. 2000. Nuclear DNA in the determination of weighting factors to estimate exergy from organism's biomass. Ecol. Model. 126, 179–190.
Forrester, JW. 1968. *Principles of Systems*, (2nd ed.), Cambridge, MA, Productivity Press. 391 pp.
Forrester JW. 1971. World Dynamics. Wright, Allen, Cambridge, MA.
Forrester JW. 1987. Nonlinearity in high-order models of social systems. Eur. J. Opt. Res. 30, 104–109.
Futuyma DJ. 1986. Evolutionary Biology. 2nd ed. Sinauer Associates, Inc. Sunderland, MA, 600 pp.
Gallopin GC. 1981. The abstract concept of environment. Int. J. Gen. Syst. 7, 139–149.
Gaston KJ. 2000. Global patterns in biodiversity. Nature 405, 220–227.
Gause GF. 1934. The struggle for Existence. Williams and Wilkins, Baltimore, MD, 163 pp.
Gibbs JW. 1902. Elementary Principles in Statistical Mechanics. Charles Scribner's Sons, New York, NY.
Gigon A, Grimm V. 1997. Stabilitätskonzepte in der Ökologie. J. Veg. Sci. 4(6), 733–736.
Giovanardi F, Poletti A, Micheli A. 1995. Indagini sulla qualita' delle acque del Lago Trasimeno-Definizione dei livelli trofici. Acqua Aria, 6 (in Italian with English summary).
Givnish TJ, Vermelj GJ. 1976. Sizes and shapes of liana leaves. Am. Nat. 110, 743–778.
Gleason HA. 1917. The structure and development of the plant association. Bull. Torrey Botanical Club 44, 463–481.
Golley FB. 1977. Ecological Succession. Dowden, Hutchinson & Ross, Stroudsburg, PA.
Goodrick RL. 1984. The wet prairies of the northern Everglades. In: Gleason PJ (ed.), Environments of South Florida: Present and Past II. Memoirs II. Miami Geological Society, Miami, FL.
Gould SJ. 1994. The evolution of life on earth. Sci. Am. 271(4), 62–69.
Gradel TE, Allenby BR. 1995. Industrial Ecology. Prentice Hall, Englewood Cliffs, NJ, 412 pp.
Grant BS. 2002. Sour grapes of wrath. Science 297, 940–941.
Grant PR. 1986. Ecology and Evolution of Darwin's Finches. Reprinted in 1999. Princeton University Press, Princeton, NY, 492 pp.
Greene B. 1999. The Elegant Universe. Superstrings, Hidden Dimensions, and the Quest for the Ultimate Theory. W. W. Norton, New York, NY, 448 pp.
Grime JF. 1979. Plant Strategies and Vegetational Processes. Wiley, Chichester, UK.
Gundersson LH, Holling CS (eds.) 2002. Panarchy. Washington, DC.
Gunderson LH, Holling CS. (eds.). 2003. Panarchy. Island Press, Washington, DC.
Gunderson LH, Holling CS, Peterson G. 2000. Resilience in ecological systems. In: Jørgensen SE, Müller F (eds.), Handbook of Ecosystem Theories and Management. Boca Raton, FL, pp. 385–394.

Haken H. 1988. Information and Self-Organization. Springer-Verlag, Berlin, Germany.

Hall CAS. 1995. Maximum Power. University Press of Colorado, Niwot, CO, 394 pp.

Hamza W, Pandolfi P, Taticchi MI. 1995. Planktonic interactions and their role in describing the trophic status of a shallow lake in Central Italy (Lake Trasimeno). Mem. Ist. It. Idrobiol. 53, 125–139.

Hannon B. 1973. The structure of ecosystems. J. Theor. Biol. 41, 535–546.

Herendeen RA. 1981. Energy intensity in ecological and economic systems. J. Theor. Biol. 91, 607–620.

Herendeen R. 1989. Energy intensity, residence time, exergy, and ascendency in dynamic ecosystems. Ecol. Model. 48, 19–44.

Herndorn AK, Taylor D. 1986. Response of a Muhlenbergia prairie to repeated burning: changes in above ground biomass. South Florida Research Center Rep. 86/05, Everglades National Park, Homestead, FL.

Heymans JJ, Baird D. 2000. Network analysis of the northern Benguela ecosystem by means of netwrk and ecopath. Ecol. Model. 131, 97–119.

Heymans JJ, Ulanowicz RE, Bondavalli C. 2002. Network analysis of the South Florida Everglades graminoid marshes and comparison with nearby cypress ecosystems. Ecol. Model. 149, 5–23.

Higashi M, Burns TP. (eds.). 1991. Theoretical Studies of Ecosystems: The Network Perspective. Cambridge University Press, New York, NY.

Higashi M, Burns TP, Patten BC. 1988. Food network unfolding: an extension of trophic dynamics for application to natural ecosystems. J. Theor. Biol. 140, 243–261.

Higashi M, Patten BC. 1986. Further aspects of the analysis of indirect effects in ecosystems. Ecol. Model. 31, 69–77.

Higashi M, Patten BC. 1989. Dominance of indirect causality in ecosystems. Am. Nat. 133, 288–302.

Ho MW, Ulanowicz R. 2005. Sustainable systems as organisms. BioSystems. 82, 39–51.

Hoffman W, Bancroft GT, Sawicki RJ. 1990. Wading bird populations and distributions in the water conservation areas of the Everglades: 1985 through 1988. Report prepared for the Environmental Sciences Division, South Florida Management Unit. Contract c396-M87-0345.

Hoffmeyer J. 1993. Signs of Meaning in the Universe. Indiana University Press, Bloomington, IN, 166 pp.

Holling CS. 1986. The resilience of terrestrial ecosystems: Local surprise and global change. In: Clark WC, Munn RE (eds.), Sustainable Development of the Biosphere. Cambridge University Press, Cambridge, MA, pp. 292–317.

Holling CS. 2004. From complex regions to complex worlds. Ecology and Society 9, Article 11.

Homer M, Kemp WM, McKellar H. 1976. Trophic analysis of an estuarine ecosystem: Salt marsh-tidal creek system near Crystal River, Florida. Unpublished data, University of Florida, Gainesville, FL.

Horn HS. 1974. Succession. In: May RM (ed.), Theoretical Ecology—Principles and Applications. Blackwell, Oxford, UK, pp. 187–204.

Howington TM, Brown MT, Wiggington M, 1997. Effect of hydrologic subsidy on self-organization of a constructed wetland in Central Florida. Ecological Engineering, 9, 137–156

Huang SL, Chen CW. 2005. Theory of urban energetics and mechanisms of urban development. Ecol. Model. 189, 49–71.

Hutchinson GE. 1948. Circular causal systems in ecology. Ann. N.Y. Acad. Sci. 50, 221–246.

Hutchinson GE. 1957. Concluding remarks. Cold Spring Harbor Symp. Quant. Biol. 22, 415–427.

Hutchinson GE. 1965. The niche: An abstractly inhabited hypervolume. In: Hutchinson GE (ed.), The Ecological Theatre and the Evolutionary Play. Yale University Press, New Haven, CT, pp. 26–78.

Huxley JS (ed). 1940. The New Systematics. Oxford University Press, Oxford, UK.

Huxley JS. 1942. Evolution: The Modern Synthesis. Allen and Unwin, Oxford, UK.

Jensen K, Schrautzer J. 1999. Consequences of abandonment for a regional fen flora and mechanism of successional change. Appl. Veg. Sci. 2, 79–88.

Jentsch A, Beierkuhnleich C, White PS. 2002. Scale, the dynamic stability of forest ecosystems, and the persistence of biodiversity. Silva Fennica 36, 393–400.

Jonasson PM, Lindegaard C. 1988. Ecosystems studies of North Atlantic ridge lakes. Verh. Int. Ver. Limnol. 23, 394–402.

Jørgensen SE. 1982. A holistic approach to ecological modelling by application of thermodynamics. In: W. Mitsch et al. (eds.), Systems and Energy, Ann Arbor, MI, 192 pp.

Jørgensen SE. 1984. Parameter estimation in toxic substance models. Ecol. Model. 22, 1–12.

Jørgensen SE. 1986. Structural dynamic model. Ecol. Model. 31, 1–9.

Jørgensen SE. 1988. Use of models as experimental tools to show that structural changes are accompanied by increased exergy. Ecol. Model. 41, 117–126.

Jørgensen SE. 1990. Ecosystem theory, ecological buffer capacity, uncertainty and complexity. Ecol. Model. 52, 125–133.

Jørgensen SE. 1992a. Parameters, ecological constraints and exergy. Ecol. Model. 62, 163–170.

Jørgensen SE. 1992b. Development of models able to account for changes in species composition. Ecol. Model. 62, 195–208.

Jørgensen SE. 1994a. Review and comparison of goal functions in system ecology. Vie Milieu 44, 11–20.

Jørgensen SE. 1994b. Fundamentals of ecological modelling. 2nd ed. Developments in Environmental Modelling, vol. 19. Elsevier, Amsterdam, 628 pp.

Jørgensen SE. 1995. Quantum mechanics and complex ecology. In: Patten BC, Jørgensen SE (eds.), Complex Ecology: The Part—Whole Relation in Ecosystems. Prentice Hall, Englewood Cliffs, NJ, pp. 34–39.

Jørgensen SE. 1997. Integration of Ecosystem Theories: A Pattern. 2nd ed. Kluwer Academic Publishers, Dordrecht, 400 pp (1st ed, 1992).

Jørgensen SE. 2000a. Principles of Pollution Abatement. Elsevier, Oxford, UK, 520 pp.

Jørgensen SE. 2000b. Application of exergy and specific exergy as ecological indicators of costal areas. Aquat. Ecosyst. Health Manag. 3, 419–430.

Jørgensen SE. 2001. Toward a consistent pattern of ecosystem theories. Sci. World 1, 71–75.

Jørgensen SE. 2002. Integration of Ecosystem Theories: A Pattern. 3rd ed. Kluwer Academic Publisher, Dordrecht, 428 pp.

Jørgensen SE. 2006. Toward a thermodynamics of biological systems. Int. J. Ecodynamics 1, 9–27.

Jørgensen SE, de Bernardi R. 1997. The application of a model with dynamic structure to simulate the effect of mass fish mortality on zooplankton structure in Lago di Annone. Hydrobiologia 356, 87–96.

Jørgensen SE, de Bernardi R. 1998. The use of structural dynamic models to explain successes and failures of biomanipulation. Hydrobiologia 359, 1–12.

Jørgensen SE, Costanza R, Fu-Liu Xu. 2004. Handbook of Ecological Indicators for Assessment of Ecosystem Health. 436 pp.

Jørgensen SE, Fath BD. 2004. Modelling the selective adaptation of Darwin's Finches. Ecol. Model. 176, 409–418.

Jørgensen SE, Ladegaard N, Debeljak M, Marques JC. 2005. Calculations of exergy for organisms. Ecol. Model. 185, 165–175.

Jørgensen SE, Marques JC. 2001. Thermodynamics and ecosystem theory, case studies from hydrobiology. Hydrobiologia, 445, 1–10.

Jørgensen SE, Marques J, Nielsen SN. 2002. Structural changes in an estuary, described by models and using exergy as orientor. Ecol. Model. 158, 233–240.

Jørgensen SE, Mejer HF. 1977. Ecological buffer capacity. Ecol. Model. 3, 39–61.

Jørgensen SE, Mejer HF. 1979. A holistic approach to ecological modelling. Ecol. Model. 7, 169–189.

Jørgensen SE, Nielsen SN, Mejer H. 1995. Emergy, environ, exergy and ecological modelling. Ecol. Model. 77, 99–109.

Jørgensen SE, Odum HT, Brown MT. 2004. Emergy and exergy stored in genetic information. Ecol. Model. 178, 11–16.

Jørgensen SE, Padisák J. 1996. Does the intermediate disturbance hypothesis comply with thermodynamics? Hydrobiologia 323, 9–21.

Jørgensen SE, Patten BC, Straskraba M. 1999. Ecosystems emerging: 3 openness. Ecol. Mod. 117, 41–64.

Jørgensen SE, Patten BC, Straškraba M. 2000. Ecosystems emerging: 4 growth. Ecol. Model. 126, 249–284.

Jørgensen SE, Svirezhev Y. 2004. Toward a Thermodynamic Theory for Ecological Systems. Elsevier, Amsterdam, 366 pp.

Kauffman SA. 1993. The Origins of Order. Oxford University Press, Oxford, UK.

Kauffman SA. 1995. At Home in the Universe: The Search for the Laws of Self Organization and Complexity. Oxford University Press, New York, NY.

Kay JJ. 1984. Self Organization in Living Systems. Ph.D. Thesis. Systems Design Engineering, University of Waterloo, Ont., Canada.

Kay JJ, Schneider ED. 1992. Thermodynamics and measures of ecological integrity. In: Proceedings of Ecological Indicators, Elsevier, Amsterdam. pp. 159–182.

Kemeny JG, Snell JL. 1960. Finite Markov Chains. D. van Nostrand Company, Inc., Princeton, NJ.

Kerekes JJ. 1990. Possible correlation of summer common loon (*Gavia immer*) population with the trophic state of a water body. Vehrh. Int. Verein. Limn. 24, 349–353.

Kettlewell HBD. 1965. Insect survival and selection for pattern. Science 148, 1290–1296.

Klipp E, Herwig R, Kowald A, Wierling C, Lehrach H. 2005. Systems Biology in Practice. Concepts, Implementation and Application. Wiley-VCH Verlag GmbH, Weinheim, Germany, 465 pp.

Kolasa J, Waltho N. 1998. A hierarchical view of habitat and its relationship to species abundance. In: Peterson DL, Parker VT (eds.), Ecological Scale: Theory and Applications. Columbia University Press, New York, NY, pp. 55–76.

Kratzer CR, Brezonik PL, 1981. A Carlson-type trophic state index for nitrogen in Florida lakes, Water Resources Bulletin, 17(4), 713–715.

Krebs CJ. 2001. Ecology. Benjamin Cummings, San Francisco, CA, 340 pp.

Krivtsov V, Corliss J, Bellinger E, Sigee D. 2000. Indirect regulation rule for consecutive stages of ecological succession. Ecol. Model. 133, 73–82.

Kullback S. 1959. Information Theory and Statistics. Peter Smith, Gloucester, MA, 399 p.

Laudenslayer WF Jr., Grenfell WE Jr. 1983. A list of amphibians, reptiles, birds and mammals of California. Outdoor Calif. Jan–Feb, 5–14.

Legendre L, Legendre P. 1984a. Ecologie Numérique. Le Traitement Multiple des Données Écologiques, vol. 1. Masson, Paris, 335 pp.

Legendre L, Legendre P. 1984b. Ecologie numerique, 2nd ed. Tome 1: Le traitement multiple des donnees ecologiques. Masson, Paris et les Presses de l'Universite du Quebec. 260 pp.

Leontief WW. 1936. Quantitative input-output relation in the economic system of the United States. Rev. Econ. Statistics 18, 105–125.

Leontief WW. 1951. The Structure of American Economy, 1919–1939: An Empirical Application of Equilibrium Analysis. Oxford University Press, New York, NY.

Leontief WW. 1966. Input–Output Economics. Oxford University Press, New York, NY.

Levine SH. 1977. Exploitation interactions and the structure of ecosystems. J. Theor. Biol. 69, 345–355.

Levine SH. 1980. Several measures of trophic structure applicable to complex food webs. J. Theor. Biol. 83, 195–207.

Levine SH. 1988. A dynamic formulation for input–output modeling of exploited ecosystems. Ecol. Model. 44, 143–151.

Li W-H, Grauer D. 1991. Fundamentals of Molecular Evolution. Sinauer, Sunderland, MA, 430 pp.

Liebig J. 1840. Chemistry in its Application to Agriculture and Physiology. Taylor and Walton, London.

Light SS, Dineen JW. 1994. Water control on the Everglades: A historical perspective. In: Davis SM, Ogden JC (eds.), Everglades: The Ecosystem and its Restoration. St. Lucie Press, Delray Beach, FL, pp. 47–84.

Likens GE, Borman FH, Pierce RS, Eaton JS, Johnson NM. 1977. The Biogeochemistry of a Forested Ecosystem. Springer-Verlag, New York, NY, 146 pp.

Lillebø AI, Flindt M, Pardal MA, Marques JC. 1999. The effect of macrofauna, meiofauna and microfauna on the degradation of *Spartina maritima* detritus from a salt marsh area. Acta Oecol. 20, 249–258.

Lima-de-Faria A. 1988. Evolution without Selection. Form and Function by Autoevolution. Elsevier, Amsterdam, 372 pp.

Lindeman RL. 1942. The trophic-dynamic aspect of ecology. Ecology 28, 399–418.

Lipton presents a scenario (available at http://www.brucelipton.com/newbiology.php).

Lodge TE. 1994. The Everglades Handbook: Understanding the Ecosystem. St. Lucie Press, Delray Beach, FL.

Loftus WF, Kushlan JA. 1987. Freshwater fishes of southern Florida. Bulletin Florida State Museum. Biol. Sci. 31, 147–344.

Loiselle SA, Bracchini L, Bonechi C, Rossi C. 2001. Modelling energy fluxes in remote wetland ecosystems with the help of remote sensing. Ecol. Model. 145, 243–261.

Lotka AJ. 1922. Contribution to the energetics of evolution. Proc. Natl. Acad. Sci. U.S.A. 8, 147–150.

Lotka AJ. 1925. Elements of Physical Biology. Williams and Wilkins, Baltimore, MD.

Lotka AJ. 1956. Elements of Mathematical Biology. Dover Publications, New York, NY, 465 pp.

Lovelock JE. 1979. Gaia: A New Look at Life on Earth. Oxford University Press, New York, NY, 157 pp.

Lowe EF, Battoe LE, Stites DL, Coveney MF. 1992. Particulate phosphorus removal via wetland filtration: an examination of potential for hypertrophic lake restoration. Environ. Man. 16, 67–74.

Lowe EF, Stites DL, Coveney MF. 1989. Potential role of marsh creation in restoration of hypereutrophic Lakes. In: Hammer DA (ed.), Constructed Wetlands for Wasterwater Treatment: Municipal, Industrial, and Agricultural. Lewis, Chelsea, MI, pp. 710–715.

Ludovisi A, Poletti A. 2003. Use of thermodynamic indices as ecological indicators of the development state of lake ecosystems. 1. Entropy production indices. Ecol. Model. 159, 203–222.

van der Maarel E. 1993. Some remarks on disturbance and its relations to diversity and stability. J. Veg. Sci. 4, 733–736.

MacArthur RH. 1955. Fluctuations of animal populations and a measure of community stability. Ecology 35, 533–536.

MacArthur RH, Pianka ER. 1966. On optimal use of a patchy environment. Am. Nat. 100, 603–609.

MacArthur RH, Wilson EO. 1967. The theory of island biogeography. Princeton University Press, Princeton, NJ.

Majerus MEN. 1998. Melanism. Evolution in Action. Oxford University Press, Oxford, UK.

Margalef R. 1977. Ecologia. Omega, Barcelona, Spain, p. 951.

Margalef RA. 1991. Networks in ecology. In: Higashi M, Burns TP (eds.), Theoretical Studies of Ecosystems: The Network Perspectives, Cambridge University Press, Cambridge, pp. 41–57.

Margalef RA. 1995. Information Theory and Complex Ecology. In: Patten BC, Jørgensen SE (eds.), Complex Ecology. Prentice-Hall PTR, New Jersey, pp. 40–50.

Margulis L. 1981. Symbiosis in Cell Evolution: Life and its Environment on the Early Earth. W.C. Freeman, San Francisco, CA, 312 pp.

Margulis L. 1991. Symbiogenesis and symbionticism. In: Margulis L, Fester R (eds.), Symbiosis as a Source of Evolutionary Innovation, MIT Press, Cambridge, MA, pp. 1–14, 346 pp.

Margulis L. 1992. Symbiosis in Cell Evolution: Microbial Communities in the Archean and Proterozoic Eons. W.H. Freeman, San Francisco, CA.

Margulis L, Sagan D. 1997. Microcosmos: Four Billion Years of Microbial Evolution. University of California Press, Berkeley, CA.

Marques JC. 1989. Amphipoda (Crustacea) Bentonicos da Costa Portuguesa: Estudo taxondmico, ecologico e biogeografico. Ph.D. Thesis. Faculdade de Ciencias e Tecnologia da Universidade de Coimbra, pp. 57–87.

Marques JC, Maranhao P, Pardal MA. 1993a. Human impact assessment on the subtidal macrobenthic community structure in the Mondego Estuary (Western Portugal). Est. Coast. Shelf Sci. 37, 403–419.

Marques JC, Nielsen SN, Pardal MA, Jørgensen SE. 2003. Impact of eutrophication and river management within a framework of ecosystem theories. Ecol. Model. 166, 147–168.

Marques JC, Pardal MA, Nielsen SN, Jørgensen SE. 1997. Analysis of the properties of exergy and biodiversity along an estuarine gradient of eutrophication. Ecol. Model. 102, 155–167.

Marques JC, Rodrigues LB, Nogueira AJA. 1993b. Intertidal macrobenthic communities structure in the Mondego estuary (Western Portugal): reference situation. Vie Millieu 43, 177–187.

Martins I, Pardal MA, Lillebø AI, Flindt MR, Marques JC. 2001. Hydrodynamics as a major factor controlling the occurrence of green macroalgal blooms in a eutrophic estuary. A case study on the influence of precipitation and river management. Est. Coast. Shelf Sci. 52, 165–177.

Maru, 1994. Piano di Gestione per il lago Trasimeno finalizzato al controllo dell'eutrofizzazione. Parte IV. Ministero dell'Ambiente e Regione Umbria (in Italian).

Matis JH, Patten BC. 1981. Environ analysis of linear compartmental systems: the static, time invariant case. Proceedings of the 42nd Session International Statistics Institute, Manila, Philippines, December 4–14.

Maturana HR, Varela FJ. 1980. Autopoiesis and Cognition: The Realization of the Living. D. Reidel, Dordrecht.

Mauersberger P. 1981. Entropie und freie Enthalpie in aquatische Ökosysteme. Acta Hydrodyn. 26, 67.

Mauersberger P. 1983. General principles in deterministic water quality modelling. In: Orlob GT (ed.), Mathematical Modeling of Water Quality: Streams, Lakes and Reservoirs (International Series on Applied Systems Analysis, 12). Wiley, New York, NY, pp. 42–115.

Mauersberger P. 1995. Entropy control of complex ecological processes. In: Patten BC, Jørgensen SE (eds.), Complex Ecology: The Part-Whole Relation in Ecosystems. Prentice-Hall Englewood Cliffs, NJ, pp. 130–165.

Maurer B, Brown JH. 1988. Distribution of energy use and biomass among species of North American terrestrial birds. Ecology 69, 1923–1932.

May RM. 1973. Stability and Complexity in Model Ecosystems. Princeton University Press, Princeton, NJ, 265 pp.

Mayr E. 1942. Systematics and the Origin of Species, Columbia University Press, Harvard University Press reprint.

Mayr E. 1982. The Growth of Biological Thought. Diversity, Evolution and Inheritance. Belknap Press, Cambridge, Ma, 974 pp.

McDonough W, Braungat M. 2002. Cradle to Cradle: Remaking the Way We Make Things. North Point Press, New York, NY, 193 pp.

Meadows DH, Meadows DL, Randers J, Behrens III WW. 1972. The Limits to Growth. Potomac Associates, New York, NY.

Mejer HF, Jørgensen SE. 1979. Energy and ecological buffer capacity. In: Jørgensen SE (ed.), State-of-the-Art of Ecological Modelling. (Environmental Sciences and Applications, 7). Proceedings of a Conference on Ecological Modelling, 28th August–2nd September 1978, Copenhagen. International Society for Ecological Modelling, Copenhagen, pp. 829–846.

Memoirs, 68, pp. 3–65. (Available at: http://newton.nap.edu/html/biomems/welsasser.html)

Minshall GW. 1988. Stream Ecosystem Theory: A Global Perspective. J. N. Am. Benthol. Soc. 7(4); Proceedings of a Symposium: Community Structure and Function in Temperate and Tropical Streams, pp. 263–288.

Minshall GW, Cummins KW, Petersen RC, Cushing CE, Bruns DA, Sedell JR, Vannote RL. 1985. Developments in stream ecosystem theory. Can. J. Fish. Aquat. Sci. 42, 1045–1055.

Mitamura O, Sijo Y. 1981. Studies on the seasonal changes of dissolved organic carbon, nitrogen, phosphorus and urea concentrations in Lake Biwa. Arch. Hydrobiol. 91, 1–14.

Morowitz HJ. 1968. Energy flow in biology. Biological Organisation as a Problem in Thermal Physics. Academic Press, New York, NY, 179 pp.

Morowitz HJ. 1992. Beginnings of Cellular Life. Yale University Press, New Haven, CT.

Moyle PB, Cech JJ Jr. 1996. Fishes: An Introduction to Ichthyology. 3rd ed. Prentice-Hall, Upper Saddle River, NJ.

Müller F. 1992. Hierarchical Approaches to Ecosystem Theory. Ecol. Model. 63, 215–242.

Müller F. 1998. Gradients, potentials and flows in ecological systems. Ecol. Model. 108, 3–21.

Müller F. 2004. Ecosystem indicators for the integrated management of landscape health and integrity. In: Joergensen SE, Costanza R (eds.), Handbook of Ecological Indicators for Assessment of Ecosystem Health. CRC Publishers, Boca Raton, FL, pp. 277–304.

Müller F. 2005. Indicating ecosystem and landscape organization. Ecol. Indicators 5, 280–294.

Müller F, Fath BD. 1998. The physical basis of ecological goal functions—an integrative discussion. In: Müller F, Leupelt M (eds.), Eco Targets, Goal Functions, and Orientors. Springer, Berlin, Germany, pp. 269–285.

Müller F, Jørgensen SE. 2000. Ecological orientors—a path to environmental applications of ecosystem theories. In: Jørgensen SE, Müller F (eds.), Handbook of Ecosystem Theories and Management. Lewis Publishers, Boca Raton, FL, pp. 561–576.

Müller F, Leupelt M. (eds.). 1998. Eco Targets, Goal Functions and Orientors. Springer Verlag, Berlin, Germany, 618 pp.

Müller FJ, Schrautzer EW, Reiche, Rinker A. (in print) Ecosystem based indicators in retrogressive successions of an agricultural landscape. Ecol. Indicators.

Murray BG. 2001. Are ecological and evolutionary theories scientific? Biol. Rev. 76, 255–289.

National Institute for Research Advancement, 1984. Data Book of World Lakes. National Institute for Research Advancement, Tokyo, Japan.

Niklas L. 2003. Soil fauna and global change. Doctoral dissertation. Dept. of Ecology and Environmental Research, SLU. Acta Universitatis agriculturae Sueciae. Silvestria vol. 270.

O'Neill RV, DeAngelis DL, Waide JB, Allen TFH. 1986. A Hierarchical Concept of Ecosystems. Princeton University Press, Princeton, NJ, 286 pp.

von Neumann J, Morgenstern O. 1944. Theory of Games and Economic Behavior. 1st ed. Princeton University Press, Princeton, NJ.

von Neumann J, Morgenstern O. 1947. Theory of Games and Economic Behavior. 2nd ed. Princeton University Press, Princeton, NJ, 625 pp., 641 pp.

Nielsen SN. 1992a. Application of maximum exergy in structural dynamic models, Ph.D. Thesis. National Environmental Research Institute, Denmark, 51 pp.

Nielsen SN. 1992b. Strategies for structural-dynamical modelling. Ecol. Model. 63, 91–102.

Nielsen SN. (in print). Towards an ecosystem semiotics. Some basic aspects for a new research programme. Ecol. Complexity.

Nielsen SN, Müller F. 2000. Emergent Properties of Ecosystems. In: Jørgensen SE, Müller F (eds.), Handbook of Ecosystem Theories and Management. Lewis Publishers, Boca Raton, FL, pp. 195–216.

Noether, A. 1918. Invariante Variationsprobleme. Nachr.v.d.Ges.d.Wiss.zu Goettingen, Math.–Phys. Kl (1918), 235–257 (English translation by MA Tavel).

Odling-Smee FJ, Laland KL, Feldman MW. 2003. Niche Construction. The Neglected Process in Evolution. Princeton University Press, Princeton, NJ, 472 pp.

Odum EP. 1959. Fundamentals of Ecology. 2nd ed.. W.B. Saunders, Philadelphia, PA, 546 pp.

Odum EP. 1969. The strategy of ecosystem development. Science 164, 262–270.

Odum EP. 1971a. Fundamentals of Ecology. 3rd ed. W.B. Saunders Co., Philadelphia, PA, 360 pp.

Odum HT. 1971b. Environment, Power, and Society. Wiley, New York, NY, 331 pp.

Odum HT. 1983. System Ecology. Wiley, New York, NY, 510 pp.

Odum HT. 1988. Self-organization, transformity, and information. Science 242, 1132–1139.

Odum HT. 1989. Experimental study of self-organization in estuarine ponds. In: Mitsch WJ, Jørgensen SE (eds.), Ecological Engineering: An Introduction to Eco-Technology. Wiley, New York, NY.

Odum HT. 1996. Environmental Accounting: Emergy and Environmental Decision Making. Wiley, New York, NY.

Odum HT, Pinkerton RC. 1955. Time's speed regulator: the optimum efficiency for maximum power output in physical and biological systems. Am. Sci. 43, 331–343.

Ostwald W. 1931. Gedanken zur Biosphäre. Wiederabdruck. BSB B. G. Teubner Verlagsgesellschaft, Leipzig, Germany, p. 1978.
Pahl-Wostl C. 1998. Ecosystem organization across a continuum of scales: A comparative analysis of lakes and rivers. In: Peterson DL, Parker VT (eds.), Ecological Scale: Theory and Applications. Columbia University Press, New York, NY, pp. 141–170.
Paine RT. 1980. Food webs: linkage, interaction strength and community infrastructure. J Anim Ecol. 49, 667–685.
Patricio J, Sala F, Pardal MA, Jørgensen MA, Marques JC. 2006. Ecological Indicators performance during a re-colonization field experiment and its compliance with ecosystem theories. Ecol. Model. 6, 46–59.
Patricio J, Ulanowicz R, Pardal MA, Marques JC. 2004. Ascendency as an ecological indicator: a case study of estuarine pulse eutrophication. Est. Coast. Shelf Sci. 60, 23–35.
Patten BC. 1978. Systems approach to the concept of environment. Ohio J. Sci. 78, 206–222.
Patten BC. 1981. Environs: the super-niches of ecosystems. Am. Zool. 21, 845–852.
Patten BC. 1982. Environs: relativistic elementary particles or ecology. Am. Nat. 119, 179–219.
Patten BC. 1983. Linearity enigma in ecology. Ecol. Model. 18, 155–170.
Patten BC. 1985. Energy cycling in the ecosystem. Ecol. Model. 28, 1–71.
Patten BC. 1991. Network ecology: Indirect determination of the life–environment relationship in ecosystems. In: Higashi M, Burns TP (eds.), Theoretical Ecosystem Ecology: The Network Perspective. Cambridge University Press, London pp. 288–351.
Patten BC. 1992. Energy, emergy and environs. Ecol. Mod. 62, 29–69.
Patten BC, 1997. Synthesis of chaos and sustainability in a non stationary linear dynamic model of the American black bear (Ursus americanus Pallas) in the Adirondack Mountains of New York. Ecol. Model. 100, 11–42.
Patten BC. 2005. Network perspectives on ecological indicators and actuators: enfolding, observability, and controllability. Ecol. Indicators (in press).
Patten BC, Auble GT. 1981. System theory of the ecological niche. Am. Nat. 117, 893–922.
Patten BC, Bosserman RW, Finn JT, Cale WG. 1976. Propagation of cause in ecosystems. Read pp. 458–471. In: Patten, BC (ed.), Systems Analysis and Simulation in Ecology, vol. 4. Academic Press, NY, pp. 457–579.
Patten BC, Higashi M, Burns TP. 1990. Trophic dynamics in ecosystem networks: significance of cycles and storages. Ecol. Model. 51, 1–28.
Patten BC. Holoecology. The Unification of Nature by Network Indirect Effects. Complexity in Ecological Systems Series, Columbia University Press, New York, NY (in preparation).
Patten BC, Jørgensen SE. 1995. Complex Ecology. The Part-Whole Relation in Ecosystems. Prentice Hall PTR, New Jersey, 706 pp.
Patten BC, Straskraba M, Jørgensen SE. 1997. Ecosystem emerging 1: conservation. Ecol. Model. 96, 221–284.
Peters RH. 1983. The Ecological Implications of Body Size. Cambridge University Press, Cambridge, MA, 329 pp.

Peterson RT. 1963. A field guide to the birds of Texas and Adjacent States. The Peterson Field Guide Series. Houghton Mifflin, Boston, MA.

Picket STA, Collings S, Armesto JJ. 1987. A hierarchical consideration of succession. Vegetatio 69, 109–114.

Picket STA, White PS. 1985. Natural disturbance and patch dynamics: An introduction. In: Picket STA, White PS (eds.), The Ecology of Disturbance and Patch Dynamics. Academic Press, New York, NY, pp. 3–13.

Pimm SL. 2002. Food Webs. University of Chicago Press, Chicago, IL.

Polis GA, 1991. Complex trophic interactions in deserts: An empirical critique of food-web theory. Am. Nat. 138, 123–155.

Polunin N (ed.). 1986. Ecosystem Theory and Application. Wiley, New York, NY.

Popper KR. 1982. The Open Universe: An Argument for Indeterminism. Rowman and Littlefield, Totowa, NJ, 185 pp.

Popper KR. 1990. A World of Propensities. Thoemmes, Bristol, 51 pp.

Post DM, Doyle MW, Sabo JL, Finlay JC. 2005. The problem of boundaries in defining ecosystems: a potential landmine for uniting geomorphology and ecology. http://www.yale.edu/post_lab/publications/pdfs/Post_et_al_2005_(Geomorph).pdf

Power ME, Rainey WE. 2000. Food webs and resource sheds: Towards spatially delimiting trophic interactions. In: Hutchings MJ, John EA, Stewart AJA (eds.), Ecological Consequences of Habitat Heterogeneity. Blackwell Scientific, Oxford, UK, pp. 291–314.

Price PW. 1992. Plant resources and insect herbivore population dynamics. In: Hunter MD, Ohgushi T, Price PW (eds.), Effects of Resource Distribution on Animal-Plant Interactions. Academic Press, San Diego, CA, pp. 139–176.

Prigogine I. 1947. L' Etude Thermodynamique des Processus Irreversibles. Desoer, Liege.

Prigogine I. 1955. Introduction to Thermodynamics of Irreversible Processes. C.C. Thomas, Springfield, IL, 115 pp.

Prigogine I. 1962. Non Equilibrium Statistical Mechanics. Wiley, New York, NY.

Prigogine I. 1978. Time, structure and fluctuations. Science 201, 777–785.

Prigogine I. 1980. From Being to Becoming: Time and Complexity in the Physical Sciences. Freeman, San Franscisco, CA, 220 pp.

Prigogine I. 1997. The End of Certainty — Time's Flow and the Laws of Nature. The Free Press, New York, NY.

Prigogine I, Stengers I. 1979. La Nouvelle Alliance, Gallimard, Paris.

Prigogine I Stengers I. 1984. Order out of Chaos: Man's New Dialogue with Nature. Bantam Books, New York, NY.

Prigogine I Stengers I. 1988. Entre le Temps et l'Eternité. Fayard, Paris.

Rafaelli D, Limia J, Hull S, Pont S. 1991. Interactions between the amphipod *Corophium volutator* and macroalgal mats on estuarine mudflats. J. Mar. Biol. Assoc. UK 71, 899–908.

Rant Z.1955. Losses in Energy Transformations (in Slovenian) Strojniski Vestnik. 1, 4–7.

Raup D, Sepkoski J. 1986. Periodic extinction of families and genera. Science 231, 833–836.

Reice SR. 1994. Nonequilibrium determinants of biological community structure. Am. Sci. 82, 424–435.

Reiche EW. 1996. WASMOD. Ein Modellsystem zur gebietsbezogenen Simulation von Wasser- und Stoffflüssen. Ecosystem 4, 143–163.

Remmert H. 1991. The Mosaic-Cycle Concept of Ecosystems. Ecological Studies 85. Springer, Berlin, Germany, 168 pp.

Reynolds CS. 1984. The Ecology of Freshwater Phytoplankton. Cambridge University Press, Cambridge, MA, 384 pp.

Reynolds CS. 1997. Vegetation Processes in the Pelagic: A Model for Ecosystem Theory. (Excellence in Ecology no. 9). Ecological Institute, Oldendorf.

Richardson H, Verbeek NA. 1986. Diet selection and optimization by north-western crows feeding in Japanese littleneck clams. Ecology, 67, 1219–1226.

Richmond B. 2001. Introduction to Systems Thinking. High Performance Systems, Inc., Winter Haven, FL, 548 pp.

Ricklefs RE. 2004. A comprehensive framework for global patterns in biodiversity. Ecol. Lett. 7, 1–15.

Riley CV. 1892. The yucca moth and yucca pollination. Third Annual Report of the Missouri Botanical Garden, pp. 99–158.

Robertson WB Jr., Kushlan JA. 1984. The southern Florida avifauna. In: Gleason PJ (ed.), Environments of Southern Florida: Present and Past II. Memoir II. Miami Geological Society, Miami, FL.

Rohde K. 1992. Latitudinal gradients in species diversity: the search for the primary cause. Oikos 65, 514–527.

Rosenzweig ML. 1992. Species diversity gradients: we know more and less than we thought. J. Mammal. 73, 715–730.

Rosenzweig ML. 1996. Species Diversity in Space and Time. Cambridge University Press, Cambridge, MA, 288 pp.

Roy K, Jablonski D, Valentine JW, Rosenberg G. 1998. Marine latitudinal diversity gradients: tests of causal hypothesis. Proc. Natl. Acad. Sci. USA 95, 3699–3702.

Rubin H. 1995. Walter M. Elsasser (Biographical Memoirs). US National Academy of Sciences.

Russell B. 1960. An Outline of Philosophy. Meridian Books, Cleveland, OH.

Rutledge RW, Basorre BL, Mulholland RJ. 1976. Ecological stability: an information theory viewpoint. J. Theor. Biol. 57, 355–371.

Sakamoto M. 1975. Trophic relation and metabolism in ecosystem. In: Mori S, Yamamoto G (eds.), Productivity of Communities in Japanese Inland Walters (JIBP Synthesis, vol. 10). University of Tokyo, Tokyo, Japan, pp. 405–410.

Sandberg J, Elmgren R, Wulff F. 2000. Carbon flows in Baltic Sea food webs—a re-evaluation using a mass balance approach. J. Mar. Sys. 25, 249–260.

Scheffer M. 1998. Ecology of Shallow Lakes. Chapman and Hall, London, UK, 357 pp.

Scheffer M, Carpenter S, Foley JA, Folke C, Walker B. 2001. Catastrophic shifts in ecosystems. Nature 413, 591–596.

Schelske C, Brezonik P. 1992. Can Lake Apopka be restored? In: Maurizi S, Poillon F (eds.), Restoration of Aquatic Ecosystems: Science, Technology, and Public Policy. National Academy Press, Washington, DC, pp. 393–398.

Schlesinger WH. 1997. Biogeochemistry. An Analysis of Global Change. 2nd ed. Academic Press, Sand Diego, CA, 680 pp.

Schmidt M. 2000. Die Blaugras-Rasen des nördlichen deutschen Mittelgebirgsraumes und ihre Kontaktgesellschaften. Dissertation in Botany 328, Berlin, Germany, 294 pp.

Schneider ED, Kay JJ. 1994. Life as a manifestation of the second law of thermodynamics. Math. Comput. Model. 19, 25–48.

Schneider ED, Sagan D. 2005. Into the Cool: Energy Flow, Thermodynamics and Life. University of Chicago Press, Chicago, IL, 362 pp.

Schramski JR, Gattie DK, Patten BC, Borrett SR, Fath BD, CR Thomas, Whipple SJ. 2006. Indirect effects and distributed control in ecosystems: distributed control in the environ networks of a seven-compartment model of nitrogen flow in the Neuse River Estuary, USA—Steady-state analysis. Ecol. Model. 194, 189–201.

Schrautzer J. 2003. Niedermoore Schleswig-Holsteins: Charakterisierung und Beurteilung ihrer Funktion im Landschaftshaushalt. Mitt. AG Geobot. Schl.-Holst. u. Hambg. (Kiel) 63, 350.

Schrautzer J, Jensen K. 1999. Quantitative und qualitative Auswirkungen von Sukzessionsprozessen auf die Flora der Niedermoorstandorte Schleswig-Holsteins. Z. Ökologie u. Natursch. 7, 219–240.

Schrödinger E. 1944. What is Life? The Physical Aspect of the Living Cell. Cambridge University Press, Cambridge, MA, 90 pp.

Schumpeter JA. 1942. Capitalism, Socialism and Democracy. Harper and Row, New York, NY.

Scienceman D. 1987. Energy and Emergy. In: Pillet G, Murota T (eds.), Environmental Economics. Roland Leimgruber, Geneva, Switzerland, pp. 279–297.

Shannon CE. 1948. A mathematical theory of communication. Bell Syst. Tech. J. 27, 379–423.

Sheley RL. 2002. Plant Invasion and Succession. Chapter 3 in Invasive Plant Management. NSF Center for Integrated Pest Management, Online Textbook. http://www.weedcenter.org/textbook/3_sheley_invasion_succession.html

Shelford VE. 1913. Animal Communities in Temperate America. University of Chicago Press, Chicago, IL, 320 pp.

Shugart HH. 1998. Terrestrial Ecosystems in Changing Environments. Cambridge University Press, New York, NY, 534 pp.

Shugart HH, West DC. 1981. Long-term dynamics of forest ecosystems. Am. Sci. 25, 25–50.

Simberloff D. 1982. A succession of paradigms in ecology: Essentials to materialism and probabilism. In: Saarinen ESA (ed.), Conceptual Issues in Ecology. D. Reichel, Dordrecht, 75–76.

Simek SL. 1995. Impact of forest land management on black bear populations in the Central Adirondacks. M.S. Thesis. State University of New York, College of Environmental Science and Forestry, Syracuse, NY.

Simon HA. 1973. The organization of complex systems. In: Pattee HH (ed.), Hierarchy Theory — The Challenge of Complex Systems. Braziller, NY, pp. 1–27.

Simpson GG. 1944. Tempo and Mode in Evolution. Columbia University Press, New York, NY.

Simpson GG. 1964. Species density of North American recent mammals. Syst. Zool. 13, 57–73.

Solbrig OT. (ed.). 1991. From Genes to Ecosystems: A Research Agenda for Biodiversity. IUBS-SCOPE-UNESCO, Cambridge, MA, 124 pp.

Sousa WP. 1984. The role of disturbance in natural communities. Ann. Rev. Ecol. Syst. 15, 353–391.
Stebbins GL. 1950. Variation and Evolution in Plants. Columbia University Press, New York, NY, 643 pp.
Stent G. 1981. Strength and weakness of the genetic approach to the development of the nervous system. Ann. Rev. Neurosci. 4, 16–194.
Stevens RD, Willig MR. 2002. Geographical ecology at the community level: perspectives on the diversity of New World bats. Ecology 83, 545–560.
Stewart KM. 1973. Detailed time variations in mean temperature and heat content in some Madison Lakes. Limnol. Oceanogr. 18, 218–226.
Straskraba M, Jørgensen SE, Patten BC. 1999. Ecosystems emerging: 2. Dissipation. Ecol. Model. 117, 3–39.
Svirezhev Yu. M. 1992. Thermodynamic Ideas in Ecology, Lecture on the School of Mathematical Ecology, ICTP, Trieste.
Svirezhev Yu. M. 1998. Thermodynamic Orientors: hows to use thermodynamic concepts in ecology. In: Muller F, Leupelt M (eds.), Eco Targets. Gola Functions and Orientors. Springer, Berlin. pp. 102–122.
Svirezhev Yu. M. 2000. Stability concepts in ecology. In: Jørgensen SE, Müller F (eds.), Handbook of Ecosystem Theories and Management. Lewis, Boca Raton, FL, pp. 361–384.
Szargut J, Morris DR, Steward FR. 1988. Exergy Analysis of Thermal, Chemical and Metallurgical Processes. Hemisphere Publishing Corporation, New York, NY, 312 pp.
Szyrmer J, Ulanowicz RE. 1987. Total flows in ecosystems. Ecol. Model. 35, 123–136.
Tansley AG. 1935. The use and abuse of vegetational concepts and terms. Ecology 16, 284–307.
Tiezzi E. 2003a. The Essence of Time. WIT Press, Southampton, UK, 125 pp.
Tiezzi E. 2003b. The End of Time. WIT Press, Southampton, UK, 202 pp.
Tiezzi E. 2005 Beauty of Science. WIT Press, Southampton, UK, 362 pp.
Tiezzi E. 2006a. Is entropy far from equilibrium a state function? Ecodynamics 1, 44–54.
Tiezzi E. 2006b. Steps Towards an Evolutionary Physics. WIT Press, Sussex, UK, 157 p
Tilley DR, Brown MT. 2006. Dynamic emergy accounting for assessing the environmental benefits of subtropical wetland stormwater management systems. Ecol. Model. 192, 327–361.
Tilly LJ. 1968. The structure and dynamics of Cone Spring. Ecol. Monogr. 38, 169–197.
Tilman D, Downing JA. 1994. Biodiversity and stability in grasslands. Nature 367, 3633–3635.
Tribus M, McIrvine EC. 1971. Energy and information. Sci. Am. 225, 179–188.
Trotter RT. 2000. Ethnography and network analysis: The study of social context in cultures and societies. In: Albrecht GL, Fitzpatrick R, Scrimshaw SC (eds.), Handbook of Studies in Health and Medicine. Sage, London, UK, pp. 210–229.
Turner FB. 1970. The ecological efficiency of consumer populations. Ecology, 51, 741–742.
Uexküll J. von 1926. Theoretical Biology. Kegan, Paul, Trench, Tubner, London.
Ulanowicz RE. 1980. An hypothesis on the development of natural communities. J. Theor. Biol. 85, 223–245.

Ulanowicz RE. 1983. Identifying the structure of cycling in ecosystems. Math. Biosci. 65, 219–237.

Ulanowicz RE. 1986a. Growth & Development: Ecosystems Phenomenology. Springer-Verlag, New York, NY, 203 pp.

Ulanowicz RE. 1986b. A phenomenological perspective of ecological development. In: Poston TM, Purdy R (eds.), Aquatic Toxicology and Environmental Fate. ASTM STP 921, vol. 9. American Society for Testing and Materials, Philadelphia, PA, USA, pp. 73–81.

Ulanowicz RE. 1995. Utricularia's secret: the advantages of positive feedback in oligotrophic environments. Ecol. Model. 79, 49–57.

Ulanowicz RE. 1996. The propensities of evolving systems. In: Khalil EL, Boulding KE (eds.), Evolution, Order and Complexity. Routledge, London, UK, pp. 217–233.

Ulanowicz RE. 1997. Ecology, the Ascendent Perspective. Columbia University Press, New York, NY.

Ulanowicz RE. 1999. http://www.glerl.noaa.gov/EcoNetwrk/. Retrieved March 22, 2006.

Ulanowicz RE. 2000. Ascendency: A measure of ecosystem performance. In: Jørgensen SE, Müller F (eds.), Handbook of Ecosystem Theories and Management. Boca Raton, FL, pp. 303–316.

Ulanowicz RE. 2002a. Ecology, a dialogue between the quick and the dead. Emergence 4, 34–52.

Ulanowicz RE. 2002b. Information theory in ecology. Comput. Chem. 25, 393–399.

Ulanowicz RE. 2004a. On the nature of ecodynamics. Ecol. Complexity 1, 341–354.

Ulanowicz RE. 2004b. Quantitative methods for ecological network analysis. Comput. Biol. Chem. 28, 321–339.

Ulanowicz RE. 2006a. Fluctuations and Order in Ecosystem Dynamics. Emerg. Complex. Org. 7(2), 14–20.

Ulanowicz RE. 2006b. Process ecology: A transactional worldview. J. Ecodynamics. 1, 103–114.

Ulanowicz RE, Abarca-Arenas LG. 1997. An informational synthesis of ecosystem structure and function. Ecol. Model. 95, 1–10.

Ulanowicz RE, Baird D. 1999. Nutrient controls on ecosystem dynamics: The Chesapeake mesohaline community. J. Mar. Syst. 19, 159–172.

Ulanowicz RE, Bondavalli C, Egnotovich MS. 1997. Network Analysis of Trophic Dynamics in South Florida ecosystems: The Cypress Wetland Ecosystem. Annual Report to USGS/BRD, Coral Gables, p. 33.

Ulanowicz RE, Jørgensen SE, Fath BD. 2006. Exergy, information and aggradation: an ecosystems reconciliation. Ecol. Model (in press).

Ulanowicz RE, Kay JJ. 1991. A package for the analysis of ecosystem flow networks. Environ. Software 6, 131–142.

Ulanowicz RE, Kemp WM. 1979. Toward a canonical trophic aggregation. Am. Nat. 114, 871–883.

Ulanowicz RE, Norden JS. 1990. Symmetrical overhead in flow networks. Int. J. Syst. Sci. 1, 429–437.

Ulanowicz RE, Wulff F. 1991. Comparing ecosystem structures: The Chesapeake Bay and the Baltic Sea. In: Cole J, Lovett G, Findlay S (eds.), Comparative Analyses of

Ecosystems: Patterns, Mechanisms and Theories. Springer-Verlag, New York, NY, pp. 140–166.

Urban DL, O'Neill RV, Shugart HH. 1987. A hierarchical perspective can help scientists understand spatial patterns. Bio. Sci. 37, 119–127.

Vannote RL, Minshall JV, Cummins KW, Seddell JR, Cushing CE. 1980. The river continuum concept. Can. J. Fish. Aquat. Sci. 37, 130–137.

Viaroli P, Azzoni R, Martoli M, Giordani G, Tajè L. 2001. Evolution of the trophic conditions and dystrophic outbreaks in the Sacca di Goro lagoon (Northen Adriatic Sea). In: Faranda FM, Guglielmo L, Spezie G (eds.), Structures and Processes in the Mediterranean Ecosystems. Springer-Verlag, Milano, Italia, pp. 443–451.

Vitousek PM. 1994. Beyond global warming: ecology and global change. Ecology 75, 1861–1876.

Vollenweider RA. 1975. Input-output models with special references to the Phosphorus loading concept in limnology. Schweiz. Z. Hydrol. 37, 53–84.

Walker B, Holling CS, Carpenter SR, Kinzig A. 2004. Resilience, Adaptability and transformability on social-ecological systems. Ecol. Soc. 9, Article 5.

Walker B, Meyers JA. 2004. Thresholds in ecological and socio-ecological systems: a developing database. Ecol. Soc. 9, Article 3.

Wall G. 1977. Exergy – a useful concept within resource accounting. Report: Physical Resources Theory. Chalmers S-412 96. Gøteborg.

Wallace AR. 1858. On the tendency of species to form varieties; and on the perpetuation of varieties and species by natural means of selection. III. On the tendency of varieties to depart indefinitely from the original type. J. Proc. Linn. Soc. London 3, 53–62.

Wallace AR. 1889. Darwinism. An Exposition of the Theory of Natural Selection with some of its Applications. MacMillan and Co., London.

Wang L, Zhang J, Ni W. 2005. Emergy evaluation of Eco-Industrial Park with Power Plant. Ecol. Model. 189, 233–240.

Warming E. 1909. Oecology of plants; an introduction to the study of plant-communities. Clarendon Press, Oxford, UK, 422 pp.

Wasserman S, Faust K. 1994. Social Network Analysis: Methods and Applications. Cambridge University Press, Cambridge, MA, 825 pp.

Watt KEF. 1968. Ecology and Resource Management. McGraw-Hill, New York, NY.

Weins JA. 1989. The Ecology of Bird Communities. Cambridge University Press, Cambridge, MA.

Wellman B. 1983. Network analysis: Some basic principles. In: Colins R (ed.), Sociological Theory. Jossey-Bass, San Francisco, CA, pp. 155–200.

Weismann A. 1885. The Continuity of the Germ-Plasm as the Foundation of a Theory of Heredity. In: Essays upon Hereditary and Kindred Biological Problems. Oxford University Press, Oxford, UK.

Welty JC. 1976. The Life of Birds. Saunders, New York, NY.

Whipple SJ. 1998. Path-based network unfolding: a solution for the problem of mixed trophic and non-trophic processes in trophic dynamic analysis. J. Theor. Biol. 190, 263–276.

Whipple SJ, Patten BC. 1993. The problem of non-trophic processes in trophic ecology: a network unfolding solution. J. Theor. Biol. Vol. 165, 393–411.

White PS, Jentsch A. 2001. The search for generality in studies of disturbance and ecosystem dynamics. Progress in Botany 62, 2–3. Checkliste für die Anwendung. In: Fränzle O, Müller F, Schröder W (eds.), Handbuch der Umweltwissenschaften, chapter III. Ecomed-Verlag, Landsberg, p. 2.3.

Whittacker RH. 1953. A consideration of climax theory: the climax as a population and pattern. Ecol. Mongr. 23, 41–78.

Whittacker RH. 1972. Recent evolution of ecological concepts in relation to eastern forests of North America. In: Egerton F (ed.), History of American Ecology. Arno Press, New York, NY, p. 347.

Whittaker RH. 1992. Stochasticism and determinism in island ecology. J. Biogeogr. 19, 587–591.

Wiley MJ, Osborne LL, Larimore RW. 1990. Longitudinal structure of an agricultural prairie river system and its relationship to current stream ecosystem theory. Can. J. Fish. Aquat. Sci. 47(2), 373–384.

Williams GC. 1966. Adaptation and Natural Selection: A Critique of some Current Evolutionary Thought. Princeton University Press, Princeton, NJ, 290 pp.

Williams RJ. 1998. Biochemical Individuality. University of Texas Press, Austin, TX, 198 pp.

Witman JD, Etter RJ, Smith F. 2004. The relationship between regional and local species diversity in marine benthic communities: A global perspective. PNAS. 101 (44), 15664-15669.

Wit, Rutger De. 2005. Do all ecosystems maximise their distance with respect to thermodynamic equilibrium? A comment on the Ecological Law of Thermodynamics (ELT) proposed by Sven Erik Jørgensen. Sci. Marina 69 (3), 427–434.

Wulff F, Field JG, Mann KH. (eds.). 1989. Network Analysis in Marine Ecology: Methods and Applications. Springer-Verlag, Berlin, Germany, 284 pp.

Zaldivar JM, Austoni M, Plus M, De Leo GA, Giordani G, Viaroli P, 2005. Ecosystem Health Assestment and Bioeconomic Analysis in Coastal Lagoon. Handbook of Ecological Indicator for Assessment of Ecosystem Health. CRC Press

Zadeh LA, Desoer CA, Linvill W, Zadeh LA, Dantzia G. 1963. Linear System Theory: The State Space Approach. McGraw-Hill Book Co., New York, NY.

Zaldívar JM, Cattaneo E, Plus M, Murray CN, Giordani G, Viaroli P. 2003. Long-term simulation of main biogeochemical events in a coastal lagoon: Sacca Di Goro (Northern Adriatic Coast, Italy). Cont. Shelf Res. 23, 1847–1875.

Index

Adaptability, 156–158, 160, 161, 166
Adaptive cycle, 114, 134, 135, 137, 158–161
Adjacency matrix, 87
Agenda 21, 1
Aggradation, 88–90, 95
Allometry, 22–26
Anabolism, 13
Anenergy, 14
Ascendency, 69–73, 77, 78, 89, 121, 122, 124, 125, 199, 210–218, 231–234, 237–240
Autocatalysis, 65, 67, 69, 76
Autocatalytic selection, 68, 76
Average mutual constraints, 71
Average mutual information, 71

Bernard cell, 120
Bifurcation, 153, 162
Biodiversity, 1, 111, 153, 180–184, 235–240
Biomanipulation, 197
Biomass packing, 127
Bottom-up, 85
Boundary simplification, 96
Buffer capacity, 111

Cardinal network hypothesis, 85, 92–101
C3-plants, 127
C4-plants, 127
CAM, 127
Catabolism, 13
Cell size, 31
Centripetality, 65, 66, 110
Chaos, 162, 163
Closed system, 8
Competitive exclusion, 189–192, see also Gause's competitive exclusion
Complexity, 155, 156
Compton effect, 41, 42
Conservation principle, 246
Constraints, 71, 113, 144
Coupling, 85, see also network

Creativity, 145–148
Creative selection, 44–46
Cycling rate, 127, 128

Darwin's finches, 130, 131, 171
Development capacity, 73, 212–214
Dissipative structure, 20–21
Distributed causation, 85
Distributed control, 93
Distribution, 153, 154
Diversity, see biodiversity

Eco-exergy, 15, 16, 75, 76, 88, 107, 111, 118, 121–127, 139–141, 157, 158, 164, 174, 175, 199, 218–222, 247–249
Eco-exergy calculations, 16, 17, 122, 123
Eco-exergy/emergy ratio, 228–231
Ecological Law of Thermodynamics, see Exergy Storage Hypothesis
Ecosystem history, 243–245
Emergence, 53, 54, 57
Emergy, 84, 98, 117, 120, 221–227
Energy circuit language, 222, 223
Energy forms, 7, 149–151
Entropy, 9, 11, 20, 21, 51, 52, 200–206
Entropy paradox, 51–52
Environ, 80, 85, 99, 101
Environ theory, 85, 92
Envirotype, 55, 56, 101
Eukaryote cells, 30, 31
Evolution, 66, 67, 76, 77
Evolutionary drive, 66, 67
Evolutionary theory, 168, 175
Exergy, 14, see also eco-exergy and Exergy storage
Exergy balance, 139–141
Exergy degradation, 137, 165
Exergy efficiency, 139–141, 165
Exergy storage, 123–129, 133, 137, 139–141, 165, 247–249
Exergy Storage Hypothesis, 123–129, 141, 173, 191, 198, 247–249

Exergy utilization, 135, 136
External variable, 14

Finn's cycling index, 90, 240
First law of thermodynamics, 9–11
Flow analysis, 87–90
Food web, 82, 83
Foraging time, 185–187

Gause's exclusion principle, 105, 189, 192
Gaya theory, 76
Genetic drift, 169
Genotype, 55, 56, 101
Goal function, 111–115
Gradients, 181
Growth forms, 112, 137–139, 141, 142, 184, 185

Heisenberg, 41, 47–49
Hierarchical level, 28
Hierarchy, 3, 27, 28, 55, 56, 149–151, 153, 249
Hierarchy theory, 27–29, 55–56, 245–247, 249
Holoevolution, 98
Homogenization, 93
Human well-being, 2
Hysteresis, 196, 197

Indeterminism, 41–42
Indicator, 111–115, 199–241
Indirect effect, 79, 87, 94, 207, 249, 250
Information, 105–110, 141, 157, 175, 199, 249
Institute of Applied Systems Analysis, 85
Integral relations, 91
Intermediate disturbance, 153, 154
Internal simplification, 93
Internal variable, 14
Irreversibility, 10, 39, 134, 153, 164, 246
Island biogeography, 176–179
Isolated system, 8, 80

K-strategists, 29, 184
Kullbach's measure of information, 121–122

Landscape, 161
Laws of thermodynamics, 9–11
Le Chatelier, 125

Leaf size, 127
Liebig's Law, 193–194
Life conditions, 103
Life expectation, 144
Limiting similarity, 188

Margalef index, 222
Markov chain, 83
Mass extinction, 148
Max. entropy, 9–11
Maximum power, 115–117, 125
Melanism, 170
Middle number system, 105
Min. entropy, 20–21, 136
Mondego estuary, 37, 232–234
Mondego river, 37
Mosaic hypothesis, 114, 115

Network analysis, 80, 82, 85, 90, 96, 101, 206–213
Network complexity, 79
Network enfolding, 97, 98
Network mutalism, 91, 95
Network synergism, 91, 94, 95
Network unfolding, 94, 96
Newtonian laws, 60, 61
Niche, ecological, 99, 187–189, 191
Non-isolated system, 8
Non-locality, 92, 93

Odum's attributes, 106, 112
Ontic openness theories, 40
Open system, definition, 8
Openness of cells, 31
Operative symbolism, 44, 46, 47
Ordered heterogeneity, 44–46
Orientors, 11–115, 151–153, 199–241
Orientor optimization, 151–153
Overhead, 73

Path analysis, 87–90
Pathway proliferation, 92
Phenomonology, 35
Phenotype, 55, 56, 101
Photosynthesis, 127
Plant species increase, 60
Probability paradox, 52, 53
Prokaryote cells, 30, 31

Index

r-strategists, 29, 184
Radiation efficiency, 139–141
RAINS, 85
Rare events, 149, 150
Redundancy, 212–214
Relative stability, 49
Relaxation, 43
Reorganization, 157, 158
Resilience, 158, 165
Resistance, 111
Retrogressive analysis, 166
Rio Declaration, 1
River continuum concept, 194–196

Sampling uncertainty, 48
Seasonal changes, 129–133
Second law of thermodynamics, 9–11, 18–20
Selection, natural, 136, 168
Self-organization, 21, 68, 87, 110, 113, 118, 144, 166
Sequence oxidation, 126
Size development, 171–173
Solar radiation, 139–141, 204, 248
Spatial distribution, 153, 154
Species diversity, 29
Species number, 104, 105, 184

Species richness, 184
Specfic exergy, 235–240
Spheres, 104
Spin relaxation, 41–42
Stabilizing selection, 169
Stability, 155, 156, 165
STELLA, 84, 172, 173, 207–209
Stochastic processes, 63
Storage analysis, 87–90
Structurally dynamic models, 128–131
Succession, 69, 112, 114, 137, 156, 257, 164, 166
Succession theory, 112–114
Sustainable, 1, 2

Top-down, 85
Total throughput, 70, 88, 90, 133, 240
Transformity, 118, 119, 225
Trophic indices, 204

Uncertainty in ecology, 47–49, 56
Unique events, 46
Utility analysis, 87–90
Utilization pattern, 129, 132

Work capacity, 141, see also exergy and eco-exergy